# Photographic and Descriptive Musculoskeletal Atlas of Bonobos

Rui Diogo • Brian Shearer
Josep M. Potau • Juan F. Pastor
Felix J. de Paz • Julia Arias-Martorell
Cassandra Turcotte • Ashley Hammond
Evie Vereecke • Marie Vanhoof
Sandra Nauwelaerts • Bernard Wood

# Photographic and Descriptive Musculoskeletal Atlas of Bonobos

## With Notes on the Weight, Attachments, Variations, and Innervation of the Muscles and Comparisons with Common Chimpanzees and Humans

Springer

Rui Diogo
Department of Anatomy
Howard University
Washington DC, District of Columbia
USA

Brian Shearer
Anthropology Center
City University of New York
Anthropology Center
New York, New York, USA

Josep M. Potau
Unity of Human Anatomy and
Embriology
University of Barcelona
Barcelona, Spain

Juan F. Pastor
Department of Anatomy and Radiology
University of Valladolid
Valladolid, Spain

Felix J. de Paz
Department of Anatomy and Radiology
University of Valladolid
Valladolid, Spain

Julia Arias-Martorell
Department of Animal Biology
University of Barcelona
Barcelona, Spain

Cassandra Turcotte
CASHP, Department of Anthropology
George Washington University
Washington, District of Columbia, USA

Ashley Hammond
CASHP, Department of Anthropology
George Washington University
Washington, District of Columbia, USA

Evie Vereecke
Development and Regeneration
Kulak Kortrijk
University of Leuven
Kortrijk, Belgium

Marie Vanhoof
Development and Regeneration
University of Leuven, Kortrijk, Belgium

Sandra Nauwelaerts
Department of Biology
University of Antwerp
Wilrijk, Belgium

Centre for Research and Conservation
Royal Zoological Society of Antwerp
Antwerp, Belgium

Bernard Wood
CASHP, Department of Anthropology
George Washington University
Washington, District of Columbia, USA

ISBN 978-3-319-85319-2 ISBN 978-3-319-54106-8 (eBook)
DOI 10.1007/978-3-319-54106-8



Printed on acid-free paper

This Springer imprint is published by Springer Nature
The registered company is Springer International Publishing AG
The registered company address is: Gewerbestrasse 11, 6330 Cham, Switzerland

# Preface

Chimpanzees, including bonobos and common chimpanzees, are our closest living relatives. However, surprisingly, information about the soft tissues of bonobos is very scarce, making it difficult to discuss and understand human evolution. This book, which is the first photographic and descriptive musculoskeletal atlas of bonobos (*Pan paniscus*), adopts the same format as the photographic atlases of other apes previously published by the various colleagues and me. These books are part of a series of monographs that set out the comparative and phylogenetic context of the gross anatomy and evolutionary history of the soft tissue morphology of modern humans and their closest relatives. This atlas, which includes detailed high-quality photographs of musculoskeletal structures from most anatomical regions of the body as well as textual information about the attachments, innervation, function and weight of the respective muscles, is based on dissections of seven bonobos, including adults, adolescents, infants and fetuses, and males and females, and on an extensive review of the literature for comparisons with common chimpanzees. It therefore provides an updated review of the anatomical variations within chimpanzees as a whole as well as an extensive list of synonyms used in the literature to designate the structures covered here. Moreover, contrary to the previous photographic atlases of apes, it also provides details on the brachial and lumbosacral plexuses. This book will therefore be of interest to students, teachers and researchers studying primatology, comparative anatomy, functional morphology, zoology and physical anthropology and to medical students, doctors and researchers who are curious about the origin, evolution, homology and variations of the musculoskeletal and limb plexuses of modern humans.

Washington, DC, USA
6 April 2017

Rui Diogo

# Acknowledgements

We gratefully acknowledge support and funding from all of the institutions and funding bodies that made this project possible: Sandra Nauwelaerts and Jeroen Stevens for organizing all the specimens and facilities used for this project, the Antwerp Zoo for providing the bonobo specimens dissected for this project and the Applied Veterinary Morphology group of the Department of Veterinary Sciences at the University of Antwerp for providing us a space to do the dissections and have our daily meetings. We also want to thank the other members of the Bonobo Morphology Initiative that are not authors of this book for their help with the dissections and the meetings and other logistical aspects, in particular to Ellen Schultz-Kornas for teaching us how to use photogrammetry, and to Heidi Vollrath for measuring and documenting the weights of each muscle of the various bonobo specimens dissected. We also acknowledge support from the several grants that have provided, in some way or another, funds that helped the authors of this book to undertake this work, and in particular from three grants that specifically included this work as part of their aims: NSF grants 1,516,557 and 1,440,624 to Rui Diogo, and grant "CGL2014-52611-C22-P; Spanish Ministerio de Economía y Competitividad" to Josep Potau and Julia Arias-Martorell.

# Contents

# About the Authors

**Rui Diogo** is a Fellow of the American Association of Anatomists, and an Associate Professor at the Howard University College of Medicine and a Resource Faculty at the Center for the Advanced Study of Hominid Paleobiology of George Washington University (USA). A multi-awarded researcher, he is the author or co-author of numerous publications, and the co-editor of the books *Catfishes* and *Gonorynchiformes and Ostariophysan Relationships*. He is the sole author or first author of the books *Morphological Evolution, Aptations, Homoplasies, Constraints and Evolutionary Trends*, *The Origin of Higher Clades, Muscles of Vertebrates*, *Photographic and Descriptive Musculoskeletal Atlas of Gorilla*, *Photographic and Descriptive Musculoskeletal Atlas of Gibbons and Siamangs (*Hylobates*), Photographic and Descriptive Musculoskeletal Atlas of Orangutans, Photographic and Descriptive Musculoskeletal Atlas of a Baby Gorilla, Photographic and Descriptive Musculoskeletal Atlas of Chimpanzees, Comparative Anatomy and Phylogeny of Primate Muscles and Human Evolution*, and *Learning and Understanding Human Anatomy and Pathology: An Evolutionary and Developmental Guide for Medical Students*.

**Brian Shearer** is a PhD candidate in physical anthropology at the Graduate Center of the City University of New York (CUNY) and the New York Consortium in Evolutionary Primatology (NYCEP). His research is primarily focused on the evolution of soft tissue complexes, comparative anatomy of the primate neuromuscular system, locomotor evolution, and the mammalian paleofauna and paleontology of the Neotropical Miocene.

**Josep Potau** is Professor at the Department of Anatomy and Embryology of the University of Barcelona (Spain) and is the Director of the University's Center for the Study of Comparative and Evolutionary Anatomy. His current research focus on the analysis of functional and anatomical adaptations associated with the evolution of different types of locomotion and of the upper limb musculature within primates. He has published several papers, book chapters and books on functional and comparative anatomy.

**Juan Pastor** is Professor at the Department of Anatomy of the University of Valladolid (Spain) and is the Director of the University's Anatomical Museum, which houses the largest comparative osteological collection in Spain. He published several papers and books on comparative anatomy and anthropology.

**Félix de Paz** is Professor at the Department of Anatomy of the University of Valladolid and is a member of the Royal Academy of Medicine and Surgery of Valladolid (Spain). He has published several papers and books on comparative anatomy and anthropology.

**Julia Arias-Martorell** is an anthropologist now undertaking postdoctoral research at the University of Kent (UK) that focuses on functional morphology and variability of the forelimbs of the hominoids related to the diverse locomotor repertoires of the members of this clade, and on the enhancement of tridimensional techniques to study evolution.

**Cassandra Turcotte** is a PhD student at George Washington University (USA) in the Department of Anthropology and the Center for Advanced Study of Human Paleobiology. Her dissertation focuses on the functional morphology of muscle attachments, implementing a combination of dissection and 3D imaging to better understand the effects of habitual activity on the mammalian bone-muscle interface and to improve methods of behavioural reconstruction using these structures.

**Ashley Hammond** is a postdoctoral scientist at George Washington University (USA) in the Department of Anthropology and the Center for Advanced Study of Human Paleobiology. She has taught human anatomy for more than 10 years and has published multiple papers on the functional anatomy of primates. Her research is focused on the evolution of the limbs and trunk of primates, and how different morphologies relate to locomotion.

**Evie Vereecke** is Professor of Anatomy at the Faculty of Medicine of the University of Leuven Campus Kulak. She obtained a PhD in 2006 with a thesis on the evolution of bipedalism. Her research focuses on the link between form and function in primate limbs and she has published several papers on the functional morphology of the hands and feet of humans, bonobos and gibbons.

**Marie Vanhoof** has a bachelor degree in Education for Biology and Mathematics and an Academic Master of Science in Biology, with a major in Anatomy, Physiology, Evolution and Morphology. She has a keen interest in integrated morphology research and human evolution.

**Sandra Nauwelaerts** has a PhD in Biology and works as a postdoctoral research associate at the University of Antwerp (Belgium), and she is also the coordinator of functional morphology research at the Center for Research and Conservation of the Antwerp Zoo. Her expertise lies in functional morphology and biomechanics of animal locomotion.

**Bernard Wood** is University Professor of Human Origins and Director of the Center for the Advanced Study of Human Paleobiology at George Washington University (USA). His edited publications include *Food Acquisition and Processing in Primates* and *Major Topics in Primate and*

*Human Evolution*. He is the author of *The Evolution of Early Man, Human Evolution, Koobi Fora Research Project – Hominid Cranial Remains (Vol. 4), Human Evolution – A Very Short Introduction*, and *Wiley-Blackwell Student Dictionary of Human Evolution*; co-author of *Comparative Anatomy and Phylogeny of Primate Muscles and Human Evolution*; and editor of the *Wiley-Blackwell Encyclopedia of Human Evolution*.

# 1 Introduction, Aims, Methodology and Materials

## 1.1 Introduction and Aims

Chimpanzees, including bonobos (*Pan paniscus*) and common chimpanzees (*Pan troglodytes*), are our closest living relatives. This photographic and descriptive musculoskeletal atlas of bonobos follows the organization used in previous photographic atlases of apes, which are part of a monograph series that is designed to help provide the comparative, phylogenetic and evolutionary context for understanding the evolutionary history of the gross anatomy of modern humans and our closest relatives.

We dissected and took high-quality photographs of 7 specimens of bonobos, including fetuses, infants, adolescents and adults, and both males and females (see Methodology and Material below), and for most of these specimens (VU PT1) we were able to record the wet weight of many of the muscles. Where there are differences between the myology (e.g., the presence/absence of a muscle or a muscle bundle, its attachments and/or its innervation) of the individuals, we provide detailed comparative notes, and use photographs to document the differences.

The atlas also includes the results of an extensive review of the literature about the musculature of both common chimpanzees and bonobos, a comprehensive review of muscle variants among individual members of these two species, and a list of the synonyms used in the literature to refer to the muscles of these primates. The data previously obtained from our dissections of numerous primates and other mammalian and non-mammalian vertebrates (e.g., Diogo 2004a, b, 2007, 2008, 2009; Diogo and Abdala 2007, 2010; Diogo et al. 2008, 2009a, b, 2010, 2012; Diogo and Wood 2011, 2012) were used to test hypotheses about the homologies among the muscles of bonobos, common chimpanzees, other apes and modern humans, and other taxa.

Although this book follows mainly the style of our previous photographic atlases of apes, some key features distinguish it from those atlases, including the fact that it also provides details on the brachial and lumbosacral plexuses. We therefore hope that this atlas will be of interest to students, teachers and researchers studying primatology, comparative anatomy, functional morphology, zoology and physical anthropology, as well as to clinicians and researchers who are interested in understanding the origin, evolution and homology of the musculoskeletal and neurovascular system of modern humans as well as the comparative context of common variants on our musculature.

## 1.2 Methodology and Material

The 7 bonobos (*Pan paniscus*) dissected for this study were made available by the Antwerp Zoo. All were fresh (frozen), except Lomela, which was formalin embalmed. The specimens are:



R. Diogo et al., *Photographic and Descriptive Musculoskeletal Atlas of Bonobos*,
DOI 10.1007/978-3-319-54106-8_1

Kidogo (ZIMS 164031), adult male, 35.000 Kg; Lomela (ZIMS 164046), adult female, 37.730 Kg; Jasiri (ZIMS 164047), 8 years-old adolescent female, 25.695 Kg; Etje (ZIMS 164040), 2 month-old infant male, 1.950 Kg; Foyo (ZIMS 164041), 8 month-old infant male, 2.700 Kg; Gabber (ZIMS 164042), 2 month-old infant female, 1.600 Kg; and Ano (ZIMS 164052), fetus female, 0.700 Kg. It should be noted that, due to poor preservation, the adult female Lomela and infant female Gabber were not dissected for most regions, and only a few notes about these animals will be made in this atlas. So, in total, we dissected in detail the heads of an adult (Kidogo), an infant (Foyo) and a fetus (Ano), and 10 upper and 10 lower limbs of representatives of four different ontogenetic stages (i.e., of adult Kidogo, adolescent Jasiri, infants E-tje and Foyo, and fetus Ano).

We took photographs of the musculoskeletal system of all of the specimens we dissected, and took notes of the weights of each muscle in each specimen, but for simplicity and utility, the muscle weights listed in this atlas are just from an adult (Kidogo), an infant (Foyo) and the fetus (Ano). In the text below, when the data are available, we therefore provide, for each muscle: (1) its weight in these three specimens. The total mass of all the striated muscles is given in parentheses immediately following the name of the muscle. In the case of paired muscles, the muscles of the left and right sides are referred to as LSB and RSB, respectively. When the muscle is part of a symmetrical structure (e.g., stylohyoideus), the weight given is that of the muscle of one side of the body; when the muscle is unpaired (e.g., arytenoideus), the weight given is that of the part of the muscle that comes from one side only; (2) the most common attachments and innervation of the muscle within bonobos, within common chimpanzees for comparison, and within the chimpanzee (*Pan*) clade as a whole, based on dissections of several common chimpanzees that various authors of the present book did for their previous photographic atlas of *P. troglodytes* (Diogo et al. 2013) and based on our literature review; (3) the function of the muscle (e.g., based on electromyographic studies (EMG) of neck and upper limb muscles or the use of the Facial Action Coding System (FACS) for facial muscles); (4) comparative notes, especially when where there are differences (e.g., regarding the presence/absence of the muscle, or of its bundles, its attachments, and/or its innervation) between the configuration usually found in the bonobos dissected by us (in these cases we often provide photographs to illustrate the differences) and/or dissected by others (such as Miller 1952); and (5) a list of the synonyms that have been used by other authors to designate that muscle.

Apart from the common chimpanzee specimens mentioned above, we have dissected numerous specimens from most vertebrate groups, including bony fish, amphibians, reptiles, monotremes, rodents, colugos, tree-shrews, and numerous primates, including modern humans (a complete list of these specimens and the terminology used to describe them is given in Diogo and Abdala 2010 and Diogo and Wood 2011, 2012). This broad comparative context proved to be crucial for generating hypotheses about the homologies among the muscular structures of hylobatids, modern humans and other primate and non-primate vertebrates, and it also informed the nomenclature proposed by Diogo et al. (2008, 2009a, b), Diogo and Abdala (2010) and Diogo and Wood (2011, 2012). This nomenclature is based on that employed in modern human anatomy (e.g., *Terminologia Anatomica* 1998), but it also takes into account the names used by researchers who have focussed on non-human mammals (e.g., Saban 1968; Jouffroy 1971). In the majority of the figures we use Latin names for the soft tissues and anglicized names for the bones. In the figures that mainly illustrate osteological structures we use Latin names, but to avoid redundancy when these names are similar to the anglicized version (e.g., processus mastoideus = mastoid process) we do not provide the latter; in those cases in which they are substantially different (e.g., incisura mandibulae and mandibular notch) we provide both the Latin names and the anglicized version.

When we describe the position, attachments and orientation of the muscles and we use the terms anterior, posterior, dorsal and ventral in the sense in which those terms are applied to pronograde tetrapods (e.g., the sternohyoideus mainly runs from the sternum, posteriorly, to the hyoid bone, anteriorly, and passes mainly ventrally to the larynx, which is, in turn, ventral to the oesophagus; the flexors of the forearm are mainly situated on the ventral side of the forearm). However, the nomenclature used in *Terminologia Anatomica* (1998) was defined on the basis of an upright posture and although most primates are not bipeds, nearly all of the osteological names and most of the myological ones used by other authors (and by us) to designate the structures of non-human primates, including common chimpanzees and bonobos, follow the *Terminologia Anatomica* nomenclature. Although this is potentially confusing we judged it to be preferable to refer to the topology of the musculoskeletal structures of non-human primates in this way because in the vast majority of primates the 'superior angle of the scapula' is actually mainly anterior, and not superior, to the 'inferior angle of the scapula', and the 'cricoarytenoideus posterior' actually lies more on the dorsal, and not on the posterior, surface of the larynx. Moreover, we think that, by keeping in mind that the actual names (both in Latin and in English) of all the osteological structures and most of the myological structures mentioned in this atlas refer to a biped posture while the actual descriptions provided here regarding the topology of these structures refer to a pronograde posture, most readers will have no difficulty in interpreting and understanding the information provided in this book.

The muscles listed below are those that are usually present in adult bonobos; muscles that are only occasionally present in adult bonobos are discussed in other parts of the atlas. In our written descriptions, we follow Edgeworth (1935), Diogo and Abdala (2010) and Diogo and Wood (2011, 2012) and divide the head and neck muscles in five main subgroups: (1) mandibular, muscles that are generally innervated by the fifth cranial nerve (CN5) and include the masticatory muscles, among others; (2) hyoid, muscles that are usually innervated by CN7 and include the facial muscles, among others; (3) branchial, muscles that are usually innervated by CN9 and CN10, and include most laryngeal and pharyngeal muscles, among others; (4) hypobranchial, muscles that include all the infrahyoid and tongue muscles, and the geniohyoideus. According to Edgeworth (1935), the hypobranchial muscles were developed primarily from the anterior myotomes of the body and then migrated into the head. Although they retain a main innervation from spinal nerves, they may also be innervated by CN11 and CN12, but they usually do not receive any branches from CN5, CN6, CN7, CN8, CN9 and CN10; and (5) extra-ocular, muscles that are usually innervated by nerves CN3, CN4 and/or CN6 in vertebrates. The head, neck and upper limb muscles are listed following the order used by Diogo et al. (2008, 2009a), Diogo and Abdala (2010) and Diogo and Wood (2011, 2012), while the lower limb muscles, as well as the other muscles of the body, are listed following the order used by Gibbs (1999). It should be emphasized that the literature review undertaken by this latter author provided a crucial basis and contribution for our own literature review.

## Bibliography

### Appendix I: Literature Including Information about the Muscles of Bonobos and Common Chimpanzees

Diogo R, Abdala V (2010) Muscles of vertebrates – comparative anatomy, evolution, homologies and development. Taylor & Francis, Oxford

Diogo R, Wood BA (2011) Soft-tissue anatomy of the primates: phylogenetic analyses based on the muscles of the head, neck, pectoral region and upper limb, with notes on the evolution of these muscles. J Anat 219:273–359

Diogo R, Wood BA (2012) Comparative anatomy and phylogeny of primate muscles and human evolution. Taylor & Francis, Oxford

Diogo R, Abdala V, Lonergan N, Wood BA (2008) From fish to modern humans – comparative anatomy, homologies and evolution of the head and neck musculature. J Anat 213:391–424

Diogo R, Abdala V, Aziz MA, Lonergan N, Wood BA (2009a) From fish to modern humans – comparative anatomy, homologies and evolution of the pectoral and forelimb musculature. J Anat 214:694–716

Diogo R, Wood BA, Aziz MA, Burrows A (2009b) On the origin, homologies and evolution of primate facial muscles, with a particular focus on hominoids and a suggested unifying nomenclature for the facial muscles of the Mammalia. J Anat 215:300–319
Edgeworth FH (1935) The cranial muscles of vertebrates. Cambridge University Press, Cambridge
Gibbs S (1999) Comparative soft tissue morphology of the extant Hominoidea, including Man. Unpublished PhD Thesis, The University of Liverpool, Liverpool
Jouffroy FK (1971) Musculature des membres. In: Grassé PP (ed) Traité de Zoologie, XVI: 3 (Mammifères). Masson et Cie, Paris, pp 1–475
Miller RA (1952) The musculature of *Pan paniscus*. Am J Anat 91:182–232
Saban R (1968) Musculature de la tête. In: Grassé PP (ed) Traité de Zoologie, XVI: 3 (Mammifères). Masson et Cie, Paris, pp 229–472

## Appendix II: Literature Cited, Not Including Information About the Muscles of Bonobos and Common Chimpanzees

Diogo R (2004a) Morphological evolution, aptations, homoplasies, constraints, and evolutionary trends: catfishes as a case study on general phylogeny and macroevolution. Science Publishers, Enfield
Diogo R (2004b) Muscles versus bones: catfishes as a case study for an analysis on the contribution of myological and osteological structures in phylogenetic reconstructions. Anim Biotechnol 54:373–391
Diogo R (2007) On the origin and evolution of higher-clades: osteology, myology, phylogeny and macroevolution of bony fishes and the rise of tetrapods. Science Publishers, Enfield
Diogo R (2008) Comparative anatomy, homologies and evolution of the mandibular, hyoid and hypobranchial muscles of bony fish and tetrapods: a new insight. Anim Biotechnol 58:123–172
Diogo R (2009) The head musculature of the Philippine colugo (Dermoptera: *Cynocephalus volans*), with a comparison to tree-shrews, primates and other mammals. J Morphol 270:14–51
Diogo R, Abdala V (2007) Comparative anatomy, homologies and evolution of the pectoral muscles of bony fish and tetrapods: a new insight. J Morphol 268:504–517
Diogo R, Potau JM, Pastor JF, de Paz FJ, Ferrero EM, Bello G, Barbosa M, Wood B (2010) Photographic and Descriptive Musculoskeletal Atlas of *Gorilla* – with notes on the attachments, variations, innervation, synonymy and weight of the muscles. Taylor & Francis, Oxford
Diogo R, Potau JM, Pastor JF, de Paz FJ, Ferrero EM, Bello G, Barbosa M, Aziz MA, Burrows AM, Arias-Martorell J, Wood B (2012) Photographic and Descriptive Musculoskeletal Atlas of Gibbons and Siamangs (*Hylobates*) – with notes on the attachments, variations, innervation, synonymy and weight of the muscles. Taylor & Francis, Oxford
Diogo R, Potau JM, Pastor JF, de Paz FJ, Ferrero EM, Bello G, Barbosa M, Aziz MA, Burrows AM, Arias-Martorell J, Wood BA (2013) Photographic and Descriptive Musculoskeletal Atlas of Chimpanzees – with notes on the attachments, variations, innervation, function and synonymy and weight of the muscles. Taylor & Francis, Oxford
Terminologia Anatomica (1998) Federative committee on anatomical terminology. Georg Thieme, Stuttgart. New York: Columbia University Press

# 2 Head and Neck Musculature

## 2.1 Mandibular Musculature

***Mylohyoideus* (*adult Kidogo 1.21 g; infant Foyo 0.39 g;* Figs. 2.5 and 2.11)**

- Attachments in bonobos: From mylohyoid line of mandible to hyoid bone, posteriorly, and to ventral midline, anteriorly, where it fuses with its counterpart through a medial raphe (our fetus and infant Koyo and adolescent Jasiri and adult Kidogo and Miller's 1952 adult bonobo).
- Usual innervation: Mylohyoid nerve of mandibular division of CN5 (Miller 1952: *P. paniscus*).
- Comparison with common chimpanzees: In *P. troglodytes* it runs from the mylohyoid line of the mandible to the hyoid bone, posteriorly, and to the ventral midline, anteriorly and, contrary to bonobos, there is usually no distinct median raphe of the mylohyoideus in common chimpanzees, according to Swindler and Wood (1973) and to our dissections. However, Göllner (1982) reported a median raphe in the two adult *P. troglodytes* dissected by him. The *intermandibularis anterior* (present in some mammals) is not present as a distinct muscle in chimpanzees.
- Synonymy: Intermandibularis (Edgeworth 1935).

***Digastricus anterior*** (adult Kidogo 1.93 g; infant Foyo 0.22 g; Figs. 2.5 and 2.11)

- Attachments in bonobos: From digastric intermediate tendon and hyoid bone to mandible, usually meeting its counterpart in the midline (our fetus and infant Koyo and adolescent Jasiri and adult Kidogo and Miller's 1952 adult bonobo).
- Usual innervation: Mylohyoid nerve of mandibular division of CN5 (Miller 1952: *P. paniscus*).
- Comparison with common chimpanzees: In *P. troglodytes* it runs from the intermediate digastric tendon, and sometimes also from the hyoid bone (e.g., Gratiolet and Alix 1866), to the mandible, usually meeting its counterpart in the midline (e.g., Vrolik 1841; Wilder 1862; Sonntag 1923; Dubrul 1958; Starck and Schneider 1960; Göllner 1982; our dissections).
- Synonymy: Anterior belly of digastricus (Miller 1952); anterior belly of biventer mandibulae (Starck and Schneider 1960).

***Tensor tympani***

- Attachments in bonobos: From auditory tube and adjacent regions of neurocranium to manubrium of malleus.
- Usual innervation: Data are not available.
- Comparison with common chimpanzees: Configuration similar to bonobos. According



R. Diogo et al., *Photographic and Descriptive Musculoskeletal Atlas of Bonobos*,
DOI 10.1007/978-3-319-54106-8_2

to Maier (2008) in common chimpanzees the chorda tympani passes above the tensor tympani (i.e., the epitensoric condition).

***Tensor veli palatini*** (adult Kidogo 0.45 g; Fig. 2.7)

- Attachments in bonobos: From Eustachian tube and adjacent regions of cranium to pterygoid hamulus and soft palate (our fetus and infant Koyo and adolescent Jasiri and adult Kidogo).
- Usual innervation: Data are not available.
- Comparison with common chimpanzees: In *P. troglodytes* it runs from the Eustachian tube and the adjacent regions of the cranium (usually the scaphoid fossa of the medial pterygoid plate: Gratiolet and Alix 1866, Sonntag 1923, 1924, Dean 1985; sometimes the apex of the petrous temporal bone, e.g., Sonntag 1923) to the pterygoid hamulus and soft palate. In modern human infants and in adult apes, including common chimpanzees, the palate lies much closer to the roof of the nasopharynx than it usually does in adult modern humans, so in the former the levator veli palatini and tensor veli palatini do not run so markedly downwards to reach the palate as they do in the latter. The *pterygotympanicus* (present, e.g., as an anomaly in modern humans) is usually missing in common chimpanzees.
- Synonymy: Peristaphylin externe (Gratiolet and Alix 1866); tensor palatini (Sonntag 1923).

***Masseter*** (adult Kidogo 31.20 g; Figs. 2.4 and 2.11)

- Attachments in bonobos: Mainly from zygomatic arch; the pars superficialis inserts mainly onto lower edge of base of mandible, while the pars profunda inserts mainly onto ascending ramus and coronoid process of mandible (our fetus and infant Koyo and adolescent Jasiri and adult Kidogo and Miller's 1952 adult bonobo).
- Usual innervation: Branch of the mandibular division of CN5 (Miller 1952: *P. paniscus*).
- Comparison with common chimpanzees: Configuration similar to bonobos. In common chimpanzees there is often a strong fascia/aponeurosis between the superficial and deep heads of the masseter (e.g., Sonntag 1923). The *zygomatico-mandibularis* is usually not present as a separate muscle in chimpanzees, but it is present as a separate bundle of the masseter in the *P. troglodytes* specimens dissected by Göllner (1982) and Gratiolet and Alix (1866) and embryo dissected by Starck (1973) and at least some of the common chimpanzees dissected by us (PFA 1016, PFA 1009, PFA 1051, HU PT1), lying superficially to the temporalis and running from the deep surface of the zygomatic process to the coronoid process of the mandible.

***Temporalis*** (adult Kidogo 59.52 g; Figs. 2.4 and 2.11)

- Attachments in bonobos: From whole fossa temporalis, the temporalis fascia and adjacent regions of cranium, to coronoid process and ramus of mandible (our fetus and infant Koyo and adolescent Jasiri and adult Kidogo and Miller's 1952 adult bonobo). In our infant Foyo there was a pars suprazygomatica (Fig. 2.11), but in our adult Kidogo and adolescent Jasiri, the pars suprazygomatica is not present as a distinct bundle (Fig. 2.4); we could not discern if a pars suprazygomatica is, or not, present in our fetus.
- Usual innervation: Branch of the mandibular division of CN5 (Miller 1952: *P. paniscus*).
- Comparison with common chimpanzees: Configuration similar to bonobos. Swindler and Wood (1973) stated that in common chimpanzees the temporalis is usually not divided into a distinct pars superficialis and a distinct pars profunda. However, Göllner (1982) dissected 4 neonate and 2 adult *P. troglodytes* specimens and found a temporalis divided into a pars superficialis and a pars profunda in those 6 specimens. In the common chimpanzees dissected by us there was in general no notable differentiation of the temporalis into a pars

superficialis and a pars profunda, but in our *P. troglodytes* specimens PFA 1077 and PFA UNC there was some differentiation between the anterior, more vertical, and deeper fibres of the temporalis and the posterior, more oblique and superficial fibres of this muscle. Regarding the pars suprazygomatica, in some neonates of *P. troglodytes* dissected by Göllner (1982) and by us, this part of the muscle is clearly differentiated, but in the adult common chimpanzees dissected by us and by other authors, including Göllner (1982), the pars suprazygomatica is usually not present as a distinct bundle.

***Pterygoideus lateralis*** (adult Kidogo 3.54 g)

- Attachments in bonobos: Two heads, from lateral pterygoid plate and adjacent regions of cranium (e.g., greater wing of sphenoid bone in Miller's 1952 adult bonobo) to capsule of temporomandibular joint and neck of mandibular condyle (our fetus and infant Koyo and adolescent Jasiri and adult Kidogo and Miller's 1952 adult bonobo).
- Usual innervation: Branch of the mandibular division of CN5 (Miller 1952: *P. paniscus*).
- Comparison with common chimpanzees: Similar to bonobos. Göllner (1982) suggested that the inferior and superior heads of the pterygoideus lateralis are not differentiated in the neonate and adult *P. troglodytes* dissected by him, but the two heads were clearly present in all the *P. troglodytes* specimens dissected by Gratiolet and Alix (1866), Sonntag (1923) and by us (including neonates and adults).
- Synonymy: Pterygoideus externus (Gratiolet and Alix 1866; Sonntag 1923; Miller 1952).

***Pterygoideus medialis*** (adult Kidogo 13.25 g)

- Attachments in bonobos: Mainly from medial surface of lateral pterygoid plate of sphenoid bone and adjacent regions of skull to the medial side of mandible (our fetus and infant Koyo and adolescent Jasiri and adult Kidogo and Miller's 1952 adult bonobo).
- Usual innervation: Branch of the mandibular division of CN5 (Miller 1952: *P. paniscus*).
- Comparison with common chimpanzees: Similar to bonobos.
- Synonymy: Pterygoideus internus (Gratiolet and Alix 1866; Sonntag 1923; Miller 1952).

## 2.2 Hyoid Musculature

***Stylohyoideus (adult Kidogo 0.62 g; infant Foyo 0.16 g;*** Figs. 2.5 and 2.11*)*

- Attachments in bonobos: From styloid process and/or adjacent regions of skull to hyoid bone (our fetus and infant Koyo and adolescent Jasiri and adult Kidogo and Miller's 1952 adult bonobo). In the adult bonobo dissected by Miller (1952), the stylohyoideus was mainly superficial to the posterior digastricus; in our fetus and infant Foyo and adult Kidogo and adolescent Jasiri, the stylohyoideus was mainly superficial to the posterior digastricus, but was pierced by this latter muscle (Figs. 2.5 and 2.11).
- Usual innervation: Branch of CN7 (Miller 1952: *P. paniscus*).
- Comparison with common chimpanzees: Similar to bonobos. A piercing of the stylohyoideus by the intermediate digastric tendon was found in the *P. troglodytes* specimens dissected by Sonntag (1923) and in most common chimpanzees dissected by us, but in our *P. troglodytes* HU PT1 specimen the stylohyoideus was mainly just superficial to the posterior digastricus. To our knowledge the *jugulohyoideus* (usually present in strepsirrhines and sometimes present in *Tarsius*) and the *stylolaryngeus* (sometimes present in orangutans) have never been reported in chimpanzees.

***Digastricus posterior*** (adult Kidogo 1.74 g; infant Foyo 0.36 g; Figs. 2.5 and 2.11)

- Attachments in bonobos: From mastoid region and adjacent regions of skull to digastric intermediate tendon, which is attached indirectly to the hyoid bone through a fibrous arch, as is the case in modern humans (our fetus and infant Koyo and adolescent Jasiri and adult Kidogo and Miller's 1952 adult bonobo).

- Usual innervation: Branch of CN7 (Miller 1952: *P. paniscus*).
- Comparison with common chimpanzees: In *P. troglodytes* it runs from the mastoid region and adjacent regions of the skull (e.g., sometimes from the occipital bone, Tyson 1699) to the digastric intermediate tendon.
- Synonymy: Posterior belly of digastricus (Miller 1952); posterior part of biventer mandibulae (Starck and Schneider 1960).

***Stapedius***

- Attachments in bonobos: Probably inserts onto the stapes, but the information provided in the literature is sparse.
- Usual innervation: Data are not available.
- Comparison with common chimpanzees: Probably similar to bonobos, but the information provided in the literature is sparse.

***Platysma myoides*** (adult Kidogo 9,53 g; Figs. 2.1, 2.2, 2.9 and 2.10)

- Attachments in bonobos: From pectoral and neck regions (no direct bony attachments) to modiolus and adjacent regions of mouth, also inserting onto mandible and fusing with its counterpart in ventral midline (our fetus and infant Koyo and adolescent Jasiri and adult Kidogo and Miller's 1952 adult bonobo). The *platysma cervicale* is often missing in bonobos, and was only present as a variant, in a very reduced form, in the adult bonobo of Miller (1952), being designated by her as a 'transversus nuchae' muscle, running from the occipital bone to the insertion of the sternocleidomastoideus (see notes just below). If one sees the configuration found in, for instance, the adolescent bonobo Jasiri (Fig. 2.1), there seem to be some vestigial fibres running ventrally from the ear region, which could correspond to a vestigial *pars auricularis of the sphincter colli profundus*, which is a well developed structure in many mammals and is sometimes designated as *auricularis inferior* in those mammals (see also notes just below).
- Usual innervation: 'Platysma' (platysma myoides *sensu* the present study) by cervical branch of CN7; 'transversus nuchae' (remaining of platysma cervicale *sensu* the present study) by posterior auricular branch of CN7 (Miller 1952: *P. paniscus*).
- Comparison with common chimpanzees: Similar to bonobos. The *platysma cervicale, sphincter colli superficialis* and *sphincter colli profundus* are usually not present as distinct muscles in common chimpanzees. Influential authors such as Owen (1830–1831) and Sonntag (1924) used 'platysma myoides' to describe the platysma complex of Asian apes such as orangutans, and this nomenclature has been followed by various researchers, including Seiler (1976), and was thus also followed in Diogo et al.'s (2009) review. However, Owen (1830–1831) stated that the 'platysma myoides' of orangutans incorporates the platysma myoides of modern humans plus the platysma cervicale of other mammals, and our recent dissections of numerous primates and comparisons with the data provided in the literature confirm that orangutans and hylobatids usually have a platysma cervicale (i.e., a part of the platysma complex that attaches onto the nuchal region). That is, juvenile and adult orangutans and hylobatids usually have a well-developed platysma cervicale similar to the muscle that is found in most other primates and that is usually markedly reduced, or even absent, in *Pan* (including the *P. troglodytes* neonates dissected by us), *Gorilla* and *Homo*. In fact, juveniles and adult members of the genus *Pan* usually do not have a well-developed platysma cervicale, often having instead a small muscle 'transversus nuchae', as is the case of Miller's (1952: see her fig. 12) bonobo as noted above, and of various common chimpanzees (e.g., Gratiolet and Alix 1866; Broca 1869; Champneys 1872; Macalister 1871; Chapman 1879; Sutton 1883; Virchow 1915; Sonntag 1923; Sullivan and Osgood 1925; Huber 1930, 1931; Loth 1931; Swindler and Wood 1973; Pellatt 1979; our dissections; e.g., fig. 1 of plate IX of Gratiolet and Alix 1866; fig. 5 of Sullivan and Osgood 1925; fig. 35 of

Huber 1930; fig. 11 of Huber 1931; fig. 785 of Edgeworth 1935). This 'transversus nuchae' seems to be a vestige of the platysma cervicale (e.g., Loth 1931; Diogo and Wood 2011, 2012). Regarding the sphincter colli superficialis, it should be noted that the 'sphincter colli' described by Burrows et al. (2006) in common chimpanzees could correspond to the sphincter colli superficialis *sensu* the present work, because they stated that this structure is superficial to the platysma myoides. However, the sphincter colli superficialis is not present as a distinct muscle in all the other *Pan* specimens described in the literature and dissected by us and, as explained above, is actually usually missing in all extant primate taxa. Regarding the sphincter colli profundus, it is often considered to be absent in *Pan*. However, Seiler (1974a, 1976) stated that the *auricularis inferior* is commonly found in *P. troglodytes*, running from the fascia glandulae parotis to the region of the ear. This statement is confusing because the auricularis inferior is often a synonym for the *mandibulo-auricularis sensu* Diogo and Wood (2011, 2012), and the mandibulo-auricularis is usually not present as a distinct, fleshy muscle in common chimpanzees, possibly corresponding to part or to all of the stylomandibular ligament, as is usually the case in modern humans. Therefore, the structure described and illustrated by Seiler in *P. troglodytes* species seems to be mainly a vestigial/ligamentous structure, and not the distinct, fleshy, mandibulo-auricularis muscle that is usually present in strepsirrhines and other mammals. As noted above, if one sees the configuration found in, for instance, the adolescent bonobo Jasiri (Fig. 2.1), there seem to be effectively some vestigial fibres running ventrally from the ear region, i.e., they could correspond to the 'auricularis inferior' of Seiler, which would thus correspond not to a vestigial mandibulo-auricularis, but instead to the vestigial pars auricularis of the sphincter colli profundus, which is well developed in many mammals and is effectively sometimes designated also as 'auricularis inferior' in those mammals (see, e.g., Jouffroy and Saban 1971).

- Synonymy: Peaucier propre plus risorius (Gratiolet and Alix 1866); platysma (MacAlister 1871); platysma + transversus nuchae (Miller 1952); tracheloplatysma (Swindler and Wood 1973).

***Occipitalis*** (Figs. 2.1 and 2.2)

- Attachments in bonobos: Mainly from superior nuchal line and mastoid process to galea aponeurotica, being fused to its counterpart at the midline (our fetus, infant Foyo, adult Kidogo and adolescent Jasiri, and Miller's 1952 adult bonobo).
- Usual innervation: Posterior auricular branch of CN7 (Miller 1952: *P. paniscus*).
- Comparison with common chimpanzees: Similar to bonobos. In hylobatids and nonhominoid primates the occipitalis is usually differentiated into a main body (or *occipitalis proprius*) and a *cervico-auriculo-occipitalis* (sensu Lightoller 1925, 1928, 1934, 1939, 1940a, b, 1942). The latter is a lateral/superficial bundle of the occipitalis that often runs anterolaterally from the occipital region to the posterior portion of the ear and that sometimes covers part of the auricularis posterior in lateral view. The 'cervico-auriculo-occipitalis' of primates such as *Macaca* was designated as the 'deep layer of the occipitalis' by Huber (1930, 1931), but it is not homologous to the 'pars profunda' found in *Pongo* by Sullivan and Osgood (1925) and by us, nor to the 'pars profunda' described in *P. troglodytes* by Burrows et al. (2006), nor to the 'pars profunda' *sensu* Seiler (1976), which corresponds to the 'occipitalis proprius' *sensu* the present study. Our dissections and the literature reviewed by us (for detailed accounts on this issue see, e.g., Lightoller 1928, 1934; Seiler 1976) confirm that the 'cervico-auriculo-occipitalis' is usually not present as a distinct structure in *Pan*, although it was apparently present as a small structure in a common chimpanzee illustrated by Seiler (1976: see his fig. 143).
- Synonymy: Part of occipito-frontalis (Owen 1868; Macalister 1871; Sutton 1883).

***Auricularis posterior*** (adult Kidogo 0.47 g; infant Foyo 0.16 g; Figs. 2.1 and 2.9)

- Attachments in bonobos: From mastoid region of temporal bone region to posterior region of ear (our fetus, infant Foyo, adult Kidogo and adolescent Jasiri, and Miller's 1952 adult bonobo).
- Usual innervation: Posterior auricular branch of CN7 (Miller 1952: *P. paniscus*).
- Comparison with common chimpanzees: Similar to bonobos. The auricularis posterior of *Pan* is usually undivided but in the *P. troglodytes* specimens dissected by authors such as MacAlister (1871) and Pellatt (1979) the muscle had two bundles.
- Synonymy: Retrahens aurem (Wilder 1862; Macalister 1871); attrahens aurem plus retrahens aurem (Sonntag 1923).

***Intrinsic facial muscles of ear***

- Attachments in bonobos: See notes below.
- Usual innervation: Branches of CN7.
- Comparison with common chimpanzees: The intrinsic facial muscles of the ear of chimpanzees have been rarely described in detail in the literature, and they were difficult to analyse in our specimens, but some authors such as Huber (e.g., 1930, 1931), Ruge (e.g., 1887a, b) and Seiler (e.g., 1974, 1976) have examined these muscles in some detail in *P. troglodytes*. According to Seiler (1976), the *helicis major* (part of 'depressor helicis' *sensu* Ruge 1887a, b), *helicis minor*, *antitragicus*, *tragicus* ('tragus' sensu Gratiolet and Alix 1866 and Swindler and Wood 1973), *obliquus auriculae* (part of 'musculus auriculae proprius posterior' *sensu* Ruge 1887a, b and of 'transversi et obliqui' *sensu* Huber 1930, 1931) and *transversus auriculae* (part of 'musculus auriculae proprius posterior' *sensu* Ruge 1887a, b and of 'transversi et obliqui' sensu Huber 1930, 1931) are usually present in common chimpanzees, as is normally the case in modern humans. He suggested that the *incisurae terminalis* (*incisurae Santorini*) and the *intercartilagineus* are usually absent in common chimpanzees, but that the *pyramidalis auriculae* (*trago-helicinus sensu* Edgeworth 1935 and Seiler 1974) is usually present in these primates. He stated that the *depressor helicis* is inconstant in common chimpanzees. Edgeworth (1935) stated that the depressor helicis is usually missing in common chimpanzees; Ruge (1887a, b) reported a 'depressor helicis' in these primates, but the structure reported by him corresponds to the helicis major and helicis minor *sensu* Seiler (1974, 1976).

***Zygomaticus major*** (adult Kidogo 1.50 g; infant Foyo 0.42 g; Figs. 2.1, 2.2, 2.9 and 2.10)

- Attachments in bonobos: From zygomatic arch/bone (not from ear, although it often originates nearer to ear than is usually the case in modern humans) to corner of mouth, passing superficial to levator anguli oris facialis and blending with the orbicularis oris (our fetus, infant Foyo, adult Kidogo and adolescent Jasiri, and Miller's 1952 adult bonobo). The *risorius* is often (e.g., illustrations of Virchow 1915; Sonntag 1923, 1924; Burrows et al. 2006; see also descriptions of Hartmann 1886; Huber 1930, 1931; Seiler 1976; found in at least one side of the body of five of the eight common chimpanzees dissected by us), but not always, present as a distinct muscle in common chimpanzees, but was not present in the bonobo specimens where we analyzed this feature in detail (our fetus, infant Foyo, adult Kidogo and adolescent Jasiri). It was also not reported in Miller's 1952 adult bonobo.
- Usual innervation: Buccal branches of CN7 (Miller 1952: *P. paniscus*).
- Function: Moves the lips to expose the canine, premolar and molar teeth (Gratiolet and Alix 1866: *P. troglodytes*); elevates the lip corners superiorly and draws them laterally, increasing the angle of the mouth (Waller et al. 2006: *P. troglodytes*).
- Comparison with common chimpanzees: Similar to bonobos. Sonntag 1923 reported a 'zygomaticus major' running from the anterior

end of the zygoma to the angle of the mouth and a 'zygomaticus minor' running from the zygomatic bone and the temporalis fascia to the angle of the mouth in *P. troglodytes*. This corresponds to the posterior and anterior heads of the zygomaticus major, respectively, of some of the common chimpanzees dissected by us. The structure that Sonntag (1923) designated as a 'strong bundle from the orbital part of the orbicularis palpebrarum', running from the orbicularis oculi to the angle of the mouth, corresponds to the zygomaticus minor *sensu* the present study. Pellatt (1979) reported a *P. troglodytes* specimen with a 'zygomaticus major' and also a 'malaris' and a 'portion of the orbicularis oculi' that run from the orbital region to the angle of the mouth; the 'portion of the orbicularis oculi' corresponds to part of the zygomaticus minor of modern humans, while the 'malaris' might either correspond to a bundle of the zygomaticus major (in such cases the zygomaticus major would have two bundles), or to part of the zygomaticus minor. The former hypothesis (i.e., that the 'malaris' is part of the zygomaticus major) was supported by Burrows et al. (2006), who reported a zygomaticus major with a deep head attached caudally to the zygomatic arch and a superficial head attached via the skin over the superolateral portion of the face; the deep head fibres are arranged more transversely, whereas the fibres of the superficial head are more oblique. In these *P. troglodytes* cases the heads fuse approximately half of the way through their courses and attach together into the lateral-most portion of the orbicularis oris muscle at the modiolus, with a small attachment into the corresponding skin. In at least some of the common chimpanzees dissected by us (e.g., HU PT1, PFA 1077 and PFA UNC) the zygomaticus major is similar to that described by Burrows et al. (2006) in that it is divided into a more anterior, oblique and superficial portion that attaches to the skin over the zygomatic arch (just posteriorly to the attachment of the zygomaticus minor) and a more posterior, horizontal and deep portion that attaches to the bony zygomatic arch (its posterior region being relatively distant to the anterior region of the ear). The *depressor tarsi* (*preorbicularis*) is usually not present as a distinct muscle in common chimpanzees.

- Synonymy: Grand zygomatique (Gratiolet and Alix 1866); part of zygomatic mass (Macalister 1871); part or totality of zygomatici (Champneys 1872); zygomaticus plus part or totality of orbito-labialis (Sullivan and Osgood 1925); zygomaticus major and probably also zygomaticus minor (Sonntag 1923); zygomaticus (Miller 1952); zygomaticus inferior (Seiler 1976); zygomaticus major and probably also part or totality of malaris (Pellatt 1979).

***Zygomaticus minor*** (adult Kidogo 0.46 g; infant Foyo 0.29 g; Figs. 2.1, 2.2, 2.9 and 2.10)

- Attachments in bonobos: Mainly from zygomatic bone and temporalis fascia and orbicularis oculi (relatively far from ear and near eye, as is usually the case in modern humans) to corner of mouth and upper lip, being mainly superficial to levator anguli oris facialis, and blending with orbicularis oris and with zygomaticus major (our fetus, infant Foyo, adult Kidogo and adolescent Jasiri, and Miller's 1952 adult bonobo). The zygomaticus minor of bonobos seems to be different to that found in most common chimpanzees because it extends more superiorly, extending to the level of the eye but also posterior to the eye, attaching to the temporalis fascia (e.g., Figs. 2.1 and 2.2).
- Usual innervation: Buccal branches of CN7 (Miller 1952: *P. paniscus*).
- Comparison with common chimpanzees: Similar to bonobos, except the difference referred to, just above.
- Synonymy: Petit zygomatique (Gratiolet and Alix 1866); part of zygomatic mass (Macalister 1871); bundle from orbital part of orbicularis palpebrarum (Sonntag 1923); zygomatic head of quadratus labii superioris (Miller 1952); zygomaticus superior (Seiler 1976); part of orbicularis oculi, and possible part or totality of malaris (Pellatt 1979).

***Frontalis*** (Figs. 2.1, 2.2, 2.9 and 2.10)

- Attachments in bonobos: From galea aponeurotica to skin of eyebrow and orbicularis oculi, being fused with its counterpart at the midline (our fetus, infant Foyo, adult Kidogo and adolescent Jasiri, and Miller's 1952 adult bonobo).
- Usual innervation: Temporal branches of CN7 (Miller 1952: *P. paniscus*).
- Function: Waller et al. (2006) reported a 'pars medialis' and a 'pars lateralis' of the frontalis in *P. troglodytes* but as they recognized this is a functional and not a morphological distinction (i.e., the former elevates the medial portion of the eyebrow while the latter elevates the lateral and mid portion of the eyebrow).
- Comparison with common chimpanzees: Similar to bonobos.
- Synonymy: Part of occipito-frontalis (Owen 1830–1831; Macalister 1871; Sutton 1883); probably corresponds to part of frontalis sensu Pellatt (1979).

***Auriculo-orbitalis*** (adult Kidogo 0.30 g; Figs. 2.1, 2.2, 2.9 and 2.10)

- Attachments in bonobos: From anterior portion of ear to fascia temporalis (our fetus, infant Foyo, adult Kidogo and adolescent Jasiri, and Miller's 1952 adult bonobo).
- Usual innervation: Temporal branches of CN7 (Miller 1952: *P. paniscus*).
- Comparison with common chimpanzees: Similar to bonobos. In *Terminologia Anatomica* (1998) the *temporoparietalis* is considered to be a muscle that is usually present in modern humans (originating mainly from the lateral part of the galea aponeurotica and passing inferiorly to insert onto the cartilage of the auricle, in an aponeurosis shared with the other auricular muscles). However, according to authors such as Loth (1931) the temporoparietalis is usually absent as a distinct muscle in modern humans. According to Diogo et al. (2008, 2009), the temporoparietalis and *auricularis anterior* derive from the auriculo-orbitalis so when the temporoparietalis is not present as a distinct muscle these authors use the name auriculo-orbitalis to designate the structure that is often designated in the literature as 'auricularis anterior': that is, according to their terminology one can only use this latter name when the temporoparietalis is present. In various primates reported by Seiler (1976) including hominoids such as *H. moloch* and *Pongo pygmaeus*, he reported both a 'pars orbito-temporalis of the frontalis' and an 'auricularis anterior' attaching posteriorly onto the ear. That is, the structure that he designated as 'auricularis anterior' is differentiated from the auriculo-orbitalis, as is the case in various other primates, but according to him, and contrary to the condition in *Pan troglodytes* and *Gorilla gorilla* (e.g., fig. 143 of Seiler 1976), in hominoids such as *H. moloch* and *P. pygmaeus* there is 'still' a connection between the main body of the auriculo-orbitalis and the ear. One would think that the temporoparietalis of taxa such as modern humans would probably correspond to those remaining fibres of the auriculo-orbitalis that did not differentiate into the 'auricularis anterior', and this was what Jouffroy and Saban (1971) suggested in their study on the facial muscles of mammals. However, in at least some, if not all, taxa this is clearly not the case. For example, in fig. 143 of Seiler (1976) the structure that he designates as 'pot', which corresponds to the remaining fibres of the auriculo-orbitalis that do not form an 'auricularis anterior', does not correspond to the structure that is usually designated as temporoparietalis in modern human anatomical atlases, which as its name indicates usually runs mainly superoinferiorly from the parietal bone to the temporal region. Therefore, in order to be as consistent as possible with our previous studies and because the 'pars orbito-temporalis of the frontalis' and the 'auricularis anterior' sensu Seiler (1976) are very likely derived from the same anlage and are often related to each other, being often even continuous, we consider these two structures as parts/bundles of the auriculo-orbitalis *sensu* the present study

(see, e.g., fig. 74 of Seiler 1976). That is, the 'auricularis anterior' *sensu* Seiler 1976 is considered to be part of the auriculo-orbitalis *sensu* the present study, except in those few primates that have a distinct temporoparietalis (in those few primates the 'auricularis anterior' *sensu* Seiler 1976 is named by us as auricularis anterior, to clearly indicate that those primates have a distinct temporoparietalis). It is however possible, and in our opinion likely, that as suggested by Jouffroy and Saban (1971), the temporoparietalis of modern humans corresponds to the 'pars orbito-temporalis of the frontalis' *sensu* Seiler (1976) and that most of the confusion related to this issue is due to erroneous descriptions of the modern human temporoparietalis in anatomical atlases (which suggest that this is mainly a vertical muscle running superoinferiorly from the parietal bone to the temporal region, a description that does not match with the usual configuration of Seiler's 'pars orbito-temporalis of the frontalis' in other primates, which mainly runs horizontally to the region of the orbit to the region of the ear). If this is the case, then most primates would have a distinct temporoparietalis and a distinct auricularis anterior because both a 'pars orbito-temporalis of the frontalis' and an 'auricularis anterior' were reported in most primates by Seiler (1976). If this is so, then the 'pars orbito-temporalis of the frontalis' and the 'auricularis anterior' of those primates should be designated as temporoparietalis and as auricularis anterior, respectively; we plan to return to this subject in the future. The right side of our *P. troglodytes* PFA UNC specimen had a thin and broad muscle that was just medial (deep) to the auricularis superior and that connected the superior portion of the ear to the galea aponeurotica, extending superiorly to the superior attachment of the auricularis superior onto this latter aponeurosis. This thin and broad muscle seems to correspond to the temporoparietalis of modern humans (i.e., in this case there was a temporoparietalis, an auricularis superior *and* an auricularis anterior). However, on the left side of that PFA UNC specimen, and on both sides of the other common chimpanzees dissected by us, we only found an auricularis superior and a single, broad muscle that extends anteriorly to, and is blended with, the frontalis. This configuration seems to us to correspond to the auriculo-orbitalis of other chimpanzees and other primates.

- Synonymy: Probably corresponds to the auricularis anterior *sensu* Gratiolet and Alix (1866), Virchow (1915), Sonntag (1923, 1924), Miller (1952), Swindler and Wood (1973), Seiler (1974), Gibbs (1999), Burrows et al. (2006) and Burrows (2008) and to the attrahens aurem *sensu* Wilder (1862) and Macalister (1871); might correspond to part of the attolens/auricularis anterior *sensu* Sonntag (1923); probably corresponds to part of frontalis *sensu* Pellatt (1979). The *zygomatico-auricularis* is not present as a distinct muscle.

***Auricularis superior*** (adult Kidogo 0.46 g; infant Foyo 0.23 g; Figs. 2.1, 2.2, 2.9 and 2.10)

- Attachments in bonobos: From superior margin of ear to galea aponeurotica (our fetus, infant Foyo, adult Kidogo and adolescent Jasiri, and Miller's 1952 adult bonobo).
- Usual innervation: Temporal branches of CN7 (Miller 1952: *P. paniscus*).
- Comparison with common chimpanzees: Similar to bonobos.
- Synonymy: Atollens aurem (Wilder 1862; Macalister 1871); temporoparietalis (Virchow 1915); part or totality of attolens aurem (Sonntag 1923); part or, more likely, totality of auricularis superior et anterior (Pellatt 1979).

***Orbicularis oculi*** (adult Kidogo 0.88 g; Figs. 2.2, 2.9 and 2.10)

- Attachments in bonobos: From a continuous bony attachment around the orbit, including lacrimal bone and frontal process of maxilla and nasal processes of frontal bone, and from medial palpebral ligament, to skin around eye and to lateral palpebral raphe, being usually divided into a pars palpebralis and a pars

orbitalis, as in modern humans (our fetus, infant Foyo, adult Kidogo and adolescent Jasiri, and Miller's 1952 adult bonobo).
- Usual innervation: Branches of CN7.
- Function: Inferior portion of pars orbitalis elevates the infraorbital triangle (or equivalent area) superiorly and medialwards; inferior portion lowers the mid and lateral portion of the eyebrows (but perhaps this is also done by the depressor supercilii instead; Waller et al. 2006: *P. troglodytes*).
- Comparison with common chimpanzees: Similar to bonobos. The orbicularis oculi of common chimpanzees is often differentiated into a pars orbitalis and a pars palpebralis, which is in turn subdivided into a pars profunda ('pars lacrimalis') and a fasciculus ciliaris (according to Sonntag 1924 the fasciculus ciliaris is not differentiated in a few common chimpanzees). Seiler (1971b, 1976) states that, within the Catarrhini, an independent muscle *infraorbitalis* is present in *Macaca*, *Pongo*, *Pan* and *Homo*; however, this structure seems to correspond to part of the orbicularis oculi and/or of the levator labii superioris alaeque nasi *sensu* the present study.
- Synonymy: Orbiculaire plus palpébral (Gratiolet and Alix 1866); orbicularis palpebrarum and tensor tarsi (Macalister 1871), the latter structure corresponding to the pars profunda, or lacrimalis, *sensu* the present study; orbicularis palpebrarum (Sutton 1883). The *zygomatico-orbicularis* is missing in chimpanzees.

***Depressor supercilii*** (adult Kidogo 0.09 g; infant Foyo 0.01 g; Figs. 2.2, 2.3, 2.9 and 2.10)

- Attachments in bonobos: From ligamentum palpebrale mediale/glabella to eyebrow (our fetus, infant Foyo, adult Kidogo and adolescent Jasiri; see notes below about Millers 1952 adult bonobo).
- Usual innervation: Branches of CN7.
- Comparison with common chimpanzees: Similar to bonobos. Authors such as Gratiolet and Alix (1866), Macalister (1871), Sonntag (1923), Miller (1952), and Pellatt (1979) did not describe a depressor supercilii in the *P. troglodytes* and *P. paniscus* specimens dissected by them, but as noted by Seiler (1976) and Burrows et al. (2006) and corroborated by our dissections, common chimpanzees do usually have this muscle.

***Corrugator supercilii*** (adult Kidogo 0.09 g; infant Foyo 0.01 g; Figs. 2.2, 2.3, 2.9 and 2.10)

- Attachments in bonobos: From medial angle of supraorbital ridge of frontal bone to deep surface of skin of eyebrow region (our fetus, infant Foyo, adult Kidogo and adolescent Jasiri and Miller's 1952 adult bonobo), being fused with the orbicularis oculi and the depressor supercilii in our fetus.
- Usual innervation: Temporal branches of CN7 (Miller 1952: *P. paniscus*).
- Comparison with common chimpanzees: Similar to bonobos.
- Synonymy: Sourcilier (Gratiolet and Alix 1866).

***Levator labii superioris*** (adult Kidogo 0.78 g; infant Foyo 0.10 g; Figs. 2.2, 2.3, 2.9 and 2.10)

- Attachments in bonobos: From infraorbital region to upper lip, being fused with the orbicularis oris (our fetus, infant Foyo, adult Kidogo and adolescent Jasiri and Miller's 1952 adult bonobo).
- Usual innervation: Buccal branches of CN7 (Miller 1952: *P. paniscus*).
- Function: Elevates the upper lip (Waller et al. 2006: *P. troglodytes*).
- Comparison with common chimpanzees: Similar to bonobos.
- Synonymy: Releveur propre de la lèvre supérieure (Gratiolet and Alix 1866); part of levator labii superioris (Macalister 1871); part or totality of maxillo-labialis (Edgeworth 1935); infraorbital head of quadratus labii superioris (Miller 1952).

***Levator labii superioris alaeque nasi*** (adult Kidogo 0.59 g; infant Foyo 0.11 g; Figs. 2.1, 2.2, 2.9 and 2.10)

- Attachments in bonobos: From region near inner angle of eye and ala of nose, frontal process of maxilla and/or lacrimal bone near ligamentum infraorbitale mediale, to upper lip, being fused with orbicularis oris and levator labii superioris (our fetus, infant Foyo, adult Kidogo and adolescent Jasiri and Miller's 1952 adult bonobo).
- Usual innervation: Buccal branches of CN7 (Miller 1952: *P. paniscus*).
- Function: Wrinkles the skin surrounding the nose (Waller et al. 2006: *P. troglodytes*).
- Comparison with common chimpanzees: Virchow (1915) and Gibbs (1999) stated that in modern humans the levator labii superioris alaeque nasi originates mainly from the frontal process of the maxilla around the medial angle of the orbital opening, while in *P. troglodytes* it takes origin from the midline of the nasal dorsum. However, as reported by authors such as Gratiolet and Alix (1986) and observed in our dissections, in common chimpanzees the muscle often also, or exclusively, originates from the inner angle of the eye, the frontal process of the maxilla and/or the lacrimal bone near the ligamentum infraorbitale mediale.
- Synonymy: Releveur commun de l'aile du nez et de la lèvre supérieure (Gratiolet and Alix 1866); part of levator labii superioris (Macalister 1871); angular head of quadratus labii superioris (Miller 1952).

***Procerus*** (adult Kidogo 0.07 g, i.e., this is weight for only one side of head; Figs. 2.2 and 2.10)

- Attachments in bonobos: From frontalis muscle and nasal bones to skin on dorsomedial region of nose, being also fused with the levator labii superioris alaeque nasi and with the procerus of the other side of the body (our fetus, infant Foyo, adult Kidogo and adolescent Jasiri and Miller's 1952 adult bonobo).
- Usual innervation: Buccal branches of CN7 (Miller 1952: *P. paniscus*).
- Function: Depresses the medial corners of the eyebrows (Waller et al. 2006: *P. troglodytes*).
- Comparison with common chimpanzees: Similar to bonobos. According to Seiler (1971a, 1976) the *depressor glabellae* is usually present as a distinct muscle in common chimpanzees, whereas the procerus is inconstant. However, the 'depressor glabellae' is often considered in the literature to be part of the procerus (see, e.g., Terminologia Anatomica 1998); in the common chimpanzees dissected by us and by authors such as Gratiolet and Alix (1866), Pellatt (1979), and Burrows et al. (2006) the procerus was present.
- Synonymy: Pyramidalis nasi (Gratiolet and Alix 1866; Sutton 1883; Sonntag 1923); depressor glabellae (Virchow 1915); naso-labialis superficialis, pyramidalis narium, frontalis pars per dorsum nasi ducta, dorsalis narium, retractor naso-labialis or levator naso-labialis vestibularis (Jouffroy and Saban 1971); procerus plus part or totality of depressor glabellae (Seiler 1971a, 1976).

***Buccinatorius*** (adult Kidogo 5.10 g; Figs. 2.3 and 2.4)

- Attachments in bonobos: From pterygomandibular raphe (thus being partially fused with the constrictor pharyngis superior muscle), infero-lateral surface of maxilla, and supero-lateral border of mandible, mainly to angle of mouth and upper and lower lips, being fused with the orbicularis oris (our fetus, infant Foyo, adult Kidogo and adolescent Jasiri and Miller's 1952 adult bonobo).
- Usual innervation: Buccal branches of CN7 (Miller 1952: *P. paniscus*).
- Comparison with common chimpanzees: Similar to bonobos.

***Nasalis*** (adult Kidogo pars alaris 0.17 g, pars transversa 0.07 g; infant Foyo pars alaris + pars transversa =0.01 g; Figs. 2.3, 2.9 and 2.10)

- Attachments in bonobos: From maxilla, deep to orbicularis oris, to lateral margin of nose (pars transversa) and lateral portion of inferior margin of ala of nose (pars alaris) (our fetus, infant Foyo, adult Kidogo and adolescent Jasiri and Miller's 1952 adult bonobo). In Miller's 1952 adult bonobo, the muscle is fused with the levator labii superioris alaeque nasi, orbicularis oris and procerus.

- Usual innervation: Buccal branches of CN7 (Miller 1952: *P. paniscus*).
- Comparison with common chimpanzees: Similar to bonobos. Seiler (1970, 1971a, 1976) describes a 'nasalis' and a *subnasalis* in various catarrhines. The 'subnasalis' and 'nasalis' *sensu* Seiler could correspond to the pars alaris and pars transversa of the nasalis of modern human anatomy, respectively (compare, e.g., fig. 141 of Seiler 1976 to plate 26 of Netter 2006). However, according to Seiler the 'subnasalis' while often present in chimpanzees is usually missing in *Homo* and *Gorilla*. This indicates that the subnasalis does not correspond to the pars alaris of the nasalis *sensu* the present study, because this latter structure is usually found in modern humans. Be that as it may, as other authors do not refer to a 'subnasalis' muscle and as we also did not find a distinct separate 'subnasalis' in the chimpanzees dissected by us, this structure probably corresponds to part of the nasalis and/or of the orbicularis oris *sensu* the present study. Seiler (1970, 1971a, 1976) also describes a *depressor septi nasi* and a *musculus nasalis impar* in a few catarrhines, but not in common chimpanzees. One hypothesis is that the 'nasalis impar' corresponds to part or to the totality of the depressor septi nasi that is illustrated in a few atlases of modern human anatomy as a vertical muscle that lies on the midline and that attaches mainly onto the inferomesial margin of the nose (compare fig. 145 of Seiler 1976 with plate 26 of Netter 2006). However, most atlases of modern human anatomy show two depressor septi nasi muscles, one in each side of the body, running obliquely (superomedially) from the upper lip to a more medial part of the inferior region of the nose. Thus, the 'nasalis impar' *sensu* Seiler (1976) corresponds to an additional midline muscle that is only inconstantly present in catarrhines, while the 'depressor septi nasi'' *sensu* Seiler (1976) is effectively similar to the depressor septi nasi shown in most atlases of modern human anatomy. Table 3 of Seiler (1979) refers to a *labialis superior profundus* and this suggests that according to him the 'depressor septi nasi' and the 'labialis superior profundus' are not the same structure, whereas Lightoller (1928a, 1934) suggests that these are one and the same structure (Seiler 1976 clearly shows both these structures in various primates, e.g., in his fig. 145 of *Gorilla*, the 'labialis superior profundus' probably corresponding to part of the orbicularis oris *sensu* the present study). Jouffroy and Saban (1971) designate the nasalis as the 'naso-labialis profundus pars anterior' and the depressor septi nasi as the 'naso-labialis profundus pars mediana'; this suggests that these authors consider these two muscles derive from the same structure. In their fig. 471, Jouffroy and Saban (1971) designate the 'labii profundus superioris' as depressor septi nasi, thus suggesting that the 'labii profundus superior' could correspond to the depressor septi nasi of modern humans as suggested by Lightoller (1928a, 1934). However, our comparisons and dissections suggest that the homologies proposed by Seiler (1976) are somewhat doubtful and that: (1) the 'nasalis' *sensu* Seiler likely corresponds to part or the totality of the nasalis *sensu* the present study; (2) the pars transversa of the nasalis of modern humans might correspond to the 'subnasalis' *sensu* Seiler (see, e.g., fig. 141 of Seiler 1976; however, Seiler stated that the 'subnasalis' is missing in modern humans), to the depressor septi nasi *sensu* Seiler 1976 (see, e.g., fig. 145 of Seiler 1976), and/or to part of the nasalis *sensu* the present study (see, e.g., fig. 145 of Seiler 1976); (3) the depressor septi nasi of modern humans corresponds to the 'depressor septi' nasi *sensu* Seiler; therefore, the 'depressor septi nasi' *sensu* Seiler does not correspond to part of the 'labialis superior profundus' *sensu* Seiler, as suggested by Lightoller (1928, 1934; see, e.g., fig. 141 of Seiler 1976), nor to the 'nasalis impar' *sensu* Seiler (in fact, Seiler stated that the 'nasalis impar' is missing in modern humans, which do usually have a depressor septi nasi *sensu* the present study).

- Synonymy: Myrtiformis plus transversus (Gratiolet and Alix 1866), which correspond respectively to the pars alaris and pars transversa *sensu* the present work; includes the compressor nasi *sensu* Macalister 1871, which corresponds to the pars transversa *sensu* the present work; part of nasal muscles (Sonntag 1923); part of naso-labialis (Miller 1952).

***Depressor septi nasi*** (adult Kidogo 0.08 g; Fig. 2.3)

- Attachments in bonobos: From maxilla, deep to orbicularis oris, to inferior region of nose (our fetus, infant Foyo, adult Kidogo and adolescent Jasiri; see notes below about Miller's 1952 adult bonobo).
- Usual innervation: Branches of CN7.
- Comparison with common chimpanzees: Authors such as Ruge (1887a, b), Macalister (1871), Sonntag (1923), Sullivan and Osgood (1925), Miller (1952) and Pellatt (1979) did not refer to the depressor septi nasi in their descriptions of chimpanzees. However Burrows et al. (2006) did find this muscle in their careful dissections of common chimpanzees, and we also found it in at least some of the common chimpanzees dissected by us, with a configuration similar to that found in the bonobos dissected by us.
- Synonymy: Part of nasolabialis (Miller 1952).

***Levator anguli oris facialis*** (adult Kidogo 0.15 g; infant Foyo 0.01 g; Fig. 2.3)

- Attachments in bonobos: From canine fossa of maxilla to angle of mouth, not being fused with the buccinatorius (our fetus, infant Foyo, adult Kidogo and adolescent Jasiri, and Miller's 1952 adult bonobo).
- Usual innervation: Buccal branches of CN7 (Miller 1952: *P. paniscus*).
- Comparison with common chimpanzees: Similar to bonobos. As proposed by Diogo et al. (2008) and Diogo and Abdala (2010), we use the name 'levator anguli oris facialis' here (and not the name 'levator anguli oris', as is usual in atlases of modern human anatomy) to distinguish this muscle from the *levator anguli oris mandibularis* (which is usually also designated as 'levator anguli oris' in the literature) found in certain reptiles, which is part of the mandibular (innervated by CN5) and not of the hyoid (innervated by CN7) musculature. The *anomalus maxillae*, *anomalus nasi* and *anomalus menti*, found as anomalies in a few modern humans, are usually missing in chimpanzees.
- Synonymy: Caninus (Gratiolet and Alix 1866; Miller 1952; Burrows 2008); part of caninus *sensu* Seiler (1976), which also includes the depressor anguli oris *sensu* the present study.

***Orbicularis oris*** (adult Kidogo 13.14 g; infant Foyo 0.76 g; Figs. 2.2, 2.9 and 2.10)

- Attachments in bonobos: From skin, fascia and adjacent regions of lips to skin and fascia of lips (our fetus, infant Foyo, adult Kidogo and adolescent Jasiri, and Miller's 1952 adult bonobo).
- Usual innervation: Buccal branches of CN7 (Miller 1952: *P. paniscus*).
- Function: Reduces the lip aperture and funnels/protrudes the lips (Waller et al. 2006: *P. troglodytes*).
- Comparison with common chimpanzees: Similar to bonobos. Huber (1931) stated that the muscles *rectus labii inferioris* and *rectus labii superioris*, although poorly differentiated, are present in *P. troglodytes*, but they are more differentiated in modern humans. He also stated that the muscles *incisivus labii superioris* (sometimes designated as *pars supralabialis of the buccinatorius*) and *incisivus labii inferioris* are found in modern humans, but he suggested they are not found in *P. troglodytes*. However, Seiler (1970, 1971a) describes a *cuspidator oris* in common chimpanzees. This structure, which was designated as *labialis superior profundus* by Seiler 1976, probably corresponds to the *incisivus labii superior sensu* Lightoller (1928, 1934, 1939) and, thus to part of the orbicularis oris *sensu* the present study. Seiler (1976) also describes a *labialis inferior profundus* in common chim-

panzees, which thus probably corresponds to the *incisivus labii inferioris sensu* Lightoller (1928, 1934, 1939) and to part of the orbicularis oris *sensu* the present study. Lightoller (1925) stated that a differentiated pars marginalis of the orbicularis oris is present in modern humans but not in *P. troglodytes*, the presence of this pars marginalis being probably related to the evolution of speech in modern humans. Pellatt (1979) also stated that the 'peripheral' and 'marginal' parts of the orbicularis oris are not clearly differentiated in common chimpanzees.
- Synonymy: Orbiculaire des lèvres (Gratiolet and Alix 1866); orbicularis oris plus incisive muscles (Sonntag 1923); orbicularis oris plus rectus labii inferioris and rectus labii superioris (Huber 1930, 1931).

***Depressor labii inferioris*** (adult Kidogo 3.23 g; infant Foyo 0.40 g; Figs. 2.1, 2.2, 2.9 and 2.10)

- Attachments in bonobos: From platysma myoides and mandible to lower lip, being fused with platysma myoides and orbicularis oris, as well as with its counterpart at the midline (our fetus, infant Foyo, adult Kidogo and adolescent Jasiri, and Miller's 1952 adult bonobo).
- Usual innervation: Mandibular branches of CN7 (Miller 1952: *P. paniscus*).
- Function: Depresses the medial portion of the lower lip (Waller et al. 2006: *P. troglodytes*).
- Comparison with common chimpanzees: Simila to bonobos. Some authors (e.g., Gratiolet and Alix 1866; Champneys 1872) have suggested that in common chimpanzees the depressor labii inferioris is not present as a distinct muscle. However, the descriptions of Hartmann (1886), Sonntag (1923), Miller (1952), Seiler (1976) and Pellatt (1979) and Burrows et al. (2006) and our own dissections show that the depressor labii inferioris is usually present as a distinct muscle in common chimpanzees. The depressor labii inferioris usually arises from the mandible and inserts onto the skin of the lower lip, being mainly blended with its counterpart as well as with the orbicularis oris, depressor anguli oris, mentalis and platysma myoides (it is mainly distinct from the platysma myoides because, in lateral view, its fibres run more obliquely than the mainly anteroposteriorly oriented anterior fibres of this latter muscle).
- Synonymy: Quadratus labii inferioris (Sonntag 1923, 1924; Miller 1952; Jouffroy and Saban 1971).

***Depressor anguli oris*** (adult Kidogo 1.07 g; infant Foyo 0.38 g; Figs. 2.1, 2.2, 2.9 and 2.10)

- Attachments in bonobos: From angle of mouth to fascia of platysma myoides and to mandible, meeting its counterpart at the midline and being fused with platysma cervicale and orbicularis oris (our fetus, infant Foyo, adult Kidogo and adolescent Jasiri, and Miller's 1952 adult bonobo).
- Usual innervation: Mandibular branches of CN7 (Miller 1952: *P. paniscus*).
- Function: Pulls the corner of the mouth downward (Waller et al. 2006: *P. troglodytes*).
- Comparison with common chimpanzees: Similar to bonobos. Authors such as Gratiolet and Alix (1866), Huber (1931) and Jouffroy and Saban (1971) stated that in some common chimpanzees the depressor anguli oris extends inferomedially to the inferior margin of the mandible to meet its counterpart in the ventral midline, thus forming a *transversus menti* (Loth 1931 suggested that this structure is present in 18% of chimpanzees).
- Synonymy: Triangularis (Gratiolet and Alix 1866; Sonntag 1923, 1924: Edgeworth 1935; Miller 1952); part of caninus *sensu* Seiler (1976), which also includes the levator anguli oris facialis.

***Mentalis*** (adult Kidogo 1.30 g; infant Foyo 0.01 g; Fig. 2.3)

- Attachments in bonobos: From mandible to skin below lower lip, being mainly deep to orbicularis oris and to depressor labii inferioris, meeting with its counterpart in the midline (our fetus, infant Foyo, adult Kidogo and adolescent Jasiri). Not described in Miller's (1952) adult bonobo, but very likely present in that specimen, as it is

present in all *Pan* dissected by us and consistently found in other primates, including modern humans.
- Usual innervation: Branches of CN7.
- Function: Pushes the skin of the chin area superiorly (Waller et al. 2006: *P. troglodytes*).
- Comparison with common chimpanzees: Similar to bonobos, although it should be noted that in common chimpanzees only sometimes there is a fusion with its counterpart in the midline (e.g., Gibbs 1999), while such a fusion was seemingly present in all bonobos dissected by us.
- Synonymy: Elevator labii inferioris proprius *sensu* Tyson (1699).

## 2.3 Branchial Musculature

***Stylopharyngeus (adult Kidogo 0.41 g; infant Foyo 0.22 g;*** Fig. 2.6)

- Attachments in bonobos: From styloid process and/or adjacent regions of cranium to the pharyngeal wall (our fetus, infant Foyo, adult Kidogo and adolescent Jasiri). We could not discern if it also inserted onto the greater horn of the hyoid bone and/or the thyroid cartilage, passing between the middle and superior pharyngeal constrictors.
- Usual innervation: Glossopharyngeal nerve CN9 (Sonntag 1923: *P. troglodytes*).
- Comparison with common chimpanzees: Similar to bonobos. The *ceratohyoideus* and *petropharyngeus* are not present as distinct muscles in common chimpanzees.

***Trapezius*** (infant Foyo 7.25 g; adolescent Jasiri 90.48 g; adult Kidogo 169.73 g; fetus 2.62 g; Figs. 2.4, 2.12 and 2.13)

- Attachments in bonobos: From skull (occipital protuberance and superior nuchal line), ligamentum nuchae and vertebral column (C1-T9 in adult bonobo of Miller 1952; C4-T9 in our fetus; C6-T10 in our adolescent Jasiri and infant Foyo; C5-T9 in our adult Kidogo; C6-T9 in infant Etje) to spine and acromion of scapula and to clavicle (distal end in our fetus; lateral 1/3 in adult bonobo of Miller 1952 and our adolescent Jasiri and adult Kidogo and infants Foyo and Etje). In our fetus there is a partial fusion with the latissimus dorsi proximally, in the region of its origin from the thoracic vertebrae; the trapezius is also fused with the trapezius of the other side of the body.
- Usual innervation: Spinal accessory nerve and a branch from the anterior division of the third cervical nerve through the brachial plexus (Miller 1952: *P. paniscus*) [*P. troglodytes*: XI and C3 in one specimen and XI, C2, C3 and C4 in another specimen (Schück 1913); spinal accessory nerve and third and fourth cervical nerves (Sonntag 1923); motor innervation from the spinal accessory nerve (Swindler and Wood 1973)].
- Function: Pars ascendens of trapezius exhibited EMG activity during arm-raising (Tuttle and Basmajian 1976: *P. troglodytes*); EMG indicated that the cranial trapezius was not involved in arm-raising, its recruitment being instead closely tied to head position (Larson et al. 1991: *P. troglodytes*).
- Comparison with common chimpanzees: In *P. troglodytes* it usually runs from skull (occipital protuberance and superior nuchal line) and vertebral column (e.g., extending posteriorly to T12 according to Gratiolet and Alix 1866 and to Schück 1913 and even 'T13' according to Sonntag 1923, but only extending to T10 according to Macalister 1871 and Swindler and Wood 1973 and to T8 according to Stewart 1936) to the spine and acromion of the scapula and to the lateral 1/3, or more than the lateral 1/3, of the clavicle (N.B. the ligamentum nuchae is usually not present as a distinct, well-defined structure in common chimpanzees). As in modern humans, in common chimpanzees the trapezius is usually differentiated into a pars descendens, a pars transversa and a pars ascendens.
- Synonymy: Cucullaris (Tyson 1699).

***Sternocleidomastoideus*** (adult Kidogo sternomastoid head 29.95 g, cleidomastoid head 12.64 g;

infant Foyo cleidomastoid + sternomastoid heads = 5.00 g; Figs. 2.4, 2.9, 2.10, 2.11 and 2.13)

- Attachments in bonobos: The caput sternomastoideum runs from sternum to mastoid region and lateral portion of superior nuchal line; the caput cleidomastoideum usually runs from medial portion of clavicle (medial 1/3 in our fetus and adult Kidogo and adolescent Jasiri; medial 1/4 on the right side of our infant Foyo) to mastoid region. Contrary to the other bonobos dissected by us, in our fetus there was no differentiation between the two heads (Fig. 2.13), probably reflecting the normal earlier developmental configuration of this muscle, while on the right side of our infant Foyo there was a variant: the muscle was not attached to the clavicle, i.e., it inserted exclusively onto the sternum (Fig. 2.12). Another variant was found seen in Miller's (1952) adult bonobo: there was a cleidomastoid head from mastoid process to sternoclavicular joint and adjacent clavicle, a sternomastoid head from manubrium to mastoid process, and in addition a 'sterno-occipital' head from manubrium to superior nuchal line, which was only well developed in one side of that adult bonobo. This *sterno-occipital* head very likely corresponds to the *cleido-occipitalis* of other mammals, which is also found as a variant in common chimpanzees (see below), supporting the idea that the configuration of the cleido-occipitalis is flexible, and that this structure might well be the result of a convergent/parallel derivation from the sternocleidomastoideus, in different taxa (e.g., sometimes it inserts onto the clavicle, as its name indicates, and sometimes it inserts onto the sternum, as seen in Miller's 1952 adult bonobo).
- Usual innervation: Spinal accessory nerve and a branch from the anterior (ventral) division of the second cervical nerve through the cervical plexus (Miller 1952: *P. paniscus*) [*P. troglodytes*: XI in one specimen and XI and C2 in another specimen dissected by Schück 1913].
- Comparison with common chimpanzees: In common chimpanzees the caput cleidomastoideum runs from the medial portion of the clavicle (medial 1/4 according to Gratiolet and Alix 1866, medial 1/3 according to Sonntag 1923 and to our dissections of the *P. troglodytes* specimens PFA 1016, PFA 1009 and PFA 1051) to the mastoid region; in a specimen dissected by Sutton 1883 there was also an insertion onto the atlas. In a few common chimpanzees there is a structure that is sometimes designated as muscle *cleido-occipitalis* (e.g., Sonntag 1923) but this structure is usually deeply blended with the sternocleidomastoideus, and does not form a distinct, separate, muscle, i.e., it is instead a bundle of the sternocleidomastoideus, which is designated in the present work as *caput cleido-occipitale*). According to Wood (1870) and to the recent work of Mustafa (2006), only about 36% and 33% of modern humans, respectively, have a caput cleido-occipitale of the sternocleidomastoideus. Among the common chimpanzees dissected we only found a well-defined caput 'cleido-occipitale' of the sternocleidomastoideus on the left side of the infant chimpanzee PFA 1077, running from about the medial 1/3 of the clavicle to the occipital region, passing posteriorly and deep to the sternomastoid head and superficially to both the accessory nerve (XI) and to the cleidomastoid head, which lies deep to this nerve. Superiorly the cleido-occipitale and sternomastoid heads were blended at their attachments onto the occipital region, but the cleidomastoid head is well-separated from them by fascia and by the accessory nerve. Inferiorly these three structures are somewhat blended with each other. On the right side of this PFA 1077 specimen the cleido-occipitale and sternomastoid heads are almost completely fused to each other, the structure formed by them thus being superficial to the accessory nerve and running from the clavicle and sternum to the occipital region of the skull. This indicates that the modern human sternomastoid head might correspond to the sternomastoid plus cleido-occipitale heads of the left side of that PFA 1077 specimen (for more discussion of this issue see Diogo and Wood 2012).

***Constrictor pharyngis medius*** (adult Kidogo 1.89 g; infant Foyo 0.12 g; Fig. 2.6)

- Attachments in bonobos: From dorsal raphe of pharyngeal wall (where it meets its counterpart) to angle between greater and lesser horns of this bone. Thus, there is a pars ceratopharyngea, but no pars chondropharyngea, unless one considers that the small portion of the muscle that inserts onto the body of the hyoid bone includes the pars chondropharyngea (our fetus and infant Foyo and adult Kidogo and adolescent Jasiri). As a variant, in our adult Kidogo the muscle was fused with the stylopharyngeus (Fig. 2.6).
- Usual innervation: Data are not available.
- Comparison with common chimpanzees: Similar to bonobos. Himmelreich (1977) shows in his fig. 11 a *P. troglodytes* specimen with a 'hyopharyngeus' (which corresponds to part or to the totality of the middle constrictor *sensu* the present study) that mainly inserts onto the greater horn of the hyoid bone. His fig. 10 of *Macaca mulatta* shows, instead, a constrictor pharyngis medius divided into a pars ceratopharyngea and a thin pars chondropharyngea going to the greater and lesser horns of the hyoid bone, as is often the case in various other primates including modern humans. This indicates that at least in Himmelreich's common chimpanzee specimen there is no pars distinct chondropharyngea (it was not possible to discern if this latter structure was or not present in the common chimpanzees dissected by us).
- Synonymy: Part or totality of hyopharyngeus (Himmelreich 1977).

***Constrictor pharyngis inferior*** (adult Kidogo 3.77 g; infant Foyo 0.07 g; Fig. 2.6)

- Attachments in bonobos: From dorsal raphe of pharyngeal wall (where it meets its counterpart) to thyroid (pars thyropharyngea) and cricoid (pars cricopharyngea) cartilages (our fetus and infant Foyo and adult Kidogo and adolescent Jasiri).
- Usual innervation: Data not available.
- Comparison with common chimpanzees: Similar to bonobos, but in some cases apart from the insertions onto the cricoid and thyroid cartilages, there is also an insertion onto the fist tracheal ring (e.g., Gratiolet and Alix 1866).
- Synonymy: Thyropharyngeus plus cricopharyngeus (Starck and Schneider 1960; Himmelreich 1977)

***Cricothyroideus*** (adult Kidogo 0.39 g; infant Foyo 0.08 g; Figs. 2.8 and 2.11)

- Attachments in bonobos: From cricoid cartilage to thyroid cartilage, specifically to its inferior and inferomedial portion as well to a small area on the lateral surface of its inferior horn (our fetus and infant Foyo and adult Kidogo and adolescent Jasiri). It is divided into a straight, or anterior, portion (pars recta) and an oblique, or posterior, portion (pars interna) in our infant Foyo and adult Kidogo (Fig. 2.8) and adolescent Jasiri, but not in our fetus, probably reflecting the earlier developmental stage of this latter specimen. In our infant Foyo, the cricothyroideus did not contact its counterpart in the midline; we could not discern if such a condition was also found, or not, in our fetus, adult Kidogo and adolescent Jasiri.
- Usual innervation: Data not available for bonobos (*P. troglodytes*: external branch of the superior laryngeal nerve (Jordan 1971a, b, c)].
- Comparison with common chimpanzees: Similar to bonobos, but the muscle does contact, sometimes, its counterpart in the midline. The cricothyroideus is usually divided into a more superficial and less oblique pars recta and a deeper and more oblique pars obliqua, as is the case in modern humans. According to Gratiolet and Alix (1866), Kohlbrügge (1896), Duckworth (192), Kelemen (1948, 1969), Starck and Schneider (1960), Saban (1968), Jordan (1971a, b, c), the pars interna is present in *P. troglodytes*, whereas Harrison (1995) stated that he only found a pars interna in one of the eight common chimpanzees dissected by him. However, in at least four of the six *P. troglodytes* specimens in which we analysed this feature in detail, there was a pars interna (i.e., PFA 1016, PFA 1009, PFA 1051 and HU

PT1, the other two specimens being PFA 1077 and PFA UNC) similar to that we found in the nonfetal bonobos dissected by us. The ***thyroideus transversus*** is usually not present in chimpanzees.
- Synonymy: Cricothyreoideus anticus (Kohlbrügge 1896).

***Constrictor pharyngis superior*** (adult Kidogo 3.46 g; infant Foyo 0.47 g; Fig. 2.7)

- Attachments in bonobos: From pterygoid plate/hamulus (pars pterygopharyngea), pterygomandibular ligament/raphe (pars buccopharyngea), mandibular region near mylohyoid line (pars mylopharyngea), and tongue (pars glossopharyngea), and often basiocciput, to dorsal raphe of pharyngeal wall, meeting its counterpart in midline (our fetus and infant Foyo and adult Kidogo and adolescent Jasiri).
- Usual innervation: Data are not available.
- Comparison with common chimpanzees: Similar to bonobos. The *pterygopharyngeus* is usually not present as a distinct muscle in chimpanzees.
- Synonymy: Cephalopharyngeus or gnathopharyngeus (Himmelreich 1977).

***Palatopharyngeus*** (infant Foyo 0.13 g)

- Attachments in bonobos: From soft palate to pharyngeal wall (our fetus and infant Foyo and adult Kidogo and adolescent Jasiri).
- Usual innervation: Data are not available.
- Comparison with common chimpanzees: Similar to bonobos. This muscle was reported but not described in detail by Sonntag (1923), and reported by Gratiolet and Alix (1866) as a muscle running from the soft palate to the thyroid cartilage. In our *P. troglodytes* specimen PFA 1016 the muscle runs mainly from the soft palate to the to the pharyngeal wall, as is usually the case in modern humans, and is seemingly divided into a main body and a *sphincter palatopharyngeus* (*palatopharyngeal sphincter*, or *Passavant's ridge*), as is often also the case in modern humans (see, e.g., pl. 65 of Netter 2006).
- Synonymy: Pharyngostaphylins (Gratiolet and Alix 1866).

***Musculus uvulae***

- Attachments in bonobos: We could not discern if the muscle was present, or not, in the bonobos dissected by us.
- Usual innervation: Data are not available.
- Comparison with common chimpanzees: This muscle was reported in *P. troglodytes* by Gratiolet and Alix (1866) as a muscle running from the posterior nasal spine of the palate (as it usually does in modern humans) to the midline, finishing in a membrane without attaching to the uvula. It was also reported by Sonntag (1923) as a muscle that does not really attach directly onto the uvula. The common chimpanzee specimens dissected by us seem to have a musculus uvulae.
- Synonymy: Azygos uvulae (Sonntag 1923, 1924); medialis veli palatini (Himmelreich 1971).

***Levator veli palatini*** (adult Kidogo 0.42 g; Fig. 2.7)

- Attachments in bonobos: More horizontal than in modern human adults, running from Eustachian region and adjacent areas of neurocranium to soft palate (our fetus and infant Foyo and adult Kidogo and adolescent Jasiri).
- Usual innervation: Data are not available.
- Comparison with common chimpanzees: Gratiolet and Alix (1866) stated that in *P. troglodytes* this muscle originates from the Eustachian tube and that it is more horizontal than in modern humans, due to the prognathism of the face of *Pan*. Sonntag (1923) wrote that in *P. troglodytes* it originates together with the tensor veli palatini from the apex of the petrous temporal bone, the Eustachian tube and the scaphoid fossa and that it then separates from the tensor veli palatini, to run downwards and forwards and

spread out between layers of the palatopharyngeus, being more horizontal than in modern humans. Dean (1985) reported an origin in *P. troglodytes* from the medial aspect of the Eustachian tube and the adjacent part of the petrous apex and an insertion onto the superior surface of the soft palate. In our *P. troglodytes* specimens the muscle is more horizontal than in modern human adults, running from the Eustachian region and adjacent areas of the neurocranium to the soft palate, being medial to the tensor veli palatini and lateral to the palatopharyngeus, with exception to the specimens PFA 1077 and PFA UNC, in which the muscle seemed to pass partially lateral to the tensor veli palatini.
- Synonymy: Péristaphylin interne (Gratiolet and Alix 1866).

***Palatoglossus*** (adult Kidogo 0.22 g; infant Foyo 0.13 g; Fig. 2.7)

- Attachments in bonobos: From soft palate to posterior portion of tongue (our fetus and infant Foyo and adult Kidogo and adolescent Jasiri).
- Usual innervation: Data are not available.
- Comparison with common chimpanzees: Similar to bonobos. As noted by Duchin (1990) and Gibbs (1999) and confirmed by our dissections, in common chimpanzees the palatoglossus mainly attaches to the posterior portion of the tongue, whereas in modern humans it attaches to both the posterior portion and the lateral surface of the tongue.
- Synonymy: Glossopalatinus (Gratiolet and Alix 1866; Himmelreich 1971).

***Salpingopharyngeus***

- Attachments in bonobos: We could not discern if the salpingopharyngeus was present, or not, as a separate muscle in the bonobos dissected by us.
- Comparison with common chimpanzees: In our *P. troglodytes* specimens PFA 1016, PFA 1009, PFA 1051 and HU PT1 the salpingopharyngeus seems to be similar to that of modern humans. In our *P. troglodytes* specimen PFA 1077 we could not find a separate muscle salpingopharyngeus, but in PFA UNC there is seemingly at least a salpingopharyngeal fold such as that found in adult modern humans, although we could not discern if this fold is associated with a separate, well-defined muscle salpingopharyngeus.
- Usual innervation: Data are not available.

***Thyroarytenoideus*** (adult Kidogo 0.51 g; infant Foyo 0.05 g; Fig. 2.8)

- Attachments in bonobos: Runs from thyroid cartilage to arytenoid cartilage (our fetus and infant Foyo and adult Kidogo and adolescent Jasiri). Our dissections indicate that in the bonobos dissected by us the thyroaryntenoideus has a pars superior and a pars inferior, as is usually the case in common chimpanzees (see below), except in our bonobo fetus, where there was seemingly only one bundle, probably reflecting the normal early developmental stage of this muscle.
- Usual innervation: Data not available for bonobos [*P. troglodytes*: inferior laryngeal nerve, from recurrent laryngeal nerve (Jordan 1971a, b, c)].
- Comparison with common chimpanzees: With respect to hominoids and other primates, there has been controversy regarding the homologies of the thyroarytenoid bundles and the presence/absence of a distinct *musculus vocalis*. Kohlbrügge (1896) dissected gorillas, common chimpanzees and orangutans as well as taxa such as *Cebus*, *Semnopithecus*, *Hylobates* and *Macaca*, and stated that he could not find a distinct attachment of the thyroarytenoideus onto a true vocal cord (such as that found in modern humans) in any of these taxa, except perhaps in *Pongo*; within all the taxa mentioned above, he found an attachment onto the cricoid cartilage in *Hylobates* and *Colobus*. Giacomini (1897) examined the larynx of a gorilla and of an *Hylobates lar* specimen and based on these results and on his previous stud-

ies, he stated that only in modern humans is there a distinct, well-developed, musculus vocalis directly connected to the vocal cord, although in at least some *Hylobates* and seemingly in some *Pan*, there are a few fibres of the thyroarytenoideus that are somewhat isolated and situated near the vocal cord (see, e.g., fig. 2 of his Plate II). Duckworth (1912) examined specimens from all of the five extant hominoid genera, as well as from cadavers belonging to *Macaca*, *Cebus*, *Semnopithecus* and *Tarsius*, and suggested that a well-developed, distinct musculus vocalis associated with the plica vocalis is also only consistently present in modern humans. But Duckworth did suggest that great apes, and particularly common chimpanzees, show a configuration that is somewhat similar to that found in modern humans, in that they have a poorly developed/differentiated musculus vocalis (see, e.g., his figs. 24 and 17). According to him, in modern humans the superior portion of the thyroarytenoideus usually includes a distinct structure that is often associated with the region of the ventriculus (this corresponds to the *musculus ventricularis laryngis sensu* Kelemen 1948, 1969), although a somewhat similar configuration is also seen in gorillas and in chimpanzees. Loth (1931) argued that in nonhominoid primates the vocal cords mainly consist of a well-developed fold(s) of mucous membrane, which are not in contact with the musculus vocalis. He stated that in hominoids such as *P. troglodytes* and *Gorilla*, the folds are smaller and the musculus vocalis does not connect to the folds; such a connection was found in most modern humans. Edgeworth (1935) defended the notion that a musculus vocalis is found in some nonhuman primates and suggested that when this structure is present the thyroarytenoideus becomes a *thryroarytenoideus lateralis* because its inferior/mesial part gives rise to the vocalis muscle. Starck and Schneider (1960) described a 'pars lateralis' as well as a 'pars medialis' of the musculus vocalis that usually goes to the vocal fold/cord in *P. troglodytes*, *Pongo* and *Gorilla*, but not in *Hylobates*; the latter corresponds to the musculus vocalis of modern humans. They did not find a pars aryepiglottica or a pars thyroepiglottica of the thyroarytenoideus in *Hylobates* and *Pongo*, but stated that other authors did report at least one of these structures in *Gorilla* and *P. troglodytes*. Saban (1968) clarified the nomenclature of the thyroarytenoideus and suggested that this muscle may be divided into the following structures: (1) a pars superior (often designated as *thyroarytenoideus superior*, *thyroarytenoideus lateralis* or *ventricularis*); (2) a pars inferior (often designated as *thyroarytenoideus inferior*, *thyroarytenoideus medialis* or musculus vocalis); (3) a *ceratoarytenoideus lateralis*; (4) a *pars intermedia* (but this name was only used by a few authors such as Starck and Schneider 1960 who stated that some primates might have a pars superior, a pars inferior, and a pars intermedia); (5) a pars thyroepiglottica; (6) a pars aryepiglottica; (7) a pars arymembranosa; and (8) a pars thyromembranosa. According to Saban (1968), the 'ceratoarytenoideus lateralis' is usually fused with (but not differentiated from, as suggested in some anatomical atlases) the cricoarytenoideus posterior, being only a distinct muscle in a few taxa and, within primates, in *P. troglodytes* (and still in this case this seems to constitute a variant/anomaly) where it is a small muscle running from the dorsal face of the inferior thyrohyoid horn to the arytenoid cartilage. Also according to Saban the pars superior and pars inferior are more superior and inferior, respectively, in apes and modern humans (in modern humans the more inferior or medial part, i.e., the pars inferior is well developed and often designated as musculus vocalis), whereas in primates such as *Macaca* and *Papio* they are more lateral/medial; in *Pongo* the pars inferior is well-developed and lies anterior to the vocal cord, but it is not associated with it. Aiello and Dean (1990) stated that in nonhuman hominoids, the pars aryepiglottica is often reduced in size or absent. Our observations and review of the literature indicate that the pars superior and pars inferior of the thyroarytenoideus are usually

present in common chimpanzees. The cerato-arytenoideus lateralis is present in at least some *P. troglodytes* according to Saban (1968) and to Macalister (1871), but was not found in the common chimpanzees dissected by Jordan (1971a, b, c) or in those dissected by us. The pars intermedia is not present in common chimpanzees according to Starck and Schneider (1960), to Jordan (1971a, b, c), and to our dissections. The *pars thyroepiglottica* is present in at least some of the common chimpanzees dissected by us and by Gratiolet and Alix (1866) and Kelemen (1948, 1969), but it was not present in the common chimpanzees examined by Jordan (1971a, b, c) and Avril (1963). The *pars aryepiglottica* is seemingly absent in our specimens and is only present in one of the 10 common chimpanzees reported by Avril (1963), being absent in the common chimpanzees examined by Jordan (1971a, b, c), but present in the *P. troglodytes* specimen of Gratiolet and Alix (1866), in a specimen of Sonntag (1923), and in the specimen of Kelemen (1948, 1969). The *pars thyromembranosa* and *pars arymembranosa* are not present in the common chimpanzees dissected by Jordan (1971a, b, c) and seemingly by Avril (1963).
- Synonymy: Thyroarytenoideus plus musculus vocalis (Duckworth 1912; Jordan 1971a, b, c).

***Musculus vocalis*** (see thyroarytenoideus above)

- Attachments in bonobos: The vocalis is likely present in *Pan*, and thus in bonobos (see thyroarytenoideus above), but we could not check if it was effectively present as a distinct structure in the bonobos dissected by us.
- Usual innervation: See thyroarytenoideus above.
- Comparison with common chimpanzees: See thyroarytenoideus above.

***Cricoarytenoideus lateralis*** (adult Kidogo 0.51 g; infant Foyo 0.01 g; Fig. 2.8)

- Attachments in bonobos: From cricoid cartilage to arytenoid cartilage (our fetus and infant Foyo and adult Kidogo and adolescent Jasiri).
- Usual innervation: Data not available for bonobos [*P. troglodytes*: inferior laryngeal nerve, from recurrent laryngeal nerve (Jordan 1971a, b, c)].
- Comparison with common chimpanzees: Similar to bonobos.

***Arytenoideus*** (adult Kidogo 0.26 g; Fig. 2.8)

- Attachments in bonobos: Unpaired, transversly-oriented muscle connecting the two contralateral arytenoid cartilages that, as in most common chimpanzees but contrary to most modern humans, does not seem to be divided into an *arytenoideus obliquus* and an *arytenoideus transversus* (our fetus and infant Foyo and adult Kidogo and adolescent Jasiri); see comments below.
- Usual innervation: Data not available for bonobos [*P. troglodytes*: inferior laryngeal nerve, from recurrent laryngeal nerve (Jordan 1971a, b, c)].
- Comparison with common chimpanzees: Both the arytenoideus obliquus and arytenoideus transversus were found in *P. troglodytes* by Sonntag (1923, 1924), Gratiolet and Alix (1986) and Avril (1963) and they were present in some of the chimpanzees dissected by us; but they were not reported in the specimens of Kohlbrügge (1896), Kelemen (1948), Körner (1884), Starck and Schneider (1960) and Jordan (1971a, b, c), so it is possible that there is a true variation within *Pan*: it is therefore not clear if the single head seemingly present in the bonobos dissected by us represents a condition always found in bonobos, or instead a condition often – but not always – found in bonobos.

***Cricoarytenoideus posterior*** (adult Kidogo 0.29 g; infant Foyo 0.04 g; Fig. 2.8)

- Attachments in bonobos: From dorsal portion of cricoid cartilage to arytenoid cartilage, not contacting with its counterpart in the midline (our fetus and infant Foyo and adult Kidogo and adolescent Jasiri). In at least our adult Kidogo there is also an attachment onto the inferior horn of the thyroid cartilage (Fig. 2.8), but still these fibres did not really form a distinct muscle *ceratocricoideus sensu* Harrison (1995) (see just below).

- Usual innervation: Data not available for bonobos [*P. troglodytes*: inferior laryngeal nerve, from recurrent laryngeal nerve (Jordan 1971a, b, c)].
- Comparison with common chimpanzees: Similar to bonobos. In 8 of the 10 common chimpanzees from which we could retrieve detailed information (i.e., one specimen shown in plate 57 of Swindler and Wood 1973; one specimen shown in fig. 2 of plate 9 of Gratiolet and Alix 1866; one specimen reported by Sonntag 1923; one specimen described by Kelemen 1948; and the six specimens dissected by us where we studied this feature in detail) there was no contact between the cricoarytenoideus posterior and its counterpart in the midline. The two exceptions are two *P. troglodytes* infants dissected by us (PFA 1077, PFA UNC), in which a few fibres ran between the two muscles. It should also be noted that Jordan (1971a, b, c) examined 10 common chimpanzees and stated that in a few specimens the cricoarytenoideus posterior did meet its counterpart in the dorsal midline, but he did not specify the number of specimens in which this occurred. There is usually no *ceratocricoideus sensu* Harrison (1995) in common chimpanzees (see just above).
- Synonymy: Cricoarytenoideus posticus (Kohlbrügge 1896).

## 2.4 Hypobranchial Musculature

***Geniohyoideus* (*adult Kidogo 1.53 g; infant Foyo 0.37 g;* Fig. 2.6)**

- Attachments in bonobos: From mandible to body and greater horn of hyoid bone, lying close to and merging with its counterpart at midline (our fetus and infant Foyo and adult Kidogo and adolescent Jasiri); Miller (1952) reported an attachment onto the body, but not onto the greater horn, of the hyoid bone, in her adult bonobo.
- Usual innervation: Ventral ramus of the first cervical nerve through the hypoglossal nerve (Miller 1952: *P. paniscus*) [*P. troglodytes*: hypoglossal nerve (Sonntag 1923)].
- Comparison with common chimpanzees: In *P. troglodytes* it runs from mandible to body, and usually also part of greater horn, of hyoid bone (e.g., Gratiolet and Alix 1866; Sonntag 1923), lying close to and often merging with its counterpart at midline (e.g., Duvernoy 1855–1856; Sonntag 1923; Edgeworth 1935 and also seen in the two common chimpanzees in which we investigated this feature in detail, i.e., in PFA 1077 and PFA UNC).

***Genioglossus*** (adult Kidogo 2.78 g; infant Foyo 0.96 g; Fig. 2.6)

- Attachments in bonobos: From mandible to tongue and hyoid bone, being well-separated by a midline fibrous septum and/or fatty tissue from its counterpart (our fetus and infant Foyo and adult Kidogo and adolescent Jasiri).
- Usual innervation: Data not available for bonobos [*P. troglodytes*: hypoglossal nerve (Sonntag 1923)].
- Comparison with common chimpanzees: Similar to bonobos. The muscles *genio-epiglotticus*, *glosso-epiglotticus*, *hyo-epiglotticus* and *genio-hyo-epiglotticus*, described by authors such as Edgeworth (1935) and Saban (1968) in some primate and nonprimate mammals, do not seem to be present as distinct structures in the common chimpanzees dissected by us (according to these authors those muscles are usually not present in catarrhine taxa).

***Intrinsic muscles of tongue***

- Attachments in bonobos: To our knowledge, there are no detailed published descriptions of these muscles in *Pan* and we could not examine them in detail in our dissections. However, the *longitudinalis superior*, *longitudinalis*

*inferior, transversus linguae* and *verticalis linguae* are consistently found in modern humans and at least some other primate and nonprimate mammals, so these four muscles are very likely also present in *Pan*. Detailed studies of the tongue and its muscles in common chimpanzees and bonobos are clearly needed.
- Usual innervation: Data are not available.

***Hyoglossus*** (adult Kidogo 1.17 g; infant Foyo 0.70 g; Fig. 2.6)

- Attachments in bonobos: According to our dissections of bonobos and common chimpanzees, the hyoglossus is usually differentiated into a *ceratoglossus* and a *chondroglossus* as is usually the case in modern humans (see, e.g., Terminologia Anatomica 1998). The ceratoglossus connects the greater horn of the hyoid bone to the tongue, while the chondroglossus mainly connects the body of the hyoid bone (the inferior horn is usually poorly developed or absent) to the tongue.
- Usual innervation: Data are not available.
- Comparison with common chimpanzees: Similar to bonobos. In a few *P. troglodytes* the hyoglossus is, however, blended with the thyrohyoideus, as a variant (e.g., Sonntag 1923)

***Styloglossus*** (adult Kidogo 0.50 g; infant Foyo 0.15 g; Fig. 2.6)

- Attachments in bonobos: From stylohyoid process and/or from adjacent regions of skull to tongue (our fetus and infant Foyo and adult Kidogo and adolescent Jasiri). A variation is found in Kidogo, where it is fused with the superior constrictor and with the stylopharyngeus (Fig. 2.6).
- Usual innervation: Data are not available.
- Comparison with common chimpanzees: Similar to bonobos. As explained by Gibbs (1999) and confirmed by our dissections, contrary to the vast majority of eutherian mammals (e.g., fig. 369 of Saban 1968) and to most primates, including other hominoids, in chimpanzees and also in modern humans (e.g., plate 59 of Netter 2006), the styloglossus runs mainly longitudinally to insert onto the tongue, but it also has a distinct oblique slip that runs anteroinferiorly at about 45° from the main body of the muscle to insert more inferiorly onto the lateral surface of the hyoglossus.

***Sternohyoideus*** (infant Foyo 0.37 g; Fig. 2.11)

- Attachments in bonobos: From sternum and adjacent regions, to hyoid bone, not contacting its counterpart in the midline for most of its length in our infant Foyo and adult Kidogo and adolescent Jasiri and in Miller's 1952 adult bonobo, but blending with the omohyoideus and sternothyroid of the same side, and also with sternohyoideus of the other side, in our fetus, probably reflecting the normal early development of this group of muscles. We could not discern if there were tendinous intersections of the sternohyoid in the adult Kidogo. Such intersections were not present in our fetus nor are they described in Miller's 1952 adult bonobo, but in the adolescent Jasiri and the infant Foyo there was a clear tendinous intersection, on both sides of the body, at about the caudal 1/3 of the muscle.
- Usual innervation: Branches from the ansa hypoglossi composed of the anterior divisions of the first three cervical nerves (Miller 1952: *P. paniscus*).
- Comparison with common chimpanzees: In common chimpanzees the sternohyoideus often has transverse tendinous intersections (e.g., Gratiolet and Alix 1866; Champneys 1872; Starck and Schneider 1960) but no intersections were found in the *P. troglodytes* specimens dissected by us.

***Omohyoideus*** (adult Kidogo 2.49 g; adolescent Jasiri 1.33 g; infant Foyo 0.14 g; Figs. 2.5, 2.11, 2.12 and 2.13)

- Attachments in bonobos: From scapula to hyoid bone, passing deep to sternocleidomastoideus, not being divided into two heads by a well-defined intermediate tendon (our fetus and infants Foyo and Etje and adult Kidogo and adolescent Jasiri). This is in fact a major difference between bonobos and common chimpanzees: actually none of these four bonobos has even a tendinous intersection, on either the right or the left sides of the body, with exception to the adolescent Jasiri, in which there was an incomplete tendinous intersection at half the length of the omohyoideus. This contrasts with Miller's 1952 description of her adult bonobo, where she described a superior head and an inferior head of the omohyoideus separated by an intermediate tendon. Be that as it may, within six specimens (our fetus and infants Foyo and Etje and adult Kidogo and adolescent Jasiri and Miller's 1952 adult bonobo), 5 out of 6, i.e., 83%, do not have an intermediate tendon, clearly contrasting with common chimpanzees, in which having such an intermediate tendon is the usual configuration (see below). In the infant Foyo there was a variation: a fusion between the omohyoideus and the cleidomastoid head of the sternomastoideus (Fig. 2.11).

- Usual innervation: Branches from the ansa hypoglossi composed of the anterior divisions of the first three cervical nerves (Miller 1952: *P. paniscus*).
- Comparison with common chimpanzees: In modern humans and most common chimpanzees, the omohyoideus usually has a well-defined intermediate tendon between the superior head and the inferior head/heads (e.g., Vrolik 1841; Gratiolet and Alix 1866; Bischoff 1870; Macalister 1871; Sonntag 1923, 1924; Ashton and Oxnard 1963; our dissections). Contrary to the situation in most other primates, in at least some specimens of *Gorilla*, *Pan* and *Homo*, the omohyoideus has three bellies (usually a superior belly, an inferomedial belly and an inferolateral belly: see figs. 7 and 8 and Diogo and Wood 2011, 2012), as a variation. This condition was reported in common chimpanzees by Gratiolet and Alix (1866), who designated one of the inferior bellies as 'cleidohyoideus', by Sonntag (1923; see, e.g., his fig. 32), and it was found in two out of the common chimpanzees dissected by us (PFA 1016 and PFA 1009).
- Synonymy: Coracohyoideus (Tyson 1699); probably corresponds to omo-hyoïdien plus cléido-hyoïdien *sensu* Gratiolet and Alix (1866).

***Sternothyroideus*** (infant Foyo 0.45 g; Figs. 2.5 and 2.11)

- Attachments in bonobos: From sternum to thyroid cartilage, the main body of the sternothyroideus usually extending anteriorly (rostrally) to the posterior portion of the main body of the thyrohyoideus (e.g., Fig. 2.5). The sternothyroideus does not contact its counterpart at the midline (e.g., Fig. 2.11) and does not have tendinous intersections (our fetus, infant Foyo, adult Kidogo and adolescent Jasiri and Miller's 1952 adult bonobo).
- Usual innervation: Branches from the ansa hypoglossi composed of the anterior divisions of the first three cervical nerves (Miller 1952: *P. paniscus*).

- Comparison with common chimpanzees: In common chimpanzees, the sternothyroideus may have transverse tendinous intersections (e.g., Macalister 1871), but most specimens do not have such intersections (Champneys 1872; our dissections). Contrary to the condition in modern humans and in most other primates, in common chimpanzees – and in bonobos, see just above – the main body of the sternothyroideus usually extends anteriorly to the posterior portion of the main body of the thyrohyoideus (e.g., Gratiolet and Alix 1866; Sonntag 1923; Starck and Schneider 1960; Swindler and Wood 1973; our dissections).

***Thyrohyoideus*** (adult Kidogo 0.72 g; infant Foyo 0.10 g; Figs. 2.5 and 2.11)

- Attachments in bonobos: From thyroid cartilage to hyoid bone, being fused with sternothyroid (our fetus, infant Foyo, adult Kidogo and adolescent Jasiri and Miller's 1952 adult bonobo).
- Usual innervation: Branch of the anterior division of the first cervical nerve through the descendens hypoglossi (i.e., branch of the hypoglossal nerve) (Miller 1952: *P. paniscus*) [*P. troglodytes*: branch of hypoglossal nerve (Sonntag 1923)].
- Comparison with common chimpanzees: Swindler and Wood (1973) stated that in *P. troglodytes* there is no pyramidal lobe of the thyroid gland and we and others found no evidence of a *levator glandulae thyroideae* muscle, which is present as an anomaly in modern humans. However, Gratiolet and Alix (1866) described a *P. troglodytes* specimen in which the cricothyroideus has three divisions: a pars recta, a pars obliqua and a 'small additional bundle' (see their plate 8) running superficially to these divisions, from the inferior margin of the thyroid cartilage to the more anterior tracheal cartilages. In view of its position and of being superficial to the cricothyroideus, this 'additional bundle' could correspond to the levator glandulae thyroideae of modern humans, but this latter muscle more often originates from the hyoid bone than from the inferior margin of the thyroid cartilage.

## 2.5 Extra-Ocular Musculature

***Muscles of eye***

- Attachments in bonobos: To our knowledge, there are no detailed published descriptions of these muscles in bonobos and common chimpanzees. The *rectus inferior, rectus superior, rectus medialis, rectus lateralis, obliquus superior, obliquus inferior, orbitalis* and *levator palpebrae superioris* are consistently found in modern humans and at least some other primate and nonprimate mammals, so these muscles are very likely also present in *Pan*. Detailed studies of the eye and the extra-ocular muscles in common chimpanzees and other apes are clearly needed.
- Usual innervation: Data are not available.

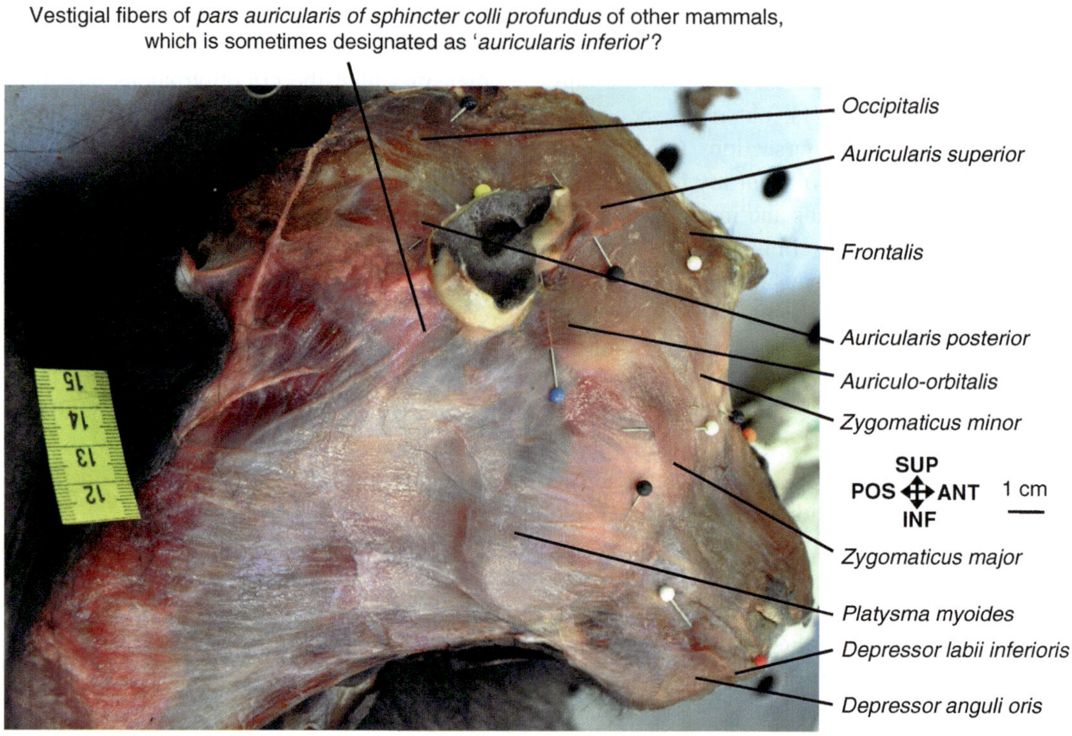


**Fig. 2.1** *Pan paniscus* (ZIMS 164047, Jasiri, adolescent female): lateral view of the right facial musculature; note that the risorius is not differentiated into a separate muscle. In this figure and the remaining figures of the atlas, the names of the muscles are in italics, and SUP, INF, ANT, POS, MED, LAT, VEN, DOR, PRO and DIS refer to superior, inferior, anterior, posterior, medial, lateral, ventral, dorsal, proximal and distal, respectively (in the sense the terms are applied to pronograde tetrapods: see Methodology and Material)

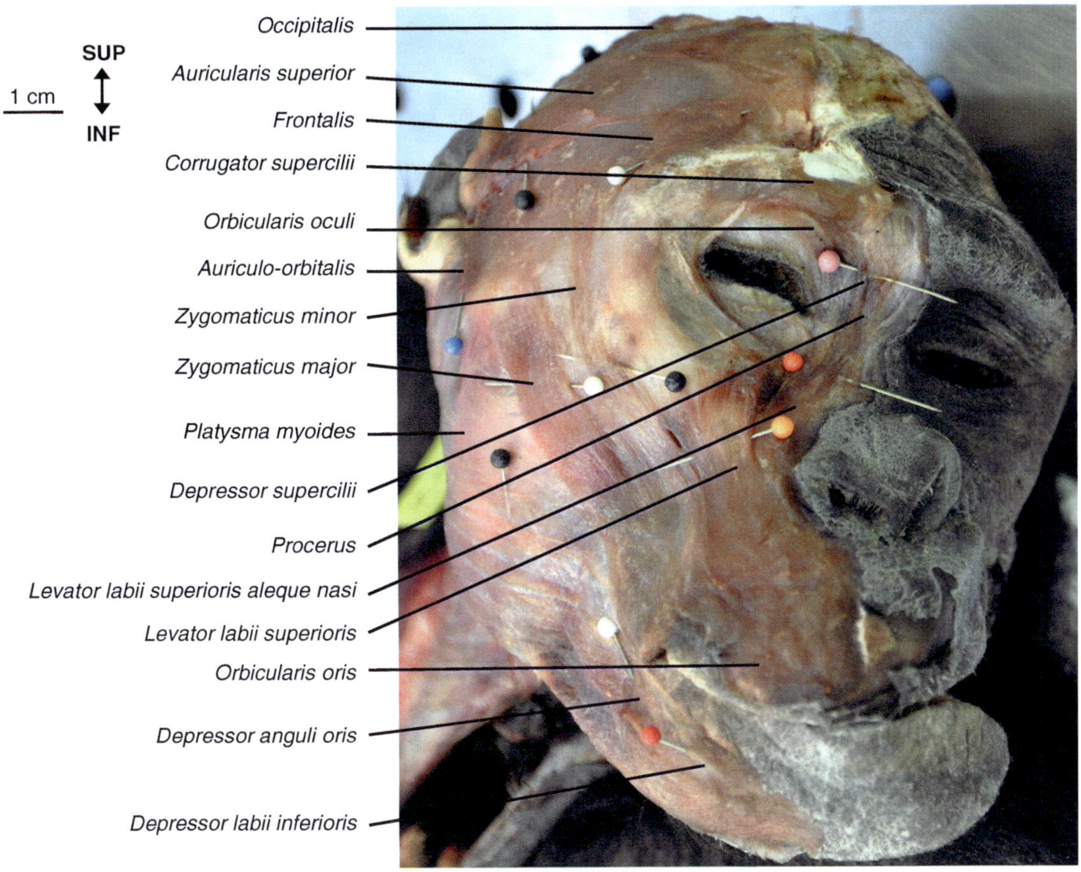


**Fig. 2.2** *Pan paniscus* (ZIMS 164047, Jasiri, adolescent female): frontal view of the right facial musculature; note that the risorius is not differentiated into a separate muscle

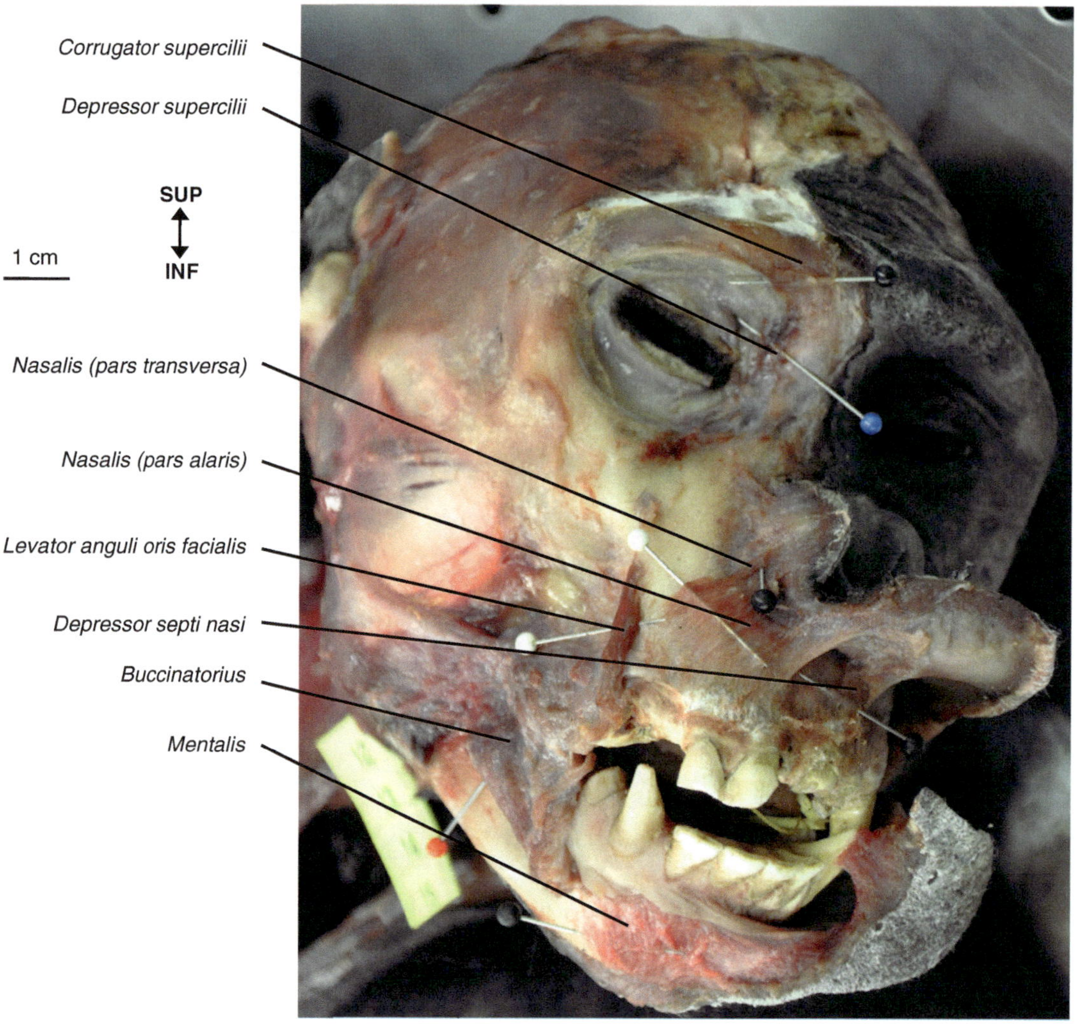


**Fig. 2.3** *Pan paniscus* (ZIMS 164047, Jasiri, adolescent female): frontal view of the deep right facial muscles

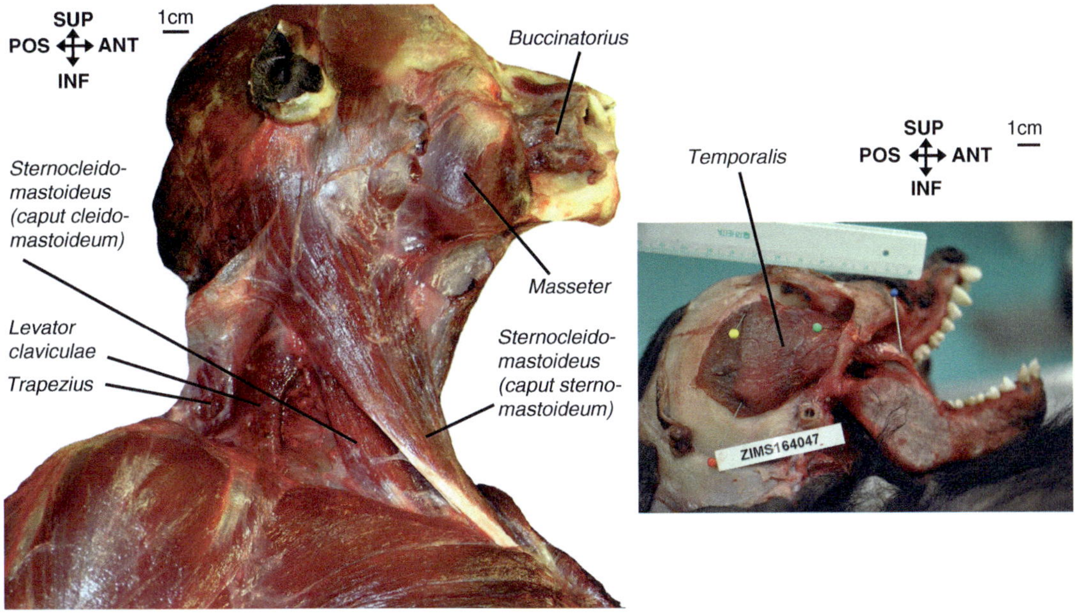


**Fig. 2.4** (*Left*) *Pan paniscus* (ZIMS 164031, Kidogo, adult male): lateral view of the right head and neck muscles after removal of most facial muscles. (*Right*) *Pan paniscus* (ZIMS 164047, Jasiri, adolescent female): lateral view of right temporalis after removal of the masseter

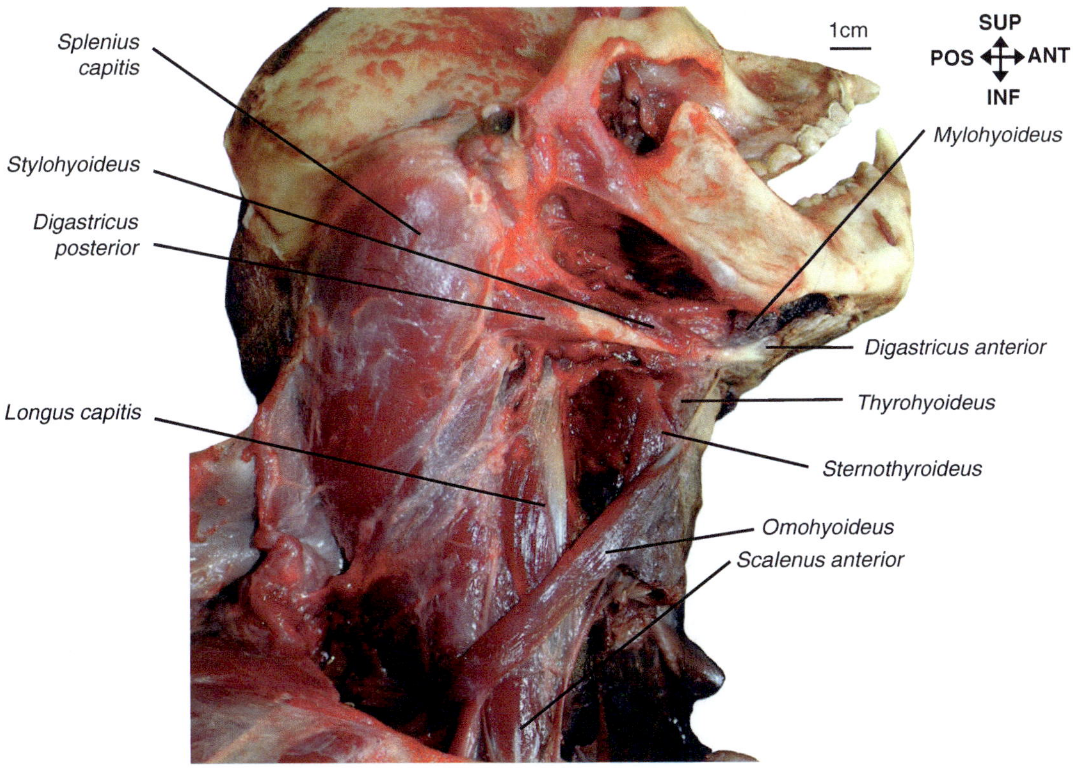


**Fig. 2.5** *Pan paniscus* (ZIMS 164031, Kidogo, adult male): lateral view of the right head and neck muscles after removal of the sternocleidomastoideus. Note that the omohyoideus has a single belly, with no intermediate tendon or tendinous intersections

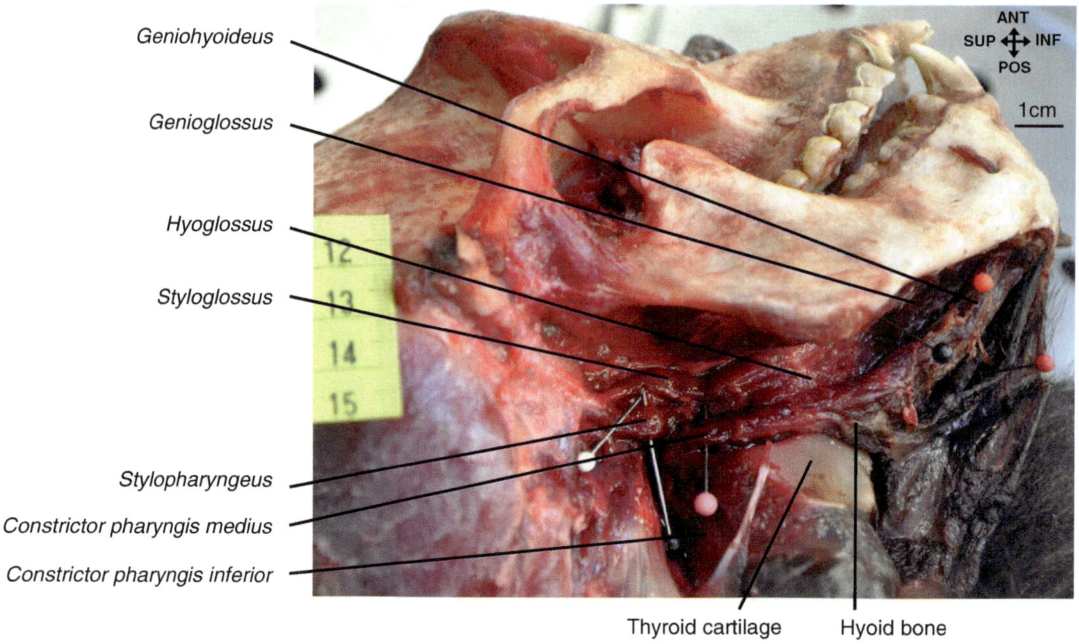


**Fig. 2.6** *Pan paniscus* (ZIMS 164031, Kidogo, adult male): lateral view of the right deep head and neck muscles after removal of the digastricus posterior and stylohyoideus and other superficial muscles

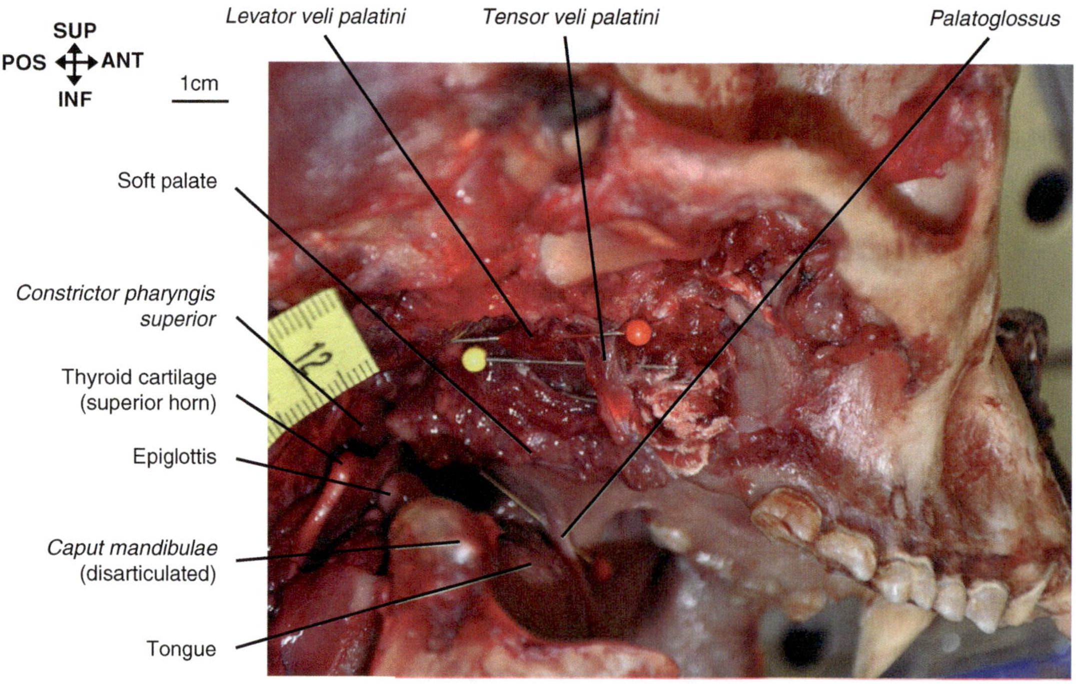


**Fig. 2.7** *Pan paniscus* (ZIMS 164031, Kidogo, adult male): lateral view of some of the deepest pharyngeal muscles after disarticulation of the right mandible

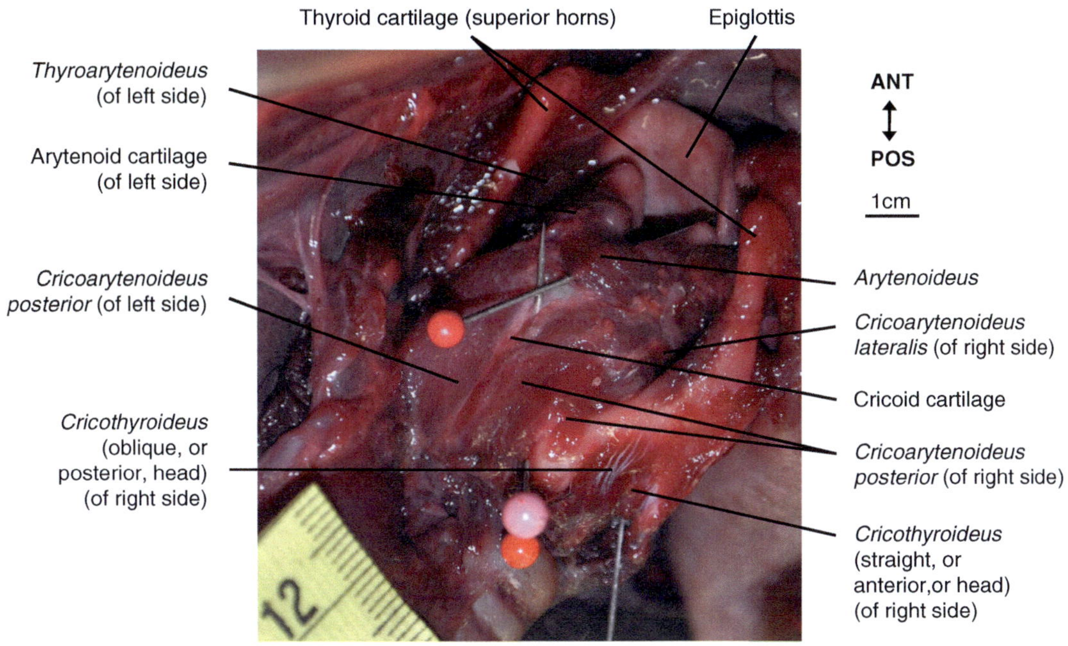


**Fig. 2.8** *Pan paniscus* (ZIMS 164031, Kidogo, adult male): dorso-lateral (postero-lateral in human anatomy) view of larynx after removal of the surrounding head and neck muscles. Note that, contrary to modern humans, there is a single arytenoideus muscle (i.e., instead of an arytenoideus transversus and an arytenoideus obliquus)

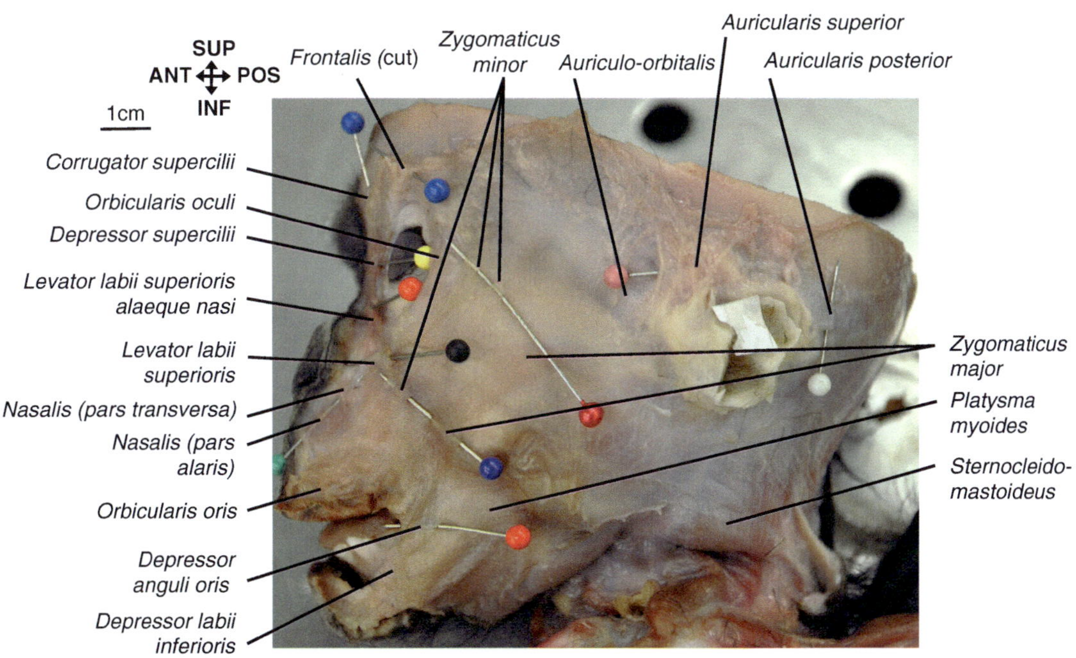


**Fig. 2.9** *Pan paniscus* (ZIMS 164041, Foyo, infant male): lateral view of the left facial musculature. Note that, contrary to modern humans, the risorius is not present as a separate muscle

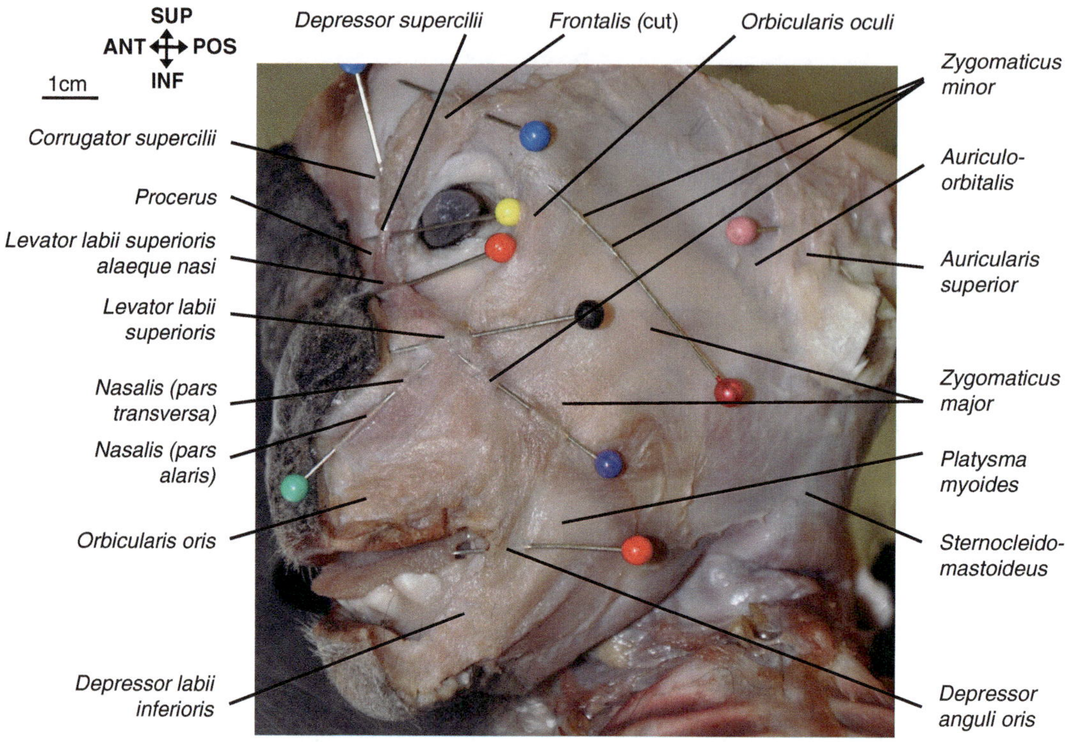


**Fig. 2.10** *Pan paniscus* (ZIMS 164041, Foyo, infant male): antero-lateral view of the left facial musculature. Note that, contrary to modern humans, the risorius is not present as a separate muscle

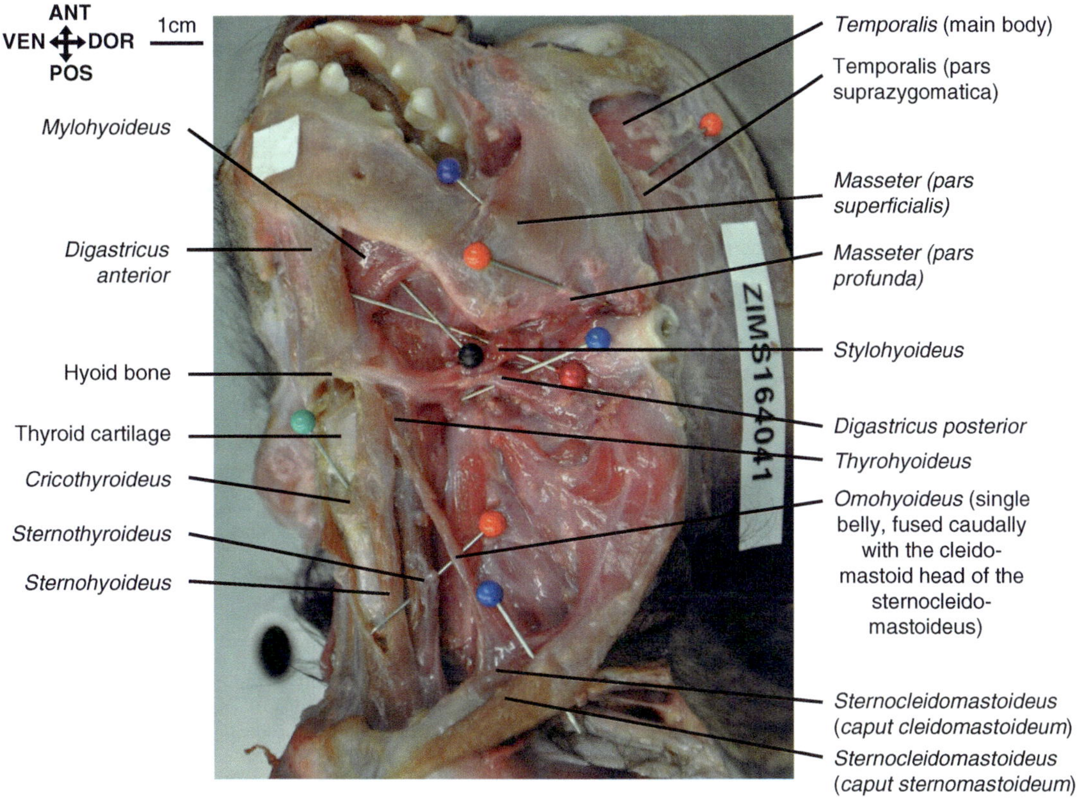


**Fig. 2.11** *Pan paniscus* (ZIMS 164041, Foyo, infant male): ventro-lateral (antero-lateral in human anatomy) view of left head and neck musculature

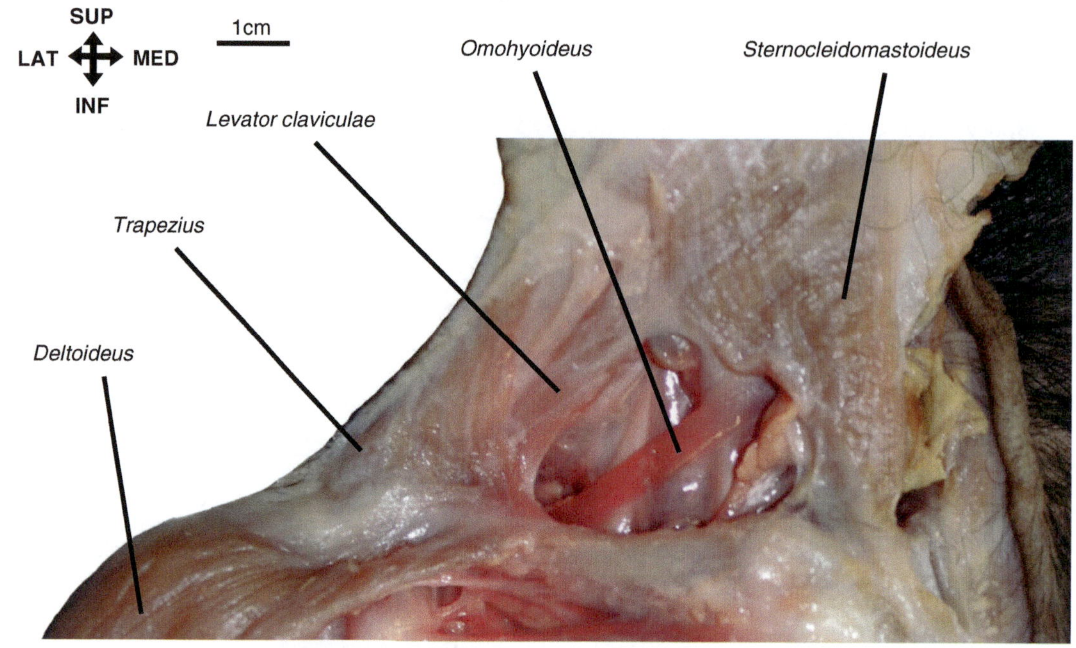


**Fig. 2.12** *Pan paniscus* (ZIMS 164041, Foyo, infant male): ventro-lateral (antero-lateral in human anatomy) view of the left head and neck musculature (as well as of the upper limb muscles deltoideus and levator claviculae), including the single belly of the omohyoideus, showing a variation of the head and neck muscles; the absence of a clavicular insertion of the sternocleidomastoideus

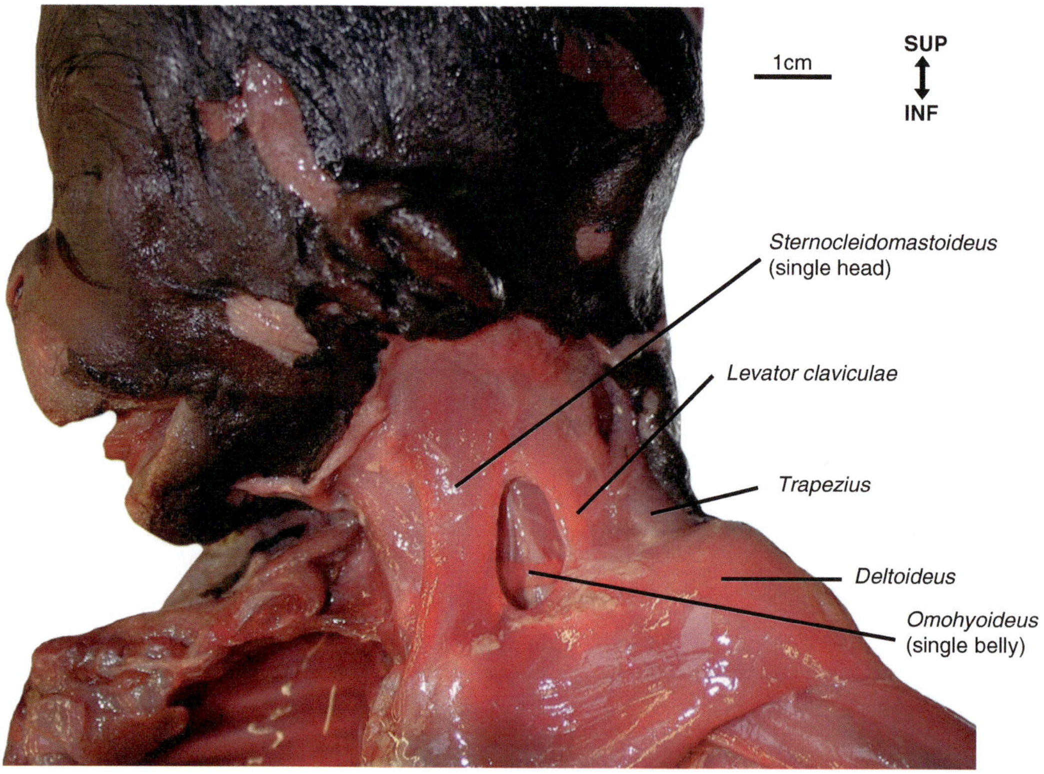


**Fig. 2.13** *Pan paniscus* (ZIMS 164052, Ano, fetus female): ventro-lateral (antero-lateral in human anatomy) view of the left head and neck musculature, as well as of the upper limb muscles levator claviculae and deltoideus

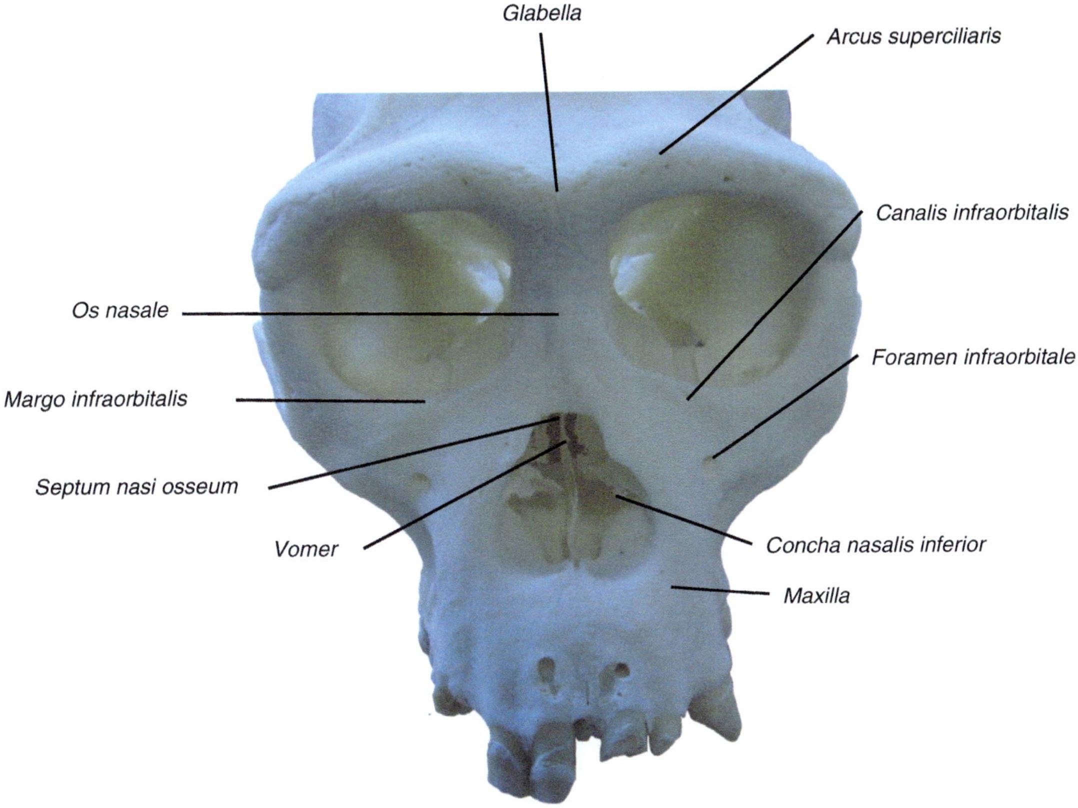


**Fig. 2.14** *Pan paniscus*: frontal view of cranium

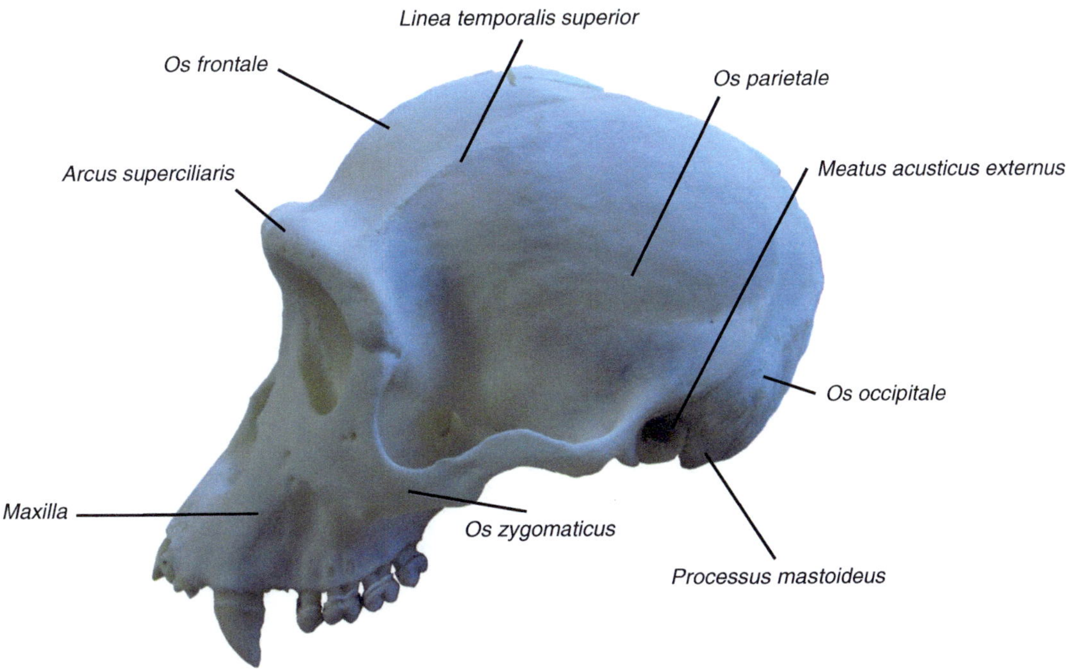


**Fig. 2.15** *Pan paniscus*: lateral view of cranium

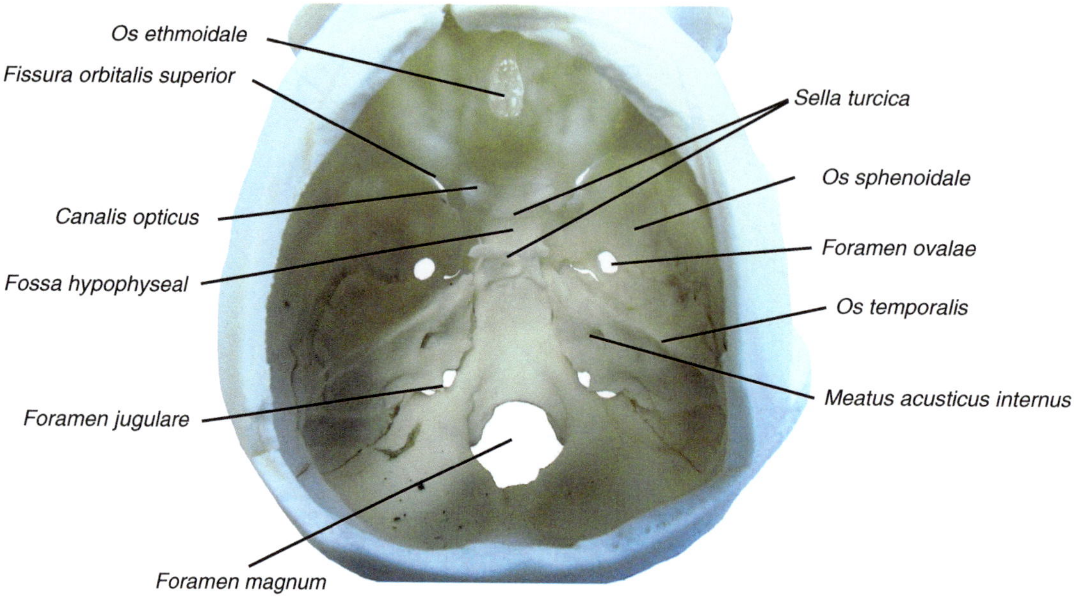


**Fig. 2.16** *Pan paniscus*: superior view of opened neurocranium

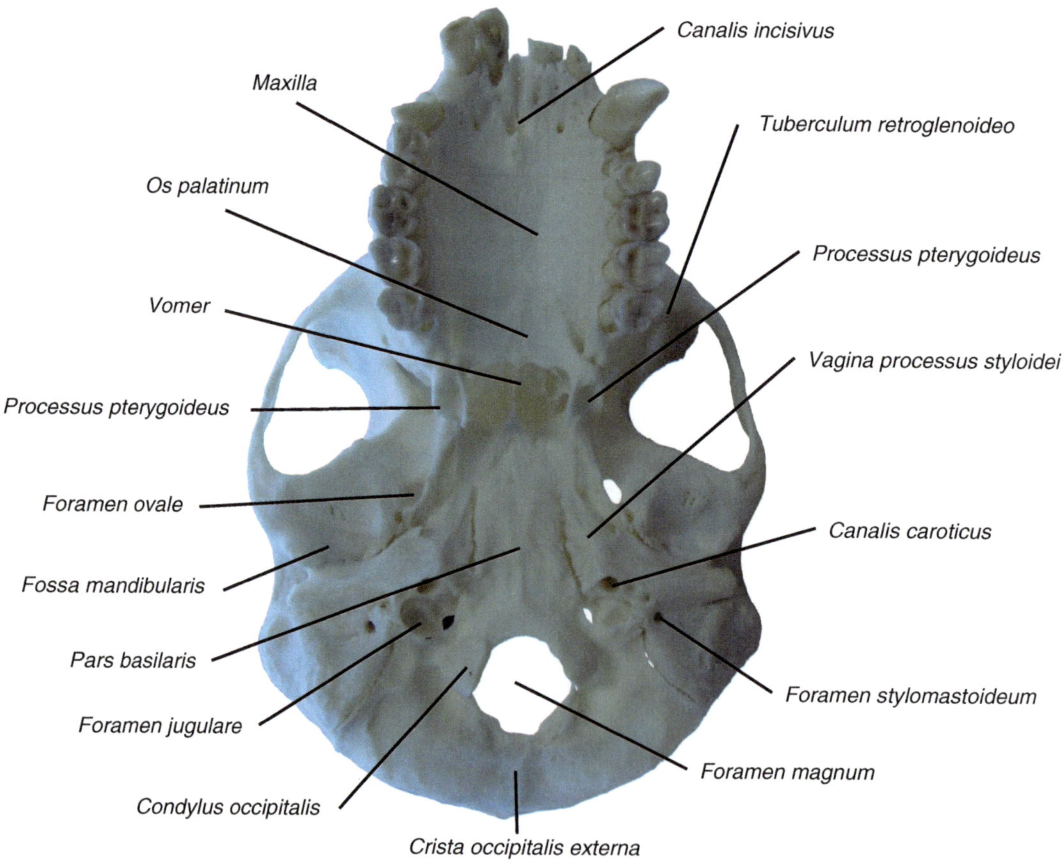


**Fig. 2.17** *Pan paniscus*: inferior view of cranium

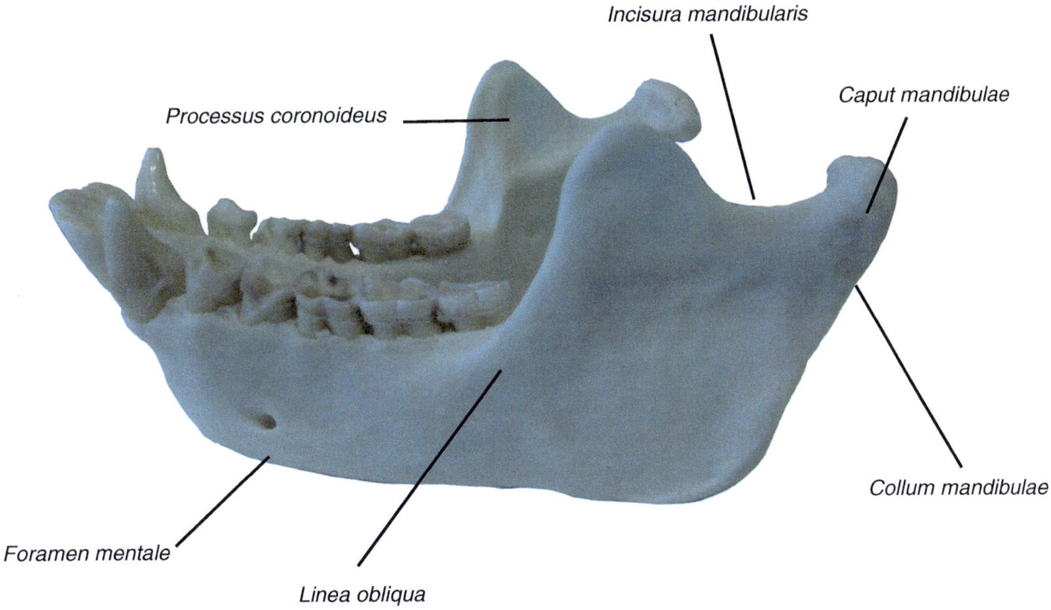


**Fig. 2.18** *Pan paniscus*: lateral view of mandible

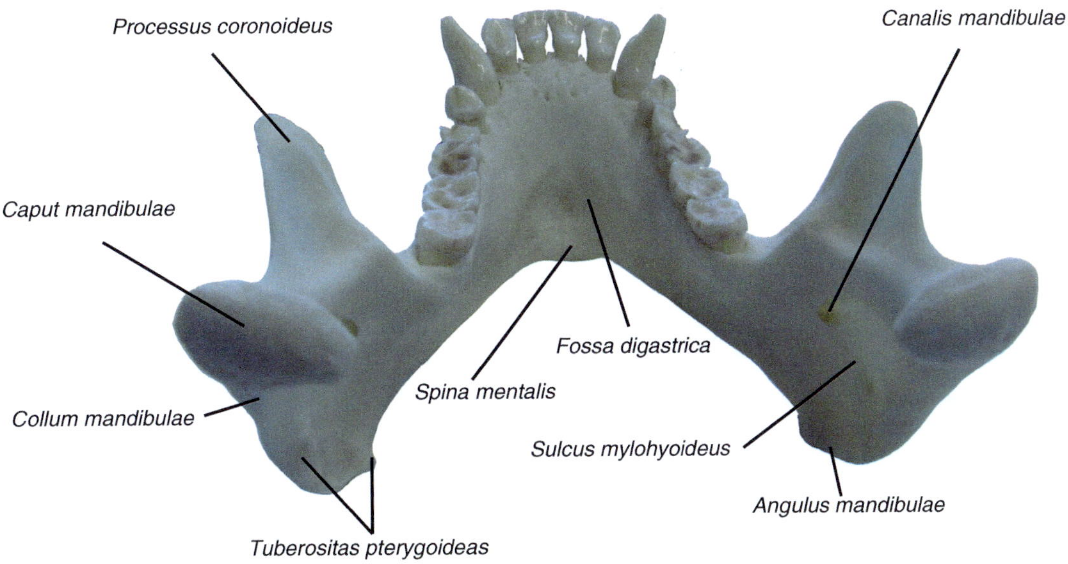


**Fig. 2.19** *Pan paniscus*: posterior view of mandible

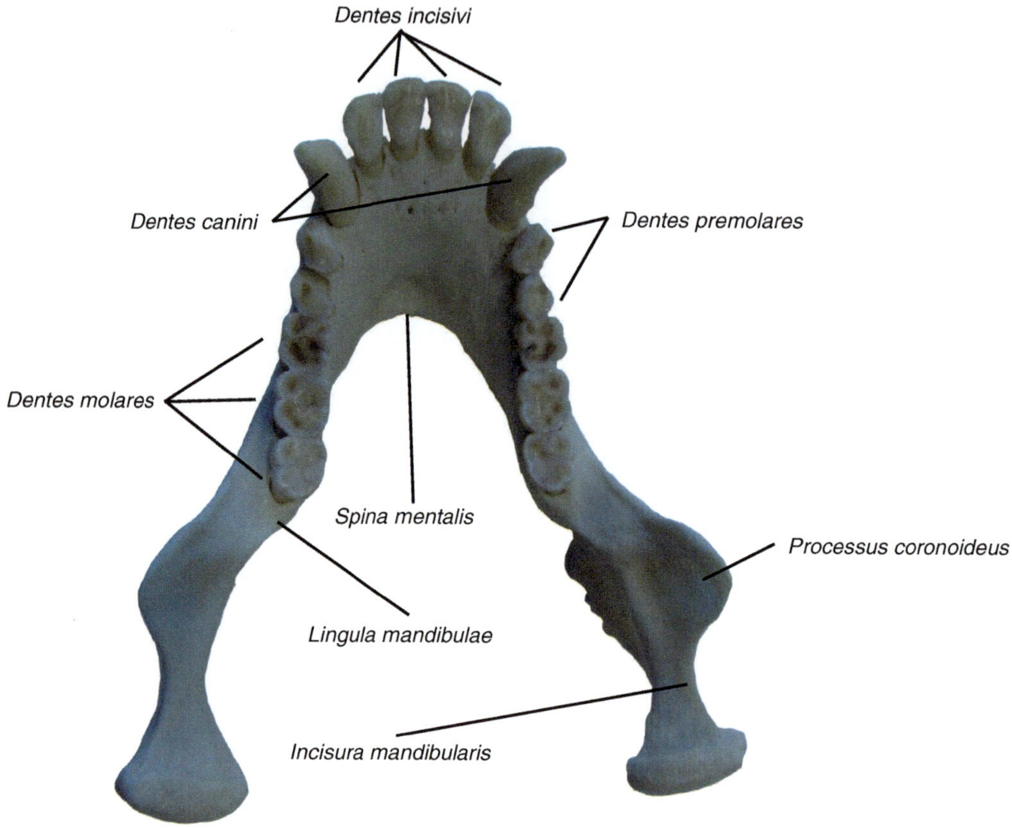


**Fig. 2.20** *Pan paniscus*: dorsal view of mandible

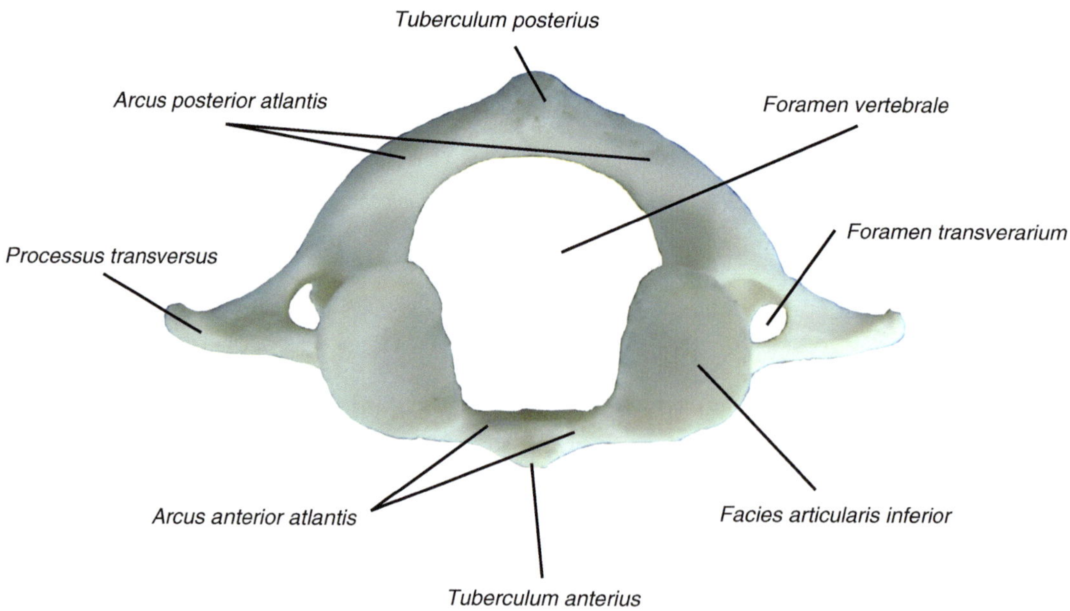


**Fig. 2.21** *Pan paniscus*: anterior (superior in human anatomy) view of atlas

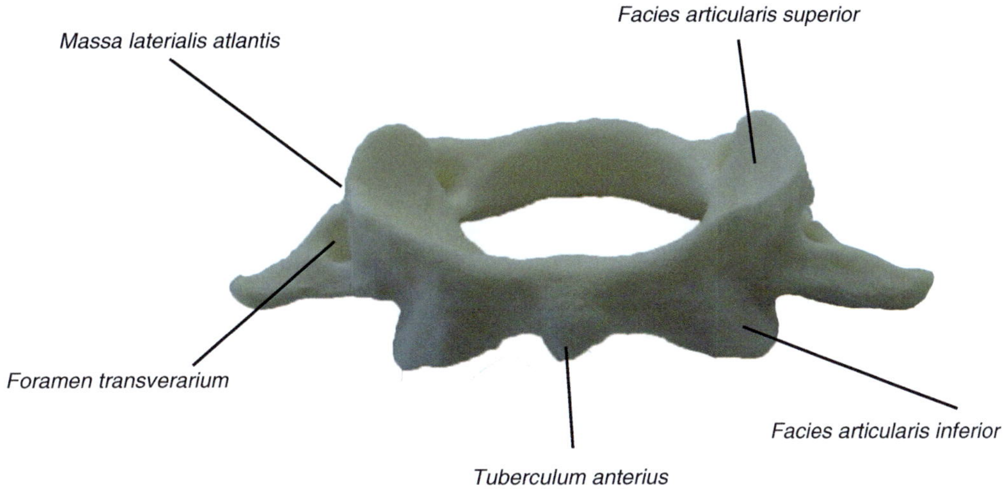


**Fig. 2.22** *Pan paniscus*: ventral (anterior in human anatomy) view of atlas

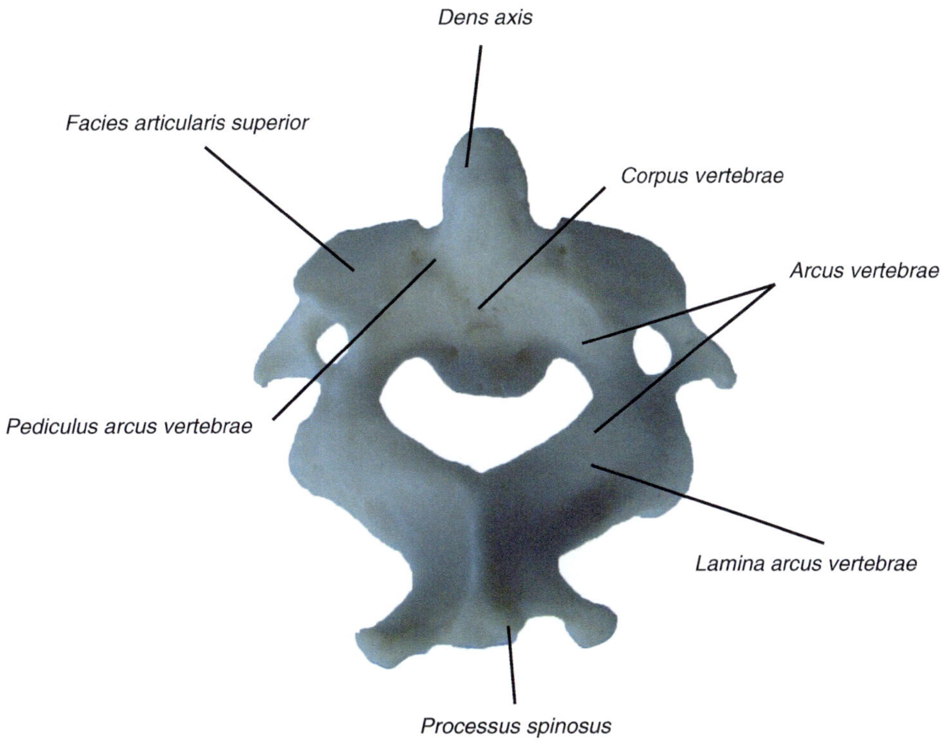


**Fig. 2.23** *Pan paniscus*: anterior (superior in human anatomy) view of axis

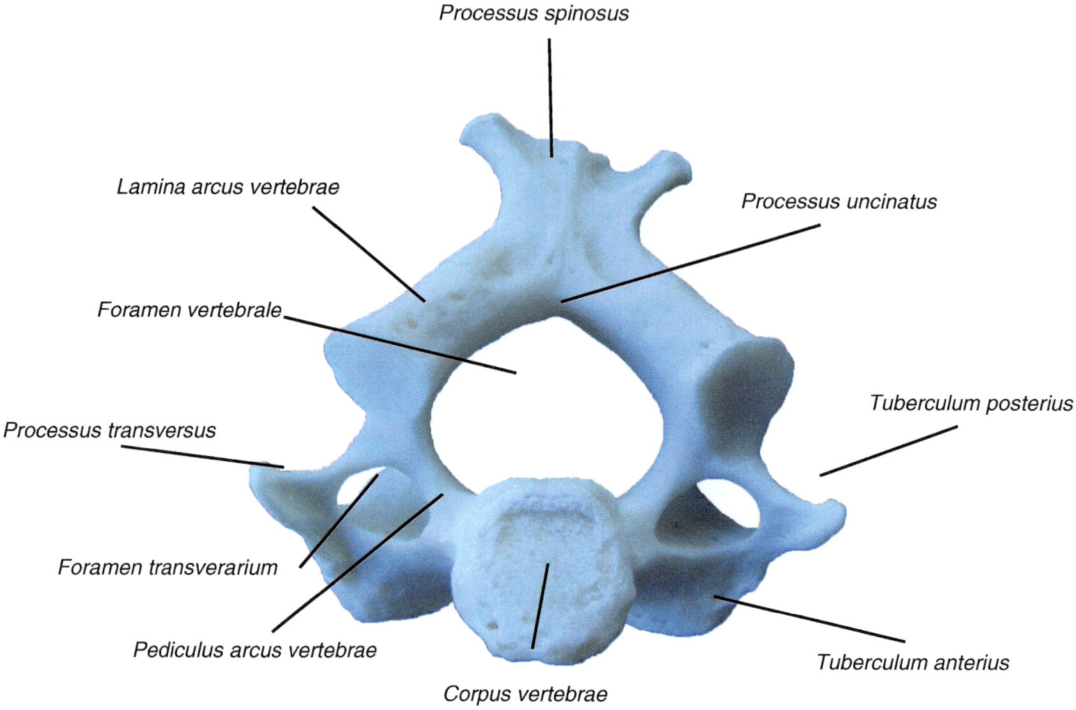


**Fig. 2.24** *Pan paniscus*: example of cervical vertebra

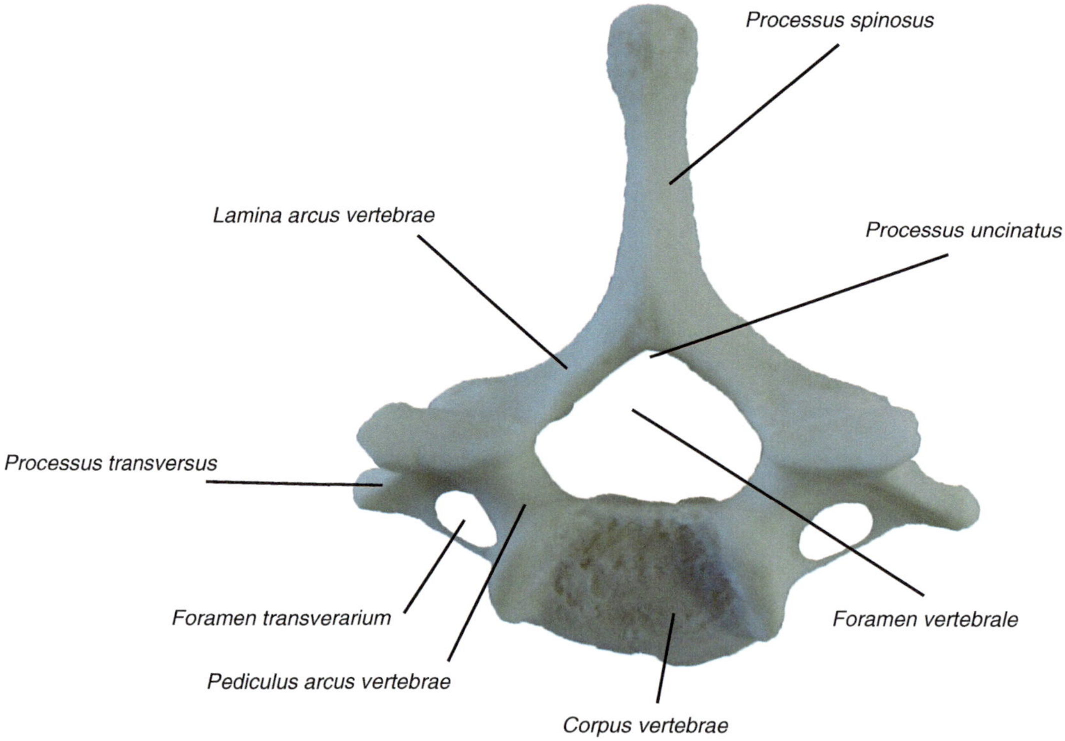


**Fig. 2.25** *Pan paniscus*: example of cervical vertebra

## Bibliography

### Appendix I: Literature Including Information about the Muscles of Bonobos and Common Chimpanzees


Aiello L, Dean C (1990) An introduction to human evolutionary anatomy. Academic Press, San Diego

Ashton EH, Oxnard CE (1963) The musculature of the primate shoulder. J Zool Soc Lond 29:553–650

Avril C (1963) Kehlkopf und Kehlsack des Schimpansen (*Pan troglodytes*). Gegenbaurs Morphol Jahrb 105: 74–129

Bischoff TLW (1870) Beitrage zur Anatomie des *Hylobates leuciscus* and zueiner vergleichenden Anatomie der Muskeln der Affen und des Menschen. Abh Bayer Akad Wiss Miinchen Math Phys Kl 10:197–297

Broca P (1869) L'ordre des primates – parallele anatomique de l'homme et des singes. Bull Soc Anthropol Paris 4:228–401

Burrows AM (2008) The facial expression musculature in primates and its evolutionary significance. Bioessays 30:212–225

Burrows AM, Waller BM, Parr LA, Bonar CJ (2006) Muscles of facial expression in the chimpanzee (*Pan troglodytes*): descriptive, comparative and phylogenetic contexts. J Anat 208:153–167

Champneys F (1872) The muscles and nerves of a Chimpanzee (*Troglodytes Niger*) and a *Cynocephalus Anubis*. J Anat Physiol 6:176–211

Chapman HC (1879) On the structure of the chimpanzee. Proc Acad Nat Sci Phila 31:52–63

Dean MC (1985) Comparative myology of the hominoid cranial base, II, the muscles of the prevertebral and upper pharyngeal region. Folia Primatol 44:40–51

Diogo R, Abdala V (2010) Muscles of vertebrates – comparative anatomy, evolution, homologies and development. Taylor & Francis, Oxford

Diogo R, Wood BA (2011) Soft-tissue anatomy of the primates: phylogenetic analyses based on the muscles of the head, neck, pectoral region and upper limb, with notes on the evolution of these muscles. J Anat 219:273–359

Diogo R, Wood BA (2012) Comparative anatomy and phylogeny of primate muscles and human evolution. Taylor & Francis, Oxford

Diogo R, Abdala V, Lonergan N, Wood BA (2008) From fish to modern humans – comparative anatomy, homologies and evolution of the head and neck musculature. J Anat 213:391–424

Diogo R, Wood BA, Aziz MA, Burrows A (2009) On the origin, homologies and evolution of primate facial muscles, with a particular focus on hominoids and a suggested unifying nomenclature for the facial muscles of the Mammalia. J Anat 215:300–319

DuBrul EL (1958) Evolution of the speech apparatus. Thomas, Springfield

Duchin LE (1990) The evolution of human speech: comparative anatomy of the oral cavity in *Pan* and *Homo*. J Hum Evol 19:687–697

Duckworth WLH (1912) On some points in the anatomy of the plica vocalis. J Anat Physiol 47:80–115

Duvernoy M (1855–1856) Des caracteres anatomiques de grands singes pseudoanthropomorphes anthropomorphes. Arch Mus Natl Hist Nat Paris 8:1–248

Edgeworth FH (1935) The cranial muscles of vertebrates. Cambridge University Press, Cambridge

Giacomini C (1897) 'Plica semilunaris' et larynx chez les singes anthropomorphes. Arch Ital Biol 28:98–119

Gibbs S (1999) Comparative soft tissue morphology of the extant Hominoidea, including Man. Unpublished PhD Thesis, The University of Liverpool, Liverpool

Göllner K (1982) Untersuchungen über die vom N. trigeminus innervierte Kiefermusculatur des Schimpansen (*Pan troglodytes*, Blumenbach 1799) und des Gorilla (*Gorilla gorilla gorilla*, Savage and Wyman 1847). Gegenbaurs Morphol Jahrb 128:851–903

Gratiolet LP, Alix PHE (1866) Recherches sur l'anatomie du *Troglodytes aubryi*. Nouv Arch Mus Hist Nat Paris 2:1–264

Harrison DFN (1995) The anatomy and physiology of the mammalian larynx. University Press, Cambridge

Hartmann R (1886) Anthropoid apes. Keegan, London

Himmelreich HA (1971) M. levator veli palatini der Säugetiere. Gegenbaurs Morphol Jahrb 116:377–400

Himmelreich HA (1977) M. cephalopharyngeus der Säugetiere. Gegenbaurs Morphol Jahrb 123:556–588

Huber E (1930) Evolution of facial musculature and cutaneous field of trigeminus – Part II. Q Rev Biol 5:389–437

Huber E (1931) Evolution of facial musculature and expression. The Johns Hopkins University Press, Baltimore

Jordan J (1971a) Studies on the structure of the organ of voice and vocalization in the chimpanzee, Part 1. Folia Morphol 30:99–117

Jordan J (1971b) Studies on the structure of the organ of voice and vocalization in the chimpanzee, Part 2. Folia Morphol 30:222–248

Jordan J (1971c) Studies on the structure of the organ of voice and vocalization in the chimpanzee, Part 3. Folia Morphol 30:323–340

Jouffroy FK, Saban R (1971) Musculature peaucière. In: Grassé PP (ed) Traité de Zoologie, XVI: 3 (Mammifères). Masson et Cie, Paris, pp 477–611

Kelemen G (1948) The anatomical basis of phonation in the chimpanzee. J Morphol 82:229–256

Kelemen G (1969) Anatomy of the larynx and the anatomical basis of vocal performance. In: Bourne GH (ed) The Chimpanzee, Anatomy, behaviour and diseases of chimpanzees, vol Vol 1. Karger, Basel, pp 35–186

Kohlbrügge JHF (1896) Der larynx und die stimmbildung der Quadrumana. Natuurk T Ned Ind 55:157–175

Körner O (1884) Beiträge zur vergleichenden Anatomie und Physiologie des Kehlkopfes der Säugethiere und des Menschen. Abh Senckenb Naturforsch Ges 13:147–261

Larson SG, Stern JT, Jungers WL (1991) EMG of serratus anterior and trapezius in the chimpanzee: scapular rotators revisited. Am J Phys Anthropol 85:71–84
Lightoller GHS (1925) Facial muscles – the modiolus and muscles surrounding the rima oris with some remarks about the panniculus adiposus. J Anat Physiol 60:1–85
Lightoller GS (1928) The facial muscles of three orang utans and two cercopithecidae. J Anat 63:19–81
Lightoller GS (1934) The facial musculature of some lesser primates and a *Tupaia*. Proc Zool Soc Lond 1934:259–309
Lightoller GS (1939) V. Probable homologues. A study of the comparative anatomy of the mandibular and hyoid arches and their musculature — Part I. Comparative myology. Trans Zool Soc Lond 24:349–382
Lightoller GS (1940a) The comparative morphology of the platysma: a comparative study of the sphincter colli profundus and the trachelo-platysma. J Anat 74:390–396
Lightoller GS (1940b) The comparative morphology of the m. caninus. J Anat 74:397–402
Lightoller GS (1942) Matrices of the facialis musculature: homologization of the musculature in monotremes with that of marsupials and placentals. J Anat 76:258–269
Loth E (1931) Anthropologie des parties molles (muscles, intestins, vaisseaux, nerfs peripheriques). Mianowski-Masson et Cie, Paris
Macalister A (1871) On some points in the myology of the chimpanzee and others of the primates. Ann Mag Nat Hist 7:341–351
Maier W (2008) Epitensoric position of the chorda tympani in Anthropoidea: a new synapomorphic character, with remarks on the fissura glaseri in Primates. In: Sargis EJ, Dagosto M (eds) Mammalian evolutionary morphology: a tribute to Frederick S. Szalay. Springer, Dordrecht, pp 339–352
Miller RA (1952) The musculature of *Pan paniscus*. Am J Anat 91:182–232
Mustafa MA (2006) Neuroanatomy. 10th Natl Congr Anat Bordum, Turkey, September 5:6–10
Owen R (1830–1831) On the anatomy of the orangutan (*Simia satyrus*, L.) Proc Zool Soc Lond 1:4–5, 9–10, 28–29, 66–72
Owen R (1868) The anatomy of vertebrates, Mammals, vol 3. Longmans, Green & Co., London
Pellatt A (1979) The facial muscles of three African primates contrasted with those of *Papio ursinus*. S Afr J Sci 75:436–440
Ruge G (1887a) Untersuchungen uber die Gesichtsmuskeln der Primaten. W. Engelmann, Leipzig
Ruge G (1887b) Die vom Facialis innervirten Muskeln des Halses, Nackens und des Schädels einen jungen *Gorilla*. Gegenbaurs Morphol Jahrb 12:459–529
Saban R (1968) Musculature de la tête. In: Grassé PP (ed) Traité de Zoologie, XVI: 3 (Mammifères). Masson et Cie, Paris, pp 229–472
Schück AC (1913) Beiträge zur Myologie der Primaten, II – 1 die gruppe sterno-cleido-mastoideus, trapezius, omo-cervicalis, 2 die gruppe levator scapulae, rhomboides, serratus anticus. Morphol Jahrb 46: 355–418
Seiler R (1970) Differences in the facial musculature of the nasal and upper-lip region in catarrhine primates and man. Z Morphol Anthropol 62:267–275
Seiler R (1971a) Facial musculature and its influence on the facial bones of catarrhine Primates, II. Morphol Jahrb 116:147–185
Seiler R (1971b) Facial musculature and its influence on the facial bones of catarrhine Primates, III. Morphol Jahrb 116:347–376
Seiler R (1974) Muscles of the external ear and their function in man, chimpanzees and *Macaca*. Morphol Jahrb 120:78–122
Seiler R (1976) Die Gesichtsmuskeln. In: Hofer H, Schultz AH, Starck D (eds) Primatologia, Handbuch der Primatenkunde, Bd. 4, Lieferung 6. Karger, Basel, pp 1–252
Seiler R (1979) Criteria of the homology and phylogeny of facial muscles in primates including man, II, Catarrhina. Morphol Jahrb 125:298–323
Sonntag CF (1923) On the anatomy, physiology, and pathology of the chimpanzee. Proc Zool Soc Lond 23:323–429
Sonntag CF (1924) The morphology and evolution of the apes and man. John Bale Sons and Danielsson, Ltd., London
Starck D (1973) The skull of the fetal chimpanzee. In: Bourne GH (ed) The Chimpanzee, vol 6. Karger, Basel, pp 1–33
Starck D, Schneider R (1960) Respirationsorgane. In: Hofer H, Schultz AH, Starck D (eds) Primatologia III/2. Karger, Basel, pp 423–587
Stewart TD (1936) The musculature of the anthropoids, I, neck and trunk. Am J Phys Anthropol 21:141–204
Sullivan WE, Osgood CW (1925) The facialis musculature of the orang, *Simia satyrus*. Anat Rec 29: 195–343
Sutton JB (1883) On some points in the anatomy of the chimpanzee (*Anthropopithecus troglodytes*). J Anat Physiol 18:66–85
Swindler DR, Wood CD (1973) An atlas of primate gross anatomy: baboon, chimpanzee and men. University of Washington Press, Seattle
Tuttle RH, Basmajian JV (1976) Electromyography of pongid shoulder muscles and hominoid evolution I – retractors of the humerus and rotators of the scapula. Yearbook Phys Anthropol 20:491–497

Tyson E (1699) Orang-Outang sive Homo sylvestris, or the anatomy of a pygmie compared to that of a monkey, an ape and a man. T. Bennett and D. Brown, London
Virchow H (1915) Gesichtsmuskeln des Schimpansen. Abh Preuss Akad Wiss 1:l–81
Vrolik W (1841) Recherches d' anatornie comparé, sur le chimpanzé. Johannes Miller, Amsterdam
Waller BM, Vick SJ, Parr LA, Bard KA, Smith Pasqualini MC, Gothard K, Fuglevand A (2006) Intramuscular electrical stimulation of facial muscles in humans and chimpanzees: Duchenne revisited and extended. Emotion 6:367–382
Wilder B (1862) Contributions to the comparative myology of the chimpanzee. Boston J Nat Hist 6:352–384

## Appendix II: Literature Cited, Not Including Information About the Muscles of Bonobos and Common Chimpanzees

Netter FH (2006) Atlas of human anatomy, 4th edn. Saunders, Philadelphia
Terminologia Anatomica (1998) Federative Committee on Anatomical Terminology. Georg Thieme, Stuttgart. New York: Columbia University Press
Wood J (1870) On a group of varieties of the muscles of the human neck, shoulder, and chest, with their transitional forms and homologies in the Mammalia. Philos Trans R Soc Lond 160:83–116

# 3 Upper Limb Musculature and Brachial Plexus

## 3.1 Upper Limb Musculature

***Serratus anterior*** (adult Kidogo 170.28 g; adolescent Jasiri 111.60 g; infant Foyo 7.21 g; fetus 1.71 g; Figs. 3.1, 3.2, 3.4, 3.19, 3.20, 3.21 and 3.30)

- Attachments in bonobos: From ribs (ribs 1–10 in adult bonobo of Miller 1952 and infant Foyo; ribs 1–11 in our fetus and adult Kidogo; ribs 1–12 in adolescent Jasiri and infant Etje) to the medial side of the scapula, being well-separated from the levator scapulae in adults. However, in the bonobo fetus, at the insertion of the levator scapulae onto the medial border of the scapula there is a fusion with the serratus anterior, reflecting their developmental closeness, although the fusion is not as extensive as is the case in many other mammals.
- Usual innervation: Long thoracic nerve from the brachial plexus (Miller 1952: *P. paniscus*) [*P. troglodytes*: long thoracic nerve, from C4, C5 and C6 (Hepburn 1892); long thoracic nerve, from C6, C6 and C7 (Schück 1913b); long thoracic nerve from the brachial plexus (Swindler and Wood 1973)].
- Function: Caudal part of muscle exhibited EMG activity during arm-raising (Tuttle and Basmajian 1976: *P. troglodytes*); EMG activity mainly related to arm-raising motions, except lowest parts of muscle, which display significant activity during suspensory postures and locomotion presumably to control the tendency of the scapula to shift cranially relative to the rib cage (Larson et al. 1991: *P. troglodytes*).
- Comparisons with common chimpanzees: In *P. troglodytes* it runs from the medial side of the scapula, being well-separated from the levator scapulae, to the ribs (e.g., ribs 1, 3 and 4–11: Wilder 1862; ribs 1–10: Gratiolet and Alix 1866, Champneys 1872; ribs 1–12: Macalister 1871; ribs 1–11: Hepburn 1892, Sonntag 1923; ribs 1–10 or 1–12: Schück 1913b; 1–11 or 1–12: Stewart 1936; extending to rib 12: Stern et al. 1980).
- Synonymy: Serratus magnus (Wilder 1862; Macalister 1871; Champneys 1872; Hepburn 1892; Beddard 1893; Sonntag 1923, 1924); grand dentelé (Gratiolet and Alix 1866; Broca 1869); pars caudalis of serratus anticus (Schück 1913b); caudal serratus anterior (Larson et al. 1991).

***Rhomboideus*** (adult Kidogo 78.51 g; adolescent Jasiri 58.28 g; infant Foyo 3.06 g; fetus 1.02 g; Figs. 3.5, 3.6, 3.19, 3.31 and 3.32)

- Attachments in bonobos: Single muscle from cervical and thoracic vertebrae (C3-T6 in adult bonobo of Miller 1952; C5-T3 in our fetus and C4-T4 in our adult Kidogo; C6-T5 in adolescent Jasiri; C6-T7 in infant Foyo;




R. Diogo et al., *Photographic and Descriptive Musculoskeletal Atlas of Bonobos*,
DOI 10.1007/978-3-319-54106-8_3

C7-T5 in infant Etje) to medial border of the scapula. In the bonobo fetus (Fig. 3.32), it is partially fused with the serratus anterior and the levator scapulae – reflecting their developmental closeness – and it is also fused with the rhomboideus of the other side of the body.

- Usual innervation: C5 and dorsal scapular nerve (Miller 1952: *P. paniscus*) [*P. troglodytes*: C4 (Hepburn 1892); C4 and C5 (Champneys 1872; Sonntag 1923); C3 and C4 (in one specimen) or C4 and C5 (in one specimen) (Schück 1913b); dorsal scapular nerve (Swindler and Wood 1973)].
- Function: The rhomboideus exhibited EMG activity during arm-raising, but there was little or no activity during hoisting and suspension (Tuttle and Basmajian 1978a: *P. troglodytes*).
- Comparison with common chimpanzees: In *P. troglodytes* the rhomboideus usually runs from the medial border of the scapula to the cervical and thoracic vertebrae (e.g., C3-T8 or C3-T7: Schück 1913b; C6-T4: Sonntag 1923, 1924; C2-T6: Stewart 1936; C4-T6: Swindler and Wood 1973; often from C2 or C3 to T6 or T7: Andrews and Groves 1976; C4-T3, T4 or T5 in our specimen PFA 1077; C6-T5 in our specimen PFA UNC). The *rhomboideus major*, *rhomboideus minor* and *rhomboideus occipitalis* are usually not present as distinct entities in bonobos and in common chimpanzees; the rhomboideus of chimpanzees probably corresponds to the rhomboideus major plus rhomboideus minor of modern humans, the rhomboideus occipitalis being missing in both modern humans and common chimpanzees (e.g., Diogo and Wood 2011, 2012). However, some exceptions are worthy of notice. Gratiolet and Alix (1866) reported a *P. troglodytes* specimen with an undivided rhomboideus originating from the first 7 vertebral spines, the 'cervical ligament', the axis and the occipital region, but the occipital origin was only from an aponeurosis, so there was no distinct rhomboideus occipitalis. Champneys (1872) described a rhomboideus major (innervated by C5) and a rhomboideus minor (innervated by C4) in the *P. troglodytes* specimen dissected by him; according to him the rhomboideus major is similar to that of modern humans, its tendon of origin being fused with the overlying rhomboideus minor. Ziegler (1964) referred to, and showed in his fig. 4, a 'rhomboideus minor' and a 'rhomboideus major' in the *P. troglodytes* specimen dissected by him, but he did not describe these structures in detail so it cannot be discerned if they are well-defined and separated and thus if they effectively correspond to the rhomboideus major and rhomboideus minor of modern humans.

***Levator scapulae*** (adolescent Jasiri 12.48 g; infant Foyo 0.82 g; fetus 0.27 g; Figs. 3.6, 3.19, 3.24 and 3.32)

- Attachments in bonobos: From the cervical vertebrae (C2-C4 in our adult Kidogo and infant Foyo, C2-C5 in adolescent Jasiri, C1-C4 in bonobo fetus and in one side of adult bonobo of Miller 1952, and C1-C3 on the other side of Miller's adult bonobo) to the anterior portion of the medial side of the scapula.
- Usual innervation: Dorsal scapular nerve (Miller 1952: *P. paniscus*) [*P. troglodytes*: C4 (Champneys 1872); one specimen by C4 while in the other it was by C3 and C4 (Schück 1913b); C3 and C5 (Sonntag 1923); C3 and C4 spinal nerves (Swindler and Wood 1973; these authors state that in modern humans it is also by C3 and C4)].
- Comparison with common chimpanzees: Sonntag (1923) described an origin in *P. troglodytes* from C1-C5 and Swindler and Wood (1973) from C1-C4, but in one side of the *P. troglodytes* specimen dissected by Stewart (1936 and in the three *P. troglodytes* specimens reported by Schück (1913b) and the *P. troglodytes* specimens described by Hepburn (1892), the origin was from C1-C3. In the *P. troglodytes* specimen described by Gratiolet and Alix (1866), the origin was from C2-C3, while in one side of the *P. troglodytes* specimen dissected by Stewart (1936) and in the *P. troglodytes* specimen analysed by Champneys (1872), it was from C1-C2 only. In our *P. troglodytes* specimens PFA 1016, PFA 1051 and

HU PT1, the levator scapulae originated from C1-C3 (one slip per vertebra).
- Synonymy: Levator anguli scapulae (Wilder 1862; Macalister 1871; Champneys 1872; Hepburn 1892, Beddard 1893; Sonntag 1923, 1924); levator scapulae plus pars cranialis of serratus anticus (Schück 1913b); l'angulaire de l'omoplate (Gratiolet and Alix 1866; Broca 1869).

***Levator claviculae*** (adult Kidogo 10.78 g; adolescent Jasiri 7.67 g; infant Foyo 0.38 g; fetus 0.17 g; Figs. 3.2, 3.18, 3.19, 3.24, 3.28, 3.30 and 3.32)

- Attachments in bonobos: From atlas to clavicle (to middle 1/3 clavicle in adult bonobo of Miller 1952; to lateral 1/3 of clavicle in our fetus and adult Kidogo and adolescent Jasiri; to acromial end of clavicle in infant Foyo), passing deep to (covered either laterally or dorsally by) the trapezius, contrary to what was stated by Miller (1952) (she stated that in her adult bonobo the levator claviculae passed lateral to the trapezius; see below).
- Usual innervation: Branch of the anterior division of C3 through the cervical plexus (Miller 1952: *P. paniscus*) [*P. troglodytes*: in one specimen it was C3 while in the other specimen of the same species it was C3 and C4 (Schück 1913b)].
- Comparison with common chimpanzees: In *P. troglodytes* it usually runs from the atlas to the clavicle, passing deep to (covered either laterally or dorsally by) the trapezius (to, e.g., the lateral 1/3 of clavicle: Gratiolet and Alix 1866; the lateral 1/2 of clavicle: Champneys 1872; a point just lateral to the middle of the clavicle: Stewart 1936). Andrews and Groves (1976) stated that in their common chimpanzee specimens the levator claviculae passes lateral to the trapezius, but the reports of most authors, e.g., Champneys (1872), Schück (1913a, b) and Ashton and Oxnard's (1963), agree with our observations in *P. troglodytes* (i.e., the levator claviculae is usually mainly deep to the trapezius). Broca (1869) stated that gorillas and common chimpanzees lack a levator claviculae, but this is clearly not usually the case in either of these taxa. Tyson (1699) stated that in the *P. troglodytes* specimen dissected by him, the levator claviculae connected C2-C3 to the clavicle, but he was probably referring to part of the levator scapulae, because in the vast majority of common chimpanzees and of other primates the muscle originates from the atlas only. Sutton (1883) stated that in the two *P. troglodytes* specimens dissected by him the levator claviculae is not present, but his statements should also be taken with caution, because he also stated that most of the facial muscles that are now recognized in common chimpanzees (see above) were missing in his two specimens.
- Synonymy: Tracheloclavicular (Wyman 1855); cléido-atloïdien or cléido-trachélien (Gratiolet and Alix 1866); omo-atlanticus (Macalister 1871); acromio-basilar or acromio-trachélien (Champneys 1872); omocervicalis, cleido-cervicalis, acromiocervicalis, levator anticus scapulae (Barnard 1875; Schück 1913b); omo-trachelian (Sonntag 1923, 1924); atlantoscapularis anterior (Ashton and Oxnard 1963).

***Atlantoscapularis posticus*** (infant Foyo 0.25 g; Fig. 3.19)

- Attachments in bonobos: The atlantoscapularis posticus (see Diogo et al. 2009) is usually not present as a distinct muscle in bonobos, but the variant bundle of the levator claviculae going to the scapula on both sides of our infant Foyo (Fig. 3.19) clearly corresponds to this muscle, supporting the idea that it does derive from the levator claviculae, as proposed by Diogo and Wood (2012) (when the atlantoscapularis posticus is present as a separate muscle, the remaining – more anterior – portion of the levator claviculae is then usually designated as atlantoscapularis anticus).
- Usual innervation: No data available, but it is probably by branch of the anterior division of C3 through the cervical plexus, as this is the innervation of the levator claviculae in bonobos according to Miller (1952).

- Comparison with common chimpanzees: The atlantoscapularis posticus is usually not present as a distinct muscle in common chimpanzees.

***Subclavius*** (adult Kidogo 1.26 g; adolescent Jasiri 2.62 g; infant Foyo 0.13 g; fetus 0.05 g; Figs. 3.4, 3.11, 3.30, 3.52 and 3.53)

- Attachments in bonobos: From first rib to clavicle, lying closely to the costocoracoid ligament, which is very well-developed and strong (our fetus and adult Kidogo and adolescent Jasiri and infants Etje and Foyo).
- Usual innervation: Nerve to subclavius, from brachial plexus (Miller 1952: *P. paniscus*) [*P. troglodytes*: nerve to subclavius, from C5 and C6 (*P. troglodytes*); nerve to subclavius, from C5 (*P. troglodytes*)].
- Comparison with common chimpanzees: In *P. troglodytes* it runs from the first rib to clavicle, lying closely to the costocoracoid ligament. The *costocoracoideus* is usually not present as a distinct muscle in common chimpanzees, but these and most other primates do usually have a very well-developed, strong costocoracoid ligament, which corresponds to the costocoracoideus muscle of primitive mammals such as monotremes (see Diogo et al. 2009). However, as is the case in modern humans, a costocoracoideus muscle may be present as an anomaly in common chimpanzees. For instance, Champneys (1872) describes a 'fibrous band from the coracoid to the sternum between the articulation of the clavicle and the first rib' in the *P. troglodytes* specimen dissected by him, which is partially fused to the subclavius muscle and the costocoracoid membrane, and which thus seems to correspond to the muscle costocoracoideus sensu the present study. Also, MacDowell (1910) describes a 'sterno-chondro-scapularis' in the *P. troglodytes* specimen dissected by him, which arises from a tendon inserted in rib 1 and inserts onto the coracoid process mesially to the attachment of the short head of the biceps brachii, the only muscular fibres of this 'sterno-chondro-scapularis' being a small group of fibres that occur at the coracoid attachment, and in a still smaller manner at the costal insertion; this 'sterno-chondro-scapularis' seemingly corresponds to a vestigial costocoracoideus sensu the present study.
- Synonymy: Sous-clavier (Gratiolet and Alix 1866).

***Pectoralis major*** (adult Kidogo 301.68 g; adolescent Jasiri 243.67 g; infant Foyo 10.81 g; fetus 2.05 g; Figs. 3.1, 3.2, 3.3, 3.20, 3.28, 3.52 and 3.53)

- Attachments in bonobos: The pars clavicularis – which is not well differentiated in our fetus – originates from clavicle (medial 1/3 in our fetus and adult Kidogo, medial 1/2 in Miller's 1952 adult bonobo and our adolescent Jasiri and infants Foyo and Etje), the pars sternocostalis from sternum and ribs (1–7 in our fetus and infant Foyo and adult Kidogo and Miller's 1952 adult bonobo; 1–8 in our adolescent Jasiri and infant Etje) and the pars abdominalis – which is not well differentiated in our fetus and adult Kidogo and our adolescent Jasiri and Miller's 1952 adult bonobo – from the abdominal musculature. All these three parts insert onto the proximal humerus, the pars sternocostalis inserting proximal to the insertion of the pars clavicularis. The *chondro-epitrochlearis* sensu MacDowell 1910 (which is present as a variant in common chimpanzees, see just below) was present in our fetus (3.29) and adult Kidogo (Fig. 3.3) and adolescent Jasiri and infants Etje and Foyo, being a deep bundle running from ribs 4–6 in our fetus and 4–5 in our adult Kidogo and 5–7 in our adolescent Jasiri and infant Etje, and 5–6 in our infant Foyo, to the proximal humerus (crest of the greater tubercle), fusing with the main body of the pectoralis major distally, and not corresponding with a *pectoralis tertius* because it is central, not caudal, and is clearly part of pectoralis major, not of the pectoralis minor/deep musculature. This 'chondro-epitrochlearis' clearly seems to correspond to the deep structure of the adult bonobo shown in fig. 3 of Miller (1952), who erroneously named this structure as 'pectoralis abdominis'. Therefore, contrary to common chimpanzees, in which the 'chondro-epitrochlearis' is present as a variant, in bonobos the 'chondro-epitrochlearis' seems to be part of the normal phenotype. In bonobos, the pectoralis major is fused with its counterpart at the midline, and with the deltoid, in our fetus and adult Kidogo, and in addition with the sternocleidomastoideus and pectoralis minor in our fetus.

- Usual innervation: Medial and lateral pectoral nerves (Miller 1952: *P. paniscus*) [*P. troglodytes*: anterior thoracic ('pectoral') nerve, which includes contributions from all of the trunks of the brachial plexus, or from the C7 and the combined trunk of C8 and T1 (Champneys 1872); medial and lateral pectoral nerves (Hepburn 1892, Swindler and Wood 1973)].
- Function: Pars sternocostalis of pectoralis major exhibited moderate or high EMG activity during hoisting behaviour and crutch-walking (Tuttle and Basmajian 1976, 1978b: *P. troglodytes*).
- Comparison with common chimpanzees: In *P. troglodytes* the pars clavicularis usually runs from the medial portion of the clavicle and often from the sternum to the proximal humerus; the pars sternocostalis runs from the sternum and ribs to the proximal humerus, inserting proximal to the insertion of the pars clavicularis; the pars abdominalis runs from the ribs and aponeurosis of the external oblique, deep to the pars sternocostalis and pars clavicularis, to the proximal humerus and to the coracoid process of the scapula. The pectoralis major is sometimes, but not usually, blended with its counterpart in the midline in *P. troglodytes* (e.g., Loth 1931). According to the literature review of Andrews and Groves (1976) in *Pan* the clavicular origin of the clavicular head is usually from the medial 1/4 to 1/2 of this bone, although it might be from the medial 2/3 in some cases (e.g., MacDowell 1910). In turn, according to the literature review of Gibbs (1999) there is no clavicular origin of the pectoralis major in 2/13 of *Pan*. Regarding the origin from the ribs, its posterior limit is often on the fourth, fifth or sixth ribs according to the literature review of Andrews and Groves (1976), although in some cases it might extend to rib 8 (e.g., Gibbs 1999). The *tensor semi-vaginae articulationis humero-scapularis sensu* Macalister (1871) and Huntington (1903), which is also designated as *sterno-humeralis*, *sterno-chondro-humeralis* or *pectoralis minimus* by these authors, is a muscle that is occasionally present in modern humans, and that is found on the left side in a *P. troglodytes* specimen described by Macalister (1871), arising from the cartilages of ribs 1–2 (for more details on this structure and its evolution and homologies, see Diogo and Wood 2012). MacDowell (1910) stated that in the *P. troglodytes* specimen dissected by him there is an 'additional muscle chondro-epitrochlearis' that lies deep (dorsal) to the sternocostal head of the pectoralis major, originates from ribs 2–4 and the fascia underlying this head of the pectoralis major, and runs parallel with this head to insert with it onto the humerus. The structure that MacDowell designates as *chondro-epitrochlearis* is most likely an additional slip of the pectoralis major, and it does not correspond to the anomaly/variant seen in modern humans referred to as *muscle chondroepitrochlearis*. In fact, in modern humans this variant/anomalous muscle usually runs from the ribs to the median intermuscular septum or onto the medial epicondyle of the humerus (hence its name 'chondroepitrochlearis'), contrary to the bonobos with a 'chondroepitrochlearis' (see above). The *sternalis*, *pectoralis quartus* and *panniculus carnosus* are usually not present as distinct muscles in common chimpanzees.
- Synonymy: Pectoralis major plus the lower portion of pectoralis minor (Hartmann 1886); pectoralis major plus pectoralis abdominis, abdominalis and/or chondroepitrochlearis quartus (MacDowell 1910; Gibbs 1999).

***Pectoralis minor*** (adult Kidogo 22.30 g; adolescent Jasiri 16.83 g; infant Foyo 1.20 g; fetus 0.35 g; Figs. 3.3, 3.11, 3.18, 3.21, 3.29, 3.30, 3.52 and 3.53)

- Attachments in bonobos: From ribs (2–4 in Miller's 1952 adult bonobo and our adolescent Jasiri and infant Etje; 1–3 in our fetus and adult Kidogo; 1–4 in our infant Foyo) to coracoid process of scapula (our adult Kidogo and fetus and adolescent Jasiri, and Miller's 1952 adult bonobo and infants Etje and Foyo). A variant is seen on both sides of the adolescent Jasiri and on the right side of our infant Foyo: the insertion onto the coracoid process is made by two separate tendons (Figs. 3.11 and 3.18).
- Usual innervation: Medial pectoral nerve (Miller 1952: *P. paniscus*) [*P. troglodytes*: anterior thoracic (pectoral) nerve, which includes elements from all three trunks of the brachial plexus, or from C7 and the combined trunk of C8 and T1 (Champneys 1872); medial pectoral nerve (Hepburn 1892; Swindler and Wood 1973).
- Comparison with common chimpanzees: In *P. troglodytes* it runs from the proximal humerus and/or the glenohumeral joint capsule and, sometimes, from the coracoid process of the scapula, to the ribs (e.g., ribs 2–4: Hepburn 1892, Sonntag 1923, Stewart 1936, and our specimens PFA 1016 and PFA UNC; ribs 2–3 and aponeurosis from ribs 4: Champneys 1872, Ziegler 1964; ribs 2–5: Gratiolet and Alix 1866; ribs 2–3: Chapman 1879, and our specimen PFA 1077; ribs 3–5: Sutton 1883, Swindler and Wood 1973, and our specimen PFA 1051; ribs 1–4: MacDowell 1910). In some of the *P. troglodytes* specimens described by Gratiolet and Alix (1866), Humphry (1867), Macalister (1871), Barnard (1875), Hartmann (1886), Beddard (1893), Lander (1918), and Andrews and Groves (1976) and dissected by us, some fibres of the pectoralis minor may insert onto the coracoid process, but even in these cases the great majority of the fibres of this muscle insert onto other structures (e.g., the capsule of the glenohumeral joint). Moreover, in most of the *P. troglodytes* specimens described (e.g., Wyman 1855; Huxley 1864; Broca 1869; Champneys 1872; Chapman 1879; Sutton 1883; Hepburn 1892; MacDowell 1910; Sonntag 1923, 1924; Ziegler 1964; Swindler and Wood 1973), there is no attachment at all onto the coracoid process. In a few *P. troglodytes* specimens, such as one specimen described by Wilder (1862), the pectoralis minor inserted onto the coracoid process on one side of the body, while on the other side of the body it inserted onto the greater tuberosity of the humerus instead. In summary, the usual condition for *P. troglodytes* is for there to be no direct attachment onto the coracoid process (e.g., Ashton and Oxnard 1963 suggest that in three-quarters of *P. troglodytes* specimens described in the literature reviewed by him there is effectively no insertion onto the coracoid process), contrary to the condition found in bonobos where there is seemingly always an insertion onto the coracoid process. The *pectoralis tertius* (*xiphihumeralis*) (see Diogo et al. 2009) is usually not present as a distinct muscle in common chimpanzees (see notes about the attachments of the pectoralis major in bonobos, above).
- Synonymy: Upper portion of pectoralis minor (Hartmann 1886).

***Infraspinatus*** (adult Kidogo 92.02 g; adolescent Jasiri 52.99 g; infant Foyo 2.84 g; fetus 0.99 g; Figs. 3.10, 3.24, 3.25, 3.27 and 3.36)

- Attachments in bonobos: From infraspinous fossa of scapula and infraspinatus fascia to greater tuberosity of the humerus (our fetus and adult Kidogo and adolescent Jasiri and Miller's 1952 adult bonobo).
- Usual innervation: Suprascapular nerve (Miller 1952: *P. paniscus*) [*P. troglodytes*: suprascapular nerve, from C5 and C6 (Hepburn 1892); suprascapular nerve (Swindler and Wood 1973)].
- Function: Infraspinatus exhibited EMG activity during arm-raising; low or nil activity was exhibited during hoisting behaviour (Tuttle and Basmajian 1978a: *P. troglodytes*). EMG study indicated that the supraspinatus, infra-

spinatus and subscapularis have 'completely different' functions; the supraspinatus acts as a more or less pure abductor of the arm, assisting the deltoideus, while the infraspinatus is more an abductor/lateral rotator and the different parts of the subscapularis have distinct functions such as abduction/medial rotation and adduction/medial rotation (Larson and Stern 1986: *P. troglodytes*).
- Comparison with common chimpanzees: In *P. troglodytes* it runs from the infraspinous fossa of the scapula and the infraspinatus fascia to the greater tuberosity of the humerus and in some cases to the capsule of the glenohumeral joint (e.g., Beddard 1893).
- Synonymy: Sous-épineux (Gratiolet and Alix 1866).

***Supraspinatus*** (adult Kidogo 46.96 g; adolescent Jasiri 34.19 g; infant Foyo 1.23 g; fetus 0.58 g; Figs. 3.6, 3.8, 3.10, 3.24, 3.25, 3.32, 3.35 and 3.36)

- Attachments in bonobos: From the supraspinous fossa of the scapula and the supraspinatus fascia to the greater tuberosity of the humerus (our fetus and adult Kidogo and Miller's 1952 adult bonobo and adolescent Jasiri and infants Foyo and Etje).
- Usual innervation: Suprascapular nerve (Miller 1952: *P. paniscus*) [*P. troglodytes*: suprascapular nerve, from C5 and C6 (Hepburn 1892); suprascapular nerve (Swindler and Wood 1973).
- Function: Supraspinatus exhibited high or moderate EMG activity during arm-raising, quiet tripedal and quadrupedal stance, and, unlike *Gorilla* and *Pongo*, also during hoisting behaviour (Tuttle and Basmajian 1978a: *P. troglodytes*). For details about EMG studies of *P. troglodytes* done by Larson and Stern (1986) see infraspinatus.
- Comparison with common chimpanzees: In *P. troglodytes* it runs from the supraspinous fossa of the scapula and the supraspinatus fascia to the greater tuberosity of the humerus and in many cases to the capsule of the glenohumeral joint, as well as, in a few cases, to the acromioclavicular ligaments (e.g., Gratiolet and Alix 1866). The muscle *scapuloclavicularis*, occasionally present in modern humans, has not been described in common chimpanzees nor was it found by us. Potau et al. (2009) studied the relative masses of the deltoideus, subscapularis, supraspinatus, infraspinatus and teres minor in common chimpanzees, orangutans and modern humans, and showed that for all of the proportional values in *Pongo pygmaeus* there is marked overlap between the ranges of values in the two species. According to Potau et al. (2009), this may indicate that the functional requirements of the glenohumeral joint are similar in a great ape with fundamentally suspension/vertical climbing locomotion and in a bipedal great ape that uses the upper extremity for essentially manipulative functions. They also showed that common chimpanzees have a proportionally larger mass of the rotator cuff versus the deltoideus, primarily because of the greater role played by the subscapularis. The latter may be an adaption to knuckle-walking because in this type of locomotion the subscapularis muscle acts as the principal stabilizer of the glenohumeral joint, compensating for the shearing force on this joint caused by the dorsal position of the scapula. Modern humans have a larger mass of the deltoideus versus that of the subscapularis; the deltoideus is the principal elevator muscle of the upper extremity in the scapular plane, but its contraction elevates the humeral head to the undersurface of the acromion. This may compress and injure the supraspinatus tendon, but is usually protected by the simultaneous contraction of the rotator cuff muscles. The smaller proportional mass of the rotator cuff musculature, particularly the subscapularis, in modern humans versus common chimpanzees, together with the larger proportional mass of the deltoid in modern humans, may explain the 1–3 mm upward movement of the humeral head during the first 30° to 60° of elevation of the upper extremity in the scapula plane in modern humans; this slight movement pinches the supraspinatus tendon

between the humeral head and acromion. This anatomical pattern in modern humans may provide the structural explanation for the tendency of certain individuals to suffer a range of subacromial syndromes (e.g., sports that require elevation of the upper extremity).
- Synonymy: Sus-épineux (Gratiolet and Alix 1866).

***Deltoideus*** (adult Kidogo 222.76 g; adolescent Jasiri 148.58 g; infant Foyo 7.38 g; fetus 2.20 g; Figs. 3.1, 3.2, 3.4, 3.5, 3.6, 3.7, 3.8, 3.9, 3.10, 3.10, 3.11, 3.18, 3.19, 3.20, 3.21, 3.23, 3.24, 3.28, 3.30, 3.31, 3.32, 3.33, 3.34, 3.35, 3.36 and 3.52)

- Attachments in bonobos: From lateral portion of clavicle (lateral 1/3 in Miller's 1952 adult bonobo and in our fetus and adolescent Jasiri and adult Kidogo and infants Foyo and Etje), the acromion (pars acromialis) and the spine of the scapula and the infraspinatus fascia (pars spinalis) to the humerus. In all bonobos dissected by us there is no clear differentiation of the deltoideus into different heads, and this muscle is fused with the pectoralis major; in the fetus in addition it is fused with the infraspinatus.
- Usual innervation: Axillary (circumflex) nerve (Miller 1952: *P. paniscus*) [same innervation in *P. troglodytes*: Hepburn 1892, Sonntag 1923, Swindler and Wood 1973].
- Function: Deltoideus exhibited moderate to high EMG activity during arm-raising – no data for hoisting behaviour (Tuttle and Basmajian 1978a: *P. troglodytes*); EMG indicated that the 'anterior and middle' portions of the deltoid are mainly related with the elevation of the arm, while the 'posterior' portion of this muscle has a 'completely different' function (e.g., humeral retraction Larson and Stern 1986: *P. troglodytes*).
- Comparison with common chimpanzees: In *P. troglodytes* it runs from the lateral portion of the clavicle (pars clavicularis – either the lateral 1/2: Gratiolet and Alix 1866, Beddard 1893, Sonntag 1923; the lateral 1/3: Ashton and Oxnard 1963, Swindler and Wood 1973; or more often the lateral 1/4: Andrews and Groves 1976), the acromion (pars acromialis) and the spine of the scapula and the infraspinatus fascia (pars spinalis) to the humerus.

***Teres minor*** (adult Kidogo 23.16 g; adolescent Jasiri 16.62 g; infant Foyo 1.08 g; fetus 0.25 g; Figs. 3.10, 3.24, 3.25 and 3.36)

- Attachments in bonobos: From infraspinatus fascia and lateral border of the scapula to greater tuberosity of the humerus (our fetus and adult Kidogo and adolescent Jasiri and infants Etje and Foyo and Miller's 1952 adult bonobo), often also inserting onto the humeral shaft just distal to this tuberosity (our adult Kidogo and adolescent Jasiri and infant Foyo) or sometimes onto the shoulder capsule (Miller's 1952 adult bonobo).
- Usual innervation: Axillary (circumflex) nerve (Miller 1952: *P. paniscus*) [same innervation in *P. troglodytes*: Champneys 1872, Hepburn 1892, Sonntag 1923, Swindler and Wood 1973].
- Function: EMG study indicated that the teres minor acted more as the 'posterior' portion of the deltoideus and, even, as the teres major, being, e.g., related with functions such as humeral retraction/adduction, as is often the case in orangutans and in modern humans (contrary to what is often stated in textbooks of human anatomy; Larson and Stern 1986: *P. troglodytes*).
- Comparison with common chimpanzees: In *P. troglodytes* it runs from the infraspinatus fascia and the lateral border of the scapula to the greater tuberosity of the humerus, and often also to the humeral shaft just distal to the same tuberosity (e.g., Gratiolet and Alix 1866; Sonntag 1923; Swindler and Wodd 1973; our *P. troglodytes* specimens, e.g., PFA 1016, PFA 1051, HU PT1).
- Synonymy: Petit rond (Gratiolet and Alix 1866).

***Subscapularis*** (adult Kidogo 157.39 g; adolescent Jasiri 102.86 g; infant Foyo 4.99 g; fetus 1.98 g; Figs. 3.4, 3.7, 3.9, 3.22, 3.33, 3.34, 3.53, 3.57 and 3.58)

- Attachments in bonobos: From subscapular fossa of scapula to lesser tuberosity of the humerus (our fetus and adult Kidogo and Miller's 1952 adult bonobo and adolescent Jasiri and infants Etje and Foyo), and often also to humeral shaft just distal to this tuberosity (our fetus and adult Kidogo and adolescent Jasiri and infants Foyo and Etje), or to the shoulder capsule (Miller's 1952 adult bonobo).
- Usual innervation: Subscapular nerves (Miller 1952: *P. paniscus*] [*P. troglodytes*: subscapular nerves (Hepburn 1892); upper and lower subscapular nerves (Swindler and Wood 1973)].
- Function: Moderate or high EMG activity during arm-raising; low or nil activity was exhibited during hoisting behaviour (Tuttle and Basmajian 1978a: *P. troglodytes*); for details of EMG study of *P. troglodytes* by Larson and Stern (1986), see infraspinatus.
- Comparison with common chimpanzees: In *P. troglodytes* it runs from the subscapular fossa of the scapula to the lesser tuberosity of the humerus, and often also to the humeral shaft just distal to this tuberosity (e.g., Sonntag 1923; our specimens, e.g., PFA 1016, PFA 1051, PFA 1077, PFA UNC). In the *P. troglodytes* specimen dissected by Ziegler (1964), the subscapularis has an accessory head arising from the superior half-inch of the axillary border of the bulk of the muscle – a few fibres of this head insert onto the capsular ligament while the rest attach along a half-inch strip of the humerus distal to the lesser tuberosity.

***Teres major*** (adult Kidogo 161.06 g; adolescent Jasiri 106.06 g; infant Foyo 4.40 g; fetus 1.25 g; Figs. 3.5, 3.6, 3.7, 3.8, 3.9, 3.10, 3.24, 3.27, 3.31, 3.32, 3.33, 3.34, 3.35 and 3.36)

- Attachments in bonobos: From lateral border and inferior angle of scapula to proximal humerus by means of a distal tendon that is often fused to the distal tendon of the latissimus dorsi (our fetus and Miller's 1952 adult bonobo; did not analyse this feature in detail in our adult Kidogo nor in adolescent Jasiri and infant Foyo). In addition, in our fetus the teres major is fused with the triceps brachii and the subscapularis (Fig. 3.33).
- Usual innervation: Subscapular nerves (Miller 1952: *P. paniscus*) [*P. troglodytes*: subscapular nerves (Champneys 1872, Hepburn 1892); lower subscapular nerve (Swindler and Wood 1973)].
- Function: EMG activity during hoisting behaviour (Tuttle and Basmajian 1976: *P. troglodytes*).
- Comparison with common chimpanzees: In *P. troglodytes* it runs from the lateral border and inferior angle of the scapula to the proximal portion of the intertubercular groove of the humerus by means of a distal tendon that is usually not fused to the distal tendon of the latissimus dorsi. According to Gibbs' (1999) review of the literature, a complete separation between the insertion tendons of the latissimus dorsi and of the teres major is found in 5/7 common chimpanzees (e.g., found by Ziegler 1964, MacDowell 1910, Stewart 1936, and other authors, but not reported by Gratiolet and Alix 1866, Sonntag 1923, Miller 1952, Hepburn 1892, Champneys 1872, and Dwight 1895, who referred to a partial fusion of the tendons). If we add Gibbs' (1999) numbers with the numbers found in our own dissections of *P. troglodytes* (no fusion of the distal tendons of the two muscles in 4 of the 7 common chimpanzees where we analysed this feature), the total numbers are: no fusion between the latissimus dorsi and the teres major in 9/15 *P. troglodytes*. Thus, contrary to the condition in hylobatids and orangutans (see below), and seemingly of bonobos according to our bonobo dissections and that of Miller (1952), there is often no fusion in *P. troglodytes* (and there is usually also no fusion in gorillas and modern humans: Diogo and Wood 2011, 2012).
- Synonymy: Grand rond (Gratiolet and Alix 1866).

***Latissimus dorsi*** (adult Kidogo 493.12 g; adolescent Jasiri 253.84 g; infant Foyo 13.94 g; fetus 2.89 g; Figs. 3.1, 3.3, 3.4, 3.5, 3.6, 3.7, 3.8, 3.22, 3.27, 3.28, 3.31, 3.32, 3.33 and 3.53)

- Attachments in bonobos: From vertebrae (T6-S1 in Payne's 2001 adult bonobo; last five thoracic plus lumbar and sacral vertebrae in Miller's 1952 adult bonobo; T6 to last lumbar vertebra in our adult Kidogo; T9 to last lumbar vertebra in adolescent Jasiri and infant Foyo), ribs (last 5 ribs in Miller's 1952 adult bonobo; T7 all the way to iliac crest in our fetus), thoracolumbar fascia and, often, directly from the iliac crest (Payne's 2001 adult bonobo, our adult Kidogo and adolescent Jasiri and infant Foyo and fetus), to the intertubercular groove of the humerus; in the adult bonobo reported by Miller (1952) the muscle also attaches onto the inferior angle of the scapula. Contrary to the other bonobos dissected by us, in our fetus the latissimus dorsi was partially blended with the caudal portion of the trapezius (Fig. 3.31), supporting the idea that in fetuses muscles are in general more blended with each other, even if they derive from completely different developmental primordia, as is the case with these two muscles (the trapezius is, in reality, a head muscle).
- Usual innervation: Thoracodorsal nerve (Miller 1952: *P. paniscus*) [*P. troglodytes*: thoracodorsal nerve (Champneys 1872, Hepburn 1892); C7, C8 (Swindler and Wood 1973).
- Function: EMG activity during hoisting behaviour (Tuttle and Basmajian 1976: *P. troglodytes*).
- Comparison with common chimpanzees: In *P. troglodytes* it runs from the vertebrae, ribs, thoracolumbar fascia and, often, directly and/or indirectly from the pelvis, to the intertubercular groove of the humerus.
- Synonymy: Grand dorsal (Gratiolet and Alix 1866; Broca 1869).

***Dorsoepitrochlearis*** (adult Kidogo 43.20 g; infant Foyo 0.66 g; fetus 0.25 g; Figs. 3.4 and 3.33)

- Attachments in bonobos: From the distal portion of the latissimus dorsi to the medial epicondyle of the humerus (Payne's 2001 adult bonobo and Miller's 1952 adult bonobo, and our adult Kidogo and fetus and adolescent Jasiri and infants Etje and Foyo).
- Usual innervation: Radial nerve (Miller 1952 and our own observations: *P. paniscus*) [*P. troglodytes*: radial nerve (Champneys 1872; Hepburn 1892; Swindler and Wood 1973); branch of radial nerve that anastomoses with a similar branch of this nerve passing into the medial side of the medial head of the triceps brachii (Ziegler 1964)].
- Function: A minor part of the action of the muscle may be extension of the forearm (Ziegler 1964: *P. troglodytes*).
- Comparison with common chimpanzees: In *P. troglodytes* it runs from the distal portion of the latissimus dorsi and occasionally from the coracoid process of the scapula (e.g., Gratiolet and Alix 1866) to the medial epicondyle of the humerus and occasionally also to the adjacent supracondylar ridge of the humerus; in a few cases the dorsoepitrochlearis attachment extends distally to attach directly onto the olecranon process of the ulna (see below). As noted by Aiello and Dean (1990), in nonhuman hominoids the dorsoepitrochlearis is usually mainly attached onto the medial epicondyle, the intermuscular septum and/or surrounding structures, but not onto the olecranon process or the olecranon fascia. Regarding common chimpanzees, Tyson (1699), Gratiolet and Alix (1866), Barnard (1875), Hartmann (1886), Beddard (1893), Dwight (1895), MacDowell (1910), Schück (1913a), Sonntag (1923), Miller (1932), Swindler and Wood (1973) and Payne (2001) found a bony insertion onto the medial epicondyle only, as we did in our *P. troglodytes* specimens, while Ashton and Oxnard (1963) refer to insertions onto the medial epicondyle and the supracondylar ridge of the humerus, and only very few authors, e.g., Vrolik (1841), Testut (1883), Grönroos (1903) and Ziegler (1964), described a bony insertion onto both the medial epicondyle and olecranon process of the ulna in *P. troglodytes*. According to the literature review done by Gibbs (1999), the

usual condition for *P. troglodytes* is effectively that in which there is a bony insertion onto the medial epicondyle only. MacDowell (1910) described a *P. troglodytes* specimen in which the dorsoepitrochlearis received a fasciculus from the coracobrachialis.

– Synonymy: Appendix of the latissimus dorsi (Tyson 1699); muscle accessoire du long dorsal (Gratiolet and Alix 1866; Broca 1869); latissimo-condylus or latissimo-epitrochlearis (Barnard 1875); latissimo-condyloideus (Chapman 1879; Hepburn 1892; MacDowell 1910; Pira 1913); latissimo-tricipitalis (Schück 1913a, b; Fick 1925).

***Triceps brachii*** (adult Kidogo 353.87 g; adolescent Jasiri 187.13 g; infant Foyo 9.98 g; fetus 2.81 g; Figs. 3.7, 3.8, 3.9, 3.10, 3.27, 3.33, 3.34, 3.35 and 3.36)

– Attachments in bonobos: From scapula (less than 1/2 of length of lateral border of our adult Kidogo and fetus; more than 1/2 of this lateral border in our adolescent Jasiri and infants Foyo and Etje) (caput longum) and from shaft of humerus (caput laterale and caput mediale) to olecranon process of ulna (Payne's 2001 adult bonobo and our adult Kidogo and fetus and adolescent Jasiri and infants Foyo and Etje).
– Usual innervation: Radial nerve (Miller 1952: *P. paniscus*) [*P. troglodytes*: radial nerve (Hepburn 1892, Sonntag 1923, Swindler and Wood 1973); radial nerve but also by a branch of the ulnar nerve in at least some specimens (Champneys 1872).
– Comparison with common chimpanzees: In *P. troglodytes* it usually runs from at least half of the length of the lateral border of the scapula (caput longum) and from the shaft of the humerus (caput laterale and caput mediale) to the olecranon process of the ulna. It should be noted that the triceps brachii of modern humans includes the *articularis cubiti*, which is listed in *Terminologia Anatomica* (1998) as a muscle that is usually present in modern humans, but this term refers to a subdivision of the triceps brachii that runs from the main body of that muscle to the posterior aspect of the capsule of the elbow joint, thus lifting the capsule away from the joint; it is not to be confused with the muscle anconeus (see below). An articularis cubiti bundle of the triceps brachii has also been described in a few common chimpanzees (e.g., Champneys 1872; Sonntag 1923), but it was given the name *subanconeus*.
– Synonymy: Triceps plus subanconeus (Champneys 1872; Sonntag 1923); vaste externe plus vaste interne plus scapulo-olécrânien, which correspond respectively to the caput laterale, caput mediale and caput longum sensu the present study (Gratiolet and Alix 1866); multiceps extensor cubiti (Barnard 1875); triceps extensor cubiti (Hepburn 1892).

***Brachialis*** (adult Kidogo 132.95 g; adolescent Jasiri 112.91 g; infant Foyo 4.30 g; fetus 1.25 g; Figs. 3.4, 3.7, 3.8, 3.9, 3.10, 3.22, 3.33 and 3.34)

– Attachments in bonobos: From the humeral shaft (i.e., distal to the surgical neck of the humerus) to the ulnar tuberosity (our fetus and adult Kidogo and adolescent Jasiri and Payne's 2001 adult bonobo and infants Foyo and Etje) or coronoid process (Miller's 1952 adult bonobo).
– Usual innervation: Musculocutaneous nerve (Miller 1952: *P. paniscus*) [*P. troglodytes*: median nerve (Gratiolet and Alix 1866; Sutton 1883; Bolk 1902); musculocutaneous nerve (Hepburn 1892; Sonntag 1923; Swindler and Wood 1973; Koizumi and Sakai 1995)].
– Comparison with common chimpanzees: In *P. troglodytes* it runs from the humeral shaft (i.e., distal to the surgical neck of the humerus) to the ulnar tuberosity. Ziegler (1964) stated that on the right side of the *P. troglodytes* specimen dissected by him, two fleshy slips left the centre of the anterior aspect of the main body of the brachialis and ran distally to join the posterior aspect of the biceps brachii near its tendon of insertion; according to him such slips are recorded as rare anomalies in modern humans. Howell and Straus (1932) stated that on both sides of the *P. troglodytes* specimen

dissected by them, the medial belly of the brachialis was partially separated from the lateral belly, the medial belly having the more extensive origin from the distal half of the humerus and its medial epicondyle, inserting in the form of a tendon onto the ulna just distal to the coronoid process; the lateral belly had a shorter origin onto the humerus near the deltoid insertion. The *P. troglodytes* infant described in our common chimpanzee atlas had a configuration similar to that shown in plate 1 of Howell and Straus 1932 (i.e., proximally the brachialis is partially divided into medial and lateral bundles, but the bundles fuse distally and insert together onto the ulnar tuberosity).
- Synonymy: Brachial antérieur (Gratiolet and Alix 1866); brachialis anticus (Champneys 1872; Hepburn 1892; Beddard 1893; Sonntag 1923).

***Biceps brachii*** (adult Kidogo 173.50 g; adolescent Jasiri 118.72 g; infant Foyo 5.34 g; fetus 1.39 g; Figs. 3.4, 3.7, 3.8, 3.9, 3.12, 3.21, 3.22, 3.23, 3.27, 3.33, 3.34 and 3.53)

- Attachments in bonobos: From proximal humerus (caput breve) and supraglenoid tubercle of scapula (caput longum), to bicipital tubercle of radius (common tendon) and sometimes (in our fetus and Miller's 1952 adult bonobo, but not in our adult Kidogo nor in our infants Foyo and Etje; we did not analyse this in detail in our adolescent Jasiri), also to fascia covering the forearm flexors (bicipital aponeurosis, which is usually not fleshy, i.e., it does not form a 'lacertus carnosus'; see notes below).
- Usual innervation: Musculocutaneous nerve (Miller 1952: *P. paniscus*) [*P. troglodytes*: median nerve (Gratiolet and Alix 1866, Sutton 1883, Bolk 1902); musculocutaneous nerve (Hepburn 1892, Sonntag 1923, Swindler and Wood 1973, Koizumi and Sakai 1995)].
- Comparison with common chimpanzees: In *P. troglodytes* it runs from the proximal humerus (caput breve) and the supraglenoid tubercle of the scapula (caput longum), to the bicipital tubercle of the radius (common tendon) and to fascia covering the forearm flexors (bicipital aponeurosis). In hominoids such as *Hylobates*, *Gorilla*, *P. troglodytes* and modern humans, the biceps brachii is usually prolonged distally by this bicipital aponeurosis ('lacertus fibrosus' in *Gorilla*, *Pan* and modern humans; often 'lacertus carnosus' in *Hylobates*) that is commonly associated with the fascia covering forearm muscles such as the pronator teres. Almost all *P. troglodytes* specimens have a bicipital aponeurosis – as reported by Gratiolet and Alix (1866), Sonntag (1923), Loth (1931), Howell and Straus (1932), Glidden and De Garis (1936), Ziegler (1964) and Swindler and Wood (1973), and corroborated by our dissections – this might be a difference with bonobos, in which only 2 of the 5 specimens (i.e., < than 50%) including in our calculations had a bicipital aponeurosis. Plate 1 of Howell and Straus (1932) shows a common chimpanzee specimen with 'accessory heads of the biceps brachii' (in addition to the short and long heads and to the bicipital aponeurosis). On the right side of that specimen they report an 'accessory head' of the long head originating from the capsule of the gleno-humeral joint and fusing distally with the main body of the long head below the head of the humerus. On the left side of the same specimen, Howell and Straus (1932) report the presence of two 'accessory heads' of the long head.
- Synonymy: Biceps flexor cubiti (Owen 1868; Hepburn 1892); scapulo-radial plus coraco-antebrachial, which correspond respectively to the caput longum and caput breve sensu the present study (Gratiolet and Alix 1866).

***Coracobrachialis*** (adult Kidogo 46.22 g; adolescent Jasiri 29.39 g; infant Foyo 1.44 g; fetus 0.30 g; Figs. 3.4, 3.7, 3.22, 3.33 and 3.34)

- Attachments in bonobos: From coracoid process of scapula (fused proximally to long head of biceps brachii in our adult Kidogo and in Miller's 1952 adult bonobo and adolescent Jasiri and infants Foyo and Etje, but all the way to half of the proximodistal length of the coracobrachialis in our fetus as seen in Fig. 3.33; this might well represent the normal earlier stage of development) to proximal portion (our adult Kidogo) or midshaft (our fetus, Fig. 3.33) or middle 1/3 (in Miller's 1952 adult bonobo) of humerus, being pierced by the musculocutaneous nerve in all bonobos dissected by us.

- Usual innervation: Musculocutaneous nerve (Miller 1952: *P. paniscus*) [*P. troglodytes*: median nerve (Gratiolet and Alix 1866, Sutton 1883, Bolk 1902); musculocutaneous nerve (Hepburn 1892, Swindler and Wood 1973); musculocutaneous nerve, but occasionally also median nerve (Sonntag 1923); musculocutaneous nerve, except on one side of a specimen (Koizumi and Sakai 1995)].
- Comparison with common chimpanzees: In *P. troglodytes* it runs from the coracoid process of the scapula to the proximal portion of the humerus, being usually (e.g., Macalister 1871; Champneys 1872; Dwight 1895; Sonntag 1923; our dissections), but not always (e.g., Tyson 1699), pierced by the musculocutaneous nerve. The observations of most authors, as well as our dissections, show that the usual condition for common chimpanzees is that the *coracobrachialis superficialis/longus* and *coracobrachialis profundus/coracobrachialis brevis* are not present as distinct structures, i.e., the coracobrachialis has a single bundle that corresponds to the *coracobrachialis medius/coracobrachialis proprius* of other mammals. A few exceptions follow. Macalister (1871) reported a *P. troglodytes* specimen that had a 'small rudiment of the coracobrachialis profundus/brevis; Parsons (1898a, b) stated that small rudiment is found in 30% of common chimpanzees and Howell and Straus (1932) stated that it is present in 4 out of 13 common chimpanzees. Howell and Straus (1932) noted, correctly, that a coracobrachialis superficialis/longus is only found in a few mammals, and stated that they found it as an anomaly on the right side of the *P. troglodytes* specimen dissected by them. According to their report, the coracobrachialis superficialis/longus originates together with the short head of the biceps brachii and continues superficial to all of the nerves and muscles of the arm, except for the dorsoepitrochlearis, to its insertion on the medial epicondyle (see their plate 1). The musculocutaneous nerve passes between it and the *coracobrachialis medius/proprius*, which originates from the coracoid process and inserts on the humerus. According to these authors, no part of the coracobrachialis can be considered as representing a coracobrachialis profundus/brevis; on the left side of their specimen, there is only a *coracobrachialis medius/proprius* that originated from the coracoid process and then passed superficial (its medial portion) and deep (its lateral portion) to the musculocutaneous nerve. A distinct *coracobrachialis superficialis/longus* running from the coracoid process of the scapula to the medial epicondyle of the humerus was also reported in one of the four *P. troglodytes* dissected by Oishi et al. 2009. The 'coracobrachialis longus' – or 'superficialis' – reported by Sonntag 1923 does not correspond to the *coracobrachialis superficialis/longus sensu* the present study, i.e., this report is most likely to be due to the use of an erroneous nomenclature, because the *coracobrachialis superficialis/longus* is a bundle that should be superficial to almost all the nerves and muscles of the arm (e.g., Diogo and Abdala 2010 and Diogo and Wood 2012).

***Pronator quadratus*** (infant Foyo 0.52 g; fetus 0.10 g; Figs. 3.14, 3.40 and 3.41)

- Attachments in bonobos: From distal ulna to distal radius.
- Usual innervation: Anterior (ventral sensu the present work) interosseous branches of the medial nerve (Miller 1952: *P. paniscus*).
- Function: According to Tuttle (1969) there is a trend towards reducing the pronator musculature in the African apes, which may be associated with the fact that they do not engage as frequently as orangutans in arboreal movements that require a wide range of supination and pronation of the forearm; further-

more, the African apes, particularly *Gorilla*, require considerable stability at the elbow joint to maintain the fully extended forearm against the compressive forces incurred during knuckle-walking.

- Comparison with common chimpanzees: In *P. troglodytes* it usually runs from the distal portion of the ulna to the distal portion of radius. In the *P. troglodytes* specimen dissected by Dwight (1895), the pronator quadratus has two layers that are only partly separable, the more superficial beginning as a tendon at the top of the ulnar origin and expanding as it passes downward and outward across the muscle; it gradually becomes muscular, and having reached the outside of the radius, runs distally to the ridge of the trapezium. The deeper, more transverse part of the pronator quadratus runs obliquely and distally towards the radius; a few radial fibres go to the anterior aspect of the wrist capsule.
- Synonymy: Carré pronateur (Gratiolet and Alix 1866).

***Flexor digitorum profundus*** (infant Foyo 16.14 g; fetus 1.93 g; Figs. 3.13, 3.38, 3.39 and 3.40)

- Attachments in bonobos: From radius, ulna and interosseous membrane and sometimes also from medial epicondyle of humerus (our fetus and infant Foyo) to distal phalanges of digits 1, 2, 3, 4 and 5, and sometimes to the proximal phalanx of digit 1 as well (Miller's 1952 adult bonobo), the tendon to digit 1 being stout and continuous to the main body of the muscle, i.e., it is similar to the tendons going to the other digits (e.g., Fig. 3.13; see notes below). This clearly contrasts with the condition in common chimpanzees, in which the tendon to digit 1 is usually vestigial (see notes below). A variation was seen on the left side of our fetus, in which the tendon to digit 1 in addition anchored by a tendinous sling that is attached to the trapezium and connected to the fascia of the thenar muscles, while the tendons to digits 2 and 3 communicate by a tendinous sling at the level of the base of MII and MIII (Figs. 3.38, 3.39 and 3.40). Another variation was found on the right upper limb of the adult Kidogo, in which there were no tendons of the flexor digitorum profundus to digits 1 and 5.
- Usual innervation: Median (usually to part of muscle going to digits 2 and 3, and to digit 1 when there is a portion of the muscle going to this digit) and ulnar (usually to part of muscle going to digits 4 and 5) (Miller 1952: *P. paniscus*) [same innervation in *P. troglodytes*: Champneys 1872, Hepburn 1892, Swindler and Wood 1973)].
- Comparison with common chimpanzees: In *P. troglodytes* it runs from the radius, ulna and interosseous membrane to the distal phalanges of digits 2, 3, 4 and 5 and, sometimes, to digit 1 through a tendon that is often thin and/or not continuous to the main body of the muscle. Among the primates dissected by us, hylobatids and modern humans are the only ones in which the flexor pollicis longus is usually present as a distinct, independent muscle (e.g., the *flexor pollicis longus* is usually not present as a distinct muscle in chimpanzees Diogo and Wood 2011, 2012). That is, when some authors state that there is a 'flexor pollicis longus' in a chimpanzee, they are either referring to the part of the flexor digitorum profundus that goes to digit 1, or to the belly of this muscle that often goes to both digits 1 and 2, and not really to a distinct muscular belly going exclusively to digit 1 as that usually is found in hylobatids and modern humans (see Synonymy below). In fact, in *Pongo*, *Gorilla* and *P. troglodytes*, the tendon to digit 1 is effectively often absent or vestigial, as corroborated in the specimens dissected by us and by others, except in a few specimens of these genera according to Straus (1942a), in which the tendon to digit 1 is said to be similar to the other tendons of the flexor digitorum profundus. According to the literature reviews of Keith (1894) and Gibbs (1999), there is no tendon of the flexor digitorum profundus to digit 1 in 10/25 common chimpanzees and in 13/43 common chimpanzees, respectively. Straus (1942b) compiled evidence from his

own dissections of hominoids and from data available on the literature and stated that within 47 *P. troglodytes* specimens the tendon of the flexor digitorum profundus to digit 1 is completely absent in 30% (14 of 47), is functionless in 22% (10.5 of 47), and is in direct functional continuity with the radial muscle belly of the flexor digitorum profundus in 48% (22 of 47); thus the flexor digitorum profundus tendon to digit 1 is entirely without function in at least half of the members of *P. troglodytes* (52% of chimpanzees), contrasting with the stout, continuous tendon often present in bonobos.

- Synonymy: Flexor communis profundus (Wyman 1855); fléchisseurs des troisièmes phalanges or fléchisseurs profonds or fléchisseurs perforants (Gratiolet and Alix 1866); fléchisseur profond des doigts (Broca 1869); flexor profundus et pollicis (Macalister 1871); flexor profundus digitorum plus flexor longus pollicis (Champneys 1872; Beddard 1893; Sonntag 1923); flexor profundus digitorum plus flexor profundus indicis plus flexor longus pollicis (Sutton 1883); flexor digitorum communis profundus (Barnard 1875); flexor digitorum profundus plus flexor pollicis longus (Swindler and Wood 1973).

***Flexor digitorum superficialis*** (infant Foyo 2.25 g; fetus 1.07 g; Figs. 3.12, 3.13, 3.26, 3.37, 3.38 and 3.45)

- Attachments in bonobos: From radius (caput radiale) and from medial epicondyle of the humerus, plus occasionally (our fetus and infant Foyo and right side of adult Kidogo) the ulna (caput humeroulnare), to the middle phalanges of digits 2–5, the muscle being often blended with the flexor digitorum profundus and often having a small, extra tendon between its tendons to digits 2 and 3.
- Usual innervation: Median nerve (Miller 1952: *P. paniscus*) [same innervation in *P. troglodytes*: Hepburn 1892, Sonntag 1923, Swindler and Wood 1973].
- Function: According to Tuttle (1969), in great apes the fasciculi of the flexor digitorum superficialis and/or flexor digitorum profundus to the individual digits, especially digit 2, frequently appear as entities in the middle of the forearm and may be traced proximally at least to this level by following the individual tendons of the flexor digitorum superficialis/profundus muscles. According to Tuttle this is probably related to the ability to flex the fingers independently. This is of considerable advantage to orangutans and chimpanzees in that it allows the animals to hold a number of small twigs with some digits while the remaining fingers reach for additional supports. If some of the twigs break, the individual does not have to open the whole hand in order to grasp replacements. Also according to Tuttle, it is probable that the flexor digitorum superficialis/profundus muscles are active to prevent extreme hyperflexion at the metacarpophalangeal joints in *Gorilla* and *Pan* during static knuckle-walking postures; this contrasts with the conclusions of Ziegler (1964) that the major danger of 'collapse' at these joints was in the direction of flexion).
- Comparison with common chimpanzees: In *P. troglodytes* it usually runs from the radius (caput radiale), the ulna and the medial epicondyle of the humerus (caput humeroulnare) to the middle phalanges of digits 2–5. Sutton (1883), Hepburn (1892), Dwight (1895), MacDowell (1910), Sonntag (1923), Loth (1931), Jouffroy (1971), Swindler and Wood (1973) and Gibbs (1999) refer to an origin of the flexor digitorum superficialis that includes the ulna, radius and the medial epicondyle of the humerus and/or the common flexor tendon, in *P. troglodytes*. Such an origin was also found in our dissections of common chimpanzees. Only a few authors report examples in *P. troglodytes* in which there is no ulnar origin (e.g., Beddard 1893) and/or no radial origin (e.g., Gratiolet and Alix 1866).
- Synonymy: Fléchisseurs des secondes phalanges or fléchisseurs superficiels or fléchisseurs perforés (Gratiolet and Alix 1866); flexor sublimis digitorum (Macalister 1871; Champneys 1872; Hepburn 1892; Beddard 1893; Dwight 1895; Sonntag 1923); flexor sublimis

digitorum plus flexor sublimis indicis (Sutton 1883); flexor digitorum sublimis (MacDowell 1910; Miller 1952).

***Palmaris longus*** (fetus 0.01 g; Figs. 3.12, 3.23 and 3.37)

- Attachments in bonobos: From medial epicondyle of humerus to palmar aponeurosis, usually to the radial side of this aponeurosis (our adult Kidogo and fetus and adolescent Jasiri), often almost reaching (our fetus) or reaching (Miller's 1952 adult bonobo) the fascia covering the abductor pollicis brevis. On the left forearm of our infant Foyo, the palmaris longus was missing; and on the right forearm of our infant Etje the palmaris longus seemed to be absent. Unfortunately, we could not check whether the muscle was present or not in the other forearms of these two infants, but still the usual condition for bonobos clearly seems to be the presence of the muscle, as in total it was present in 7 forearms (both sides of our fetus and adult Kidogo and of Miller's 1952 adult bonobo, and right side of adolescent Jasiri, which is the only side in which we examined this feature in this specimen), and only absent in two (left side of infant Foyo and right side of infant Etje, i.e., presence in 78% of the cases, not very different from the 65% in common chimpanzees according to the numbers of Gibbs (1999; see below).
- Usual innervation: Median nerve (Miller 1952: *P. paniscus*) [same innervation in *P. troglodytes*: Hepburn 1892, Ziegler 1964, Swindler and Wood 1973)].
- Comparison with common chimpanzees: In *P. troglodytes* it usually runs from the medial epicondyle of the humerus to the palmar aponeurosis. According to Ziegler 1964, the branch of the median nerve that innervates the palmaris longus in the *P. troglodytes* specimen dissected by him pierces the flexor digitorum superficialis before entering the palmaris longus; this might suggest that the palmaris longus of at least some primates such as common chimpanzees derives phylogenetically from the flexor digitorum superficialis (for more details, see Diogo and Abdala 2010 and Diogo and Wood 2012). In the *P. troglodytes* specimens described by Tyson (1699; 1 specimen), Vrolik (1841; 1 specimen), Wilder (1862; 1 specimen), Gratiolet and Alix (1866; 1 specimen), Humphry (1867; 2 specimens), Macalister (1871; 1 specimen), Champneys (1872; 1 specimen), Chapman (1879; 1 specimen), Hepburn (1892; 1 specimen), Beddard (1893; 1 specimen), Dwight (1895; 1 specimen), MacDowell (1910; 1 specimen), Ribbing and Hermansson (1912; 1 specimen), Ziegler (1964; 1 specimen), Swindler and Wood (1973; 1 specimen), Sarmiento (1994; 2 specimens), Ogihara et al. (2005; 1 specimen), Oishi et al. (2009; 4 specimens), and Kikuchi (2010; 1 specimen) the muscle was present, but it was absent in the specimen dissected by Sonntag (1923), in 1 of the 3 specimens reported by Sonntag (1924), in 1 of the 4 upper limbs dissected by Carlson (2006), and in 1 out of the 6 specimens dissected by us in which we could discern this feature in detail. The palmaris longus is only present in 9 out of 12 common chimpanzee upper limbs according to Keith's (1899) review, in 95% of the *P. troglodytes* cases according to Loth's (1931) review, in 15.5 out of 17 *Pan* cases according to Sarmiento's (1994) review, and in 19 out of 28 *Pan* cases according to Gibbs' (1999) review. So using the numbers of Gibbs (1999) but excluding the two limbs of the adult bonobo reported by Miller (1952), which were counted by Gibbs (1999), the total is 17 out of 26 cases for *P. troglodytes*, so 65%, a frequency that is not very different from that noted above for bonobos (78%), which is in turn not so different from the 80–85% of cases in which the muscle is said to be present, in modern humans.
- Synonymy: Fléchisseur des premières phalanges or palmaire grêle (Gratiolet and Alix 1866).

***Flexor carpi ulnaris*** (infant Foyo 2.28 g; fetus 0.33 g; Figs. 3.12, 3.23, 3.26 and 3.37)

- Attachments in bonobos: From ulna (caput ulnare) and medial epicondyle of humerus

(caput humerale) to pisiform, and sometimes also to base of MV (Miller's 1952 adult bonobo).
- Usual innervation: Ulnar nerve (Miller 1952: *P. paniscus*) [same innervation in *P. troglodytes*: Champneys 1872, Hepburn 1892, Sonntag 1923, Swindler and Wood 1973].
- Comparison with common chimpanzees: A bony origin from the ulna and humerus is usually present in common chimpanzees and was found in the *P. troglodytes* specimens dissected by us and described by Gratiolet and Alix (1866), Hepburn (1892), Beddard (1893), Sonntag (1923) and Swindler and Wood (1973). In common chimpanzees, the insertion is normally onto the pisiform, and sometimes also to the base of MV (e.g., Gratiolet and Alix 1866) and/or to the hamate (e.g., our *P. troglodytes* specimen PFA 1077).
- Synonymy: Métacarpien palmaire du cinquième doigt or cubital antérieur or cubital palmaire (Gratiolet and Alix 1866; Jouffroy and Lessertisseur 1959).

***Epitrochleoanconeus*** (fetus 0.01 g)

- Attachments in bonobos: From medial epicondyle (our fetus) or medial supracondylar ridge (one side of Miller's 1952 adult bonobo, see just below) of humerus to olecranon process of ulna (our fetus); absent on one side of the body of the adult specimen dissected by Miller 1952 (present on the other side, being fused with the flexor carpi ulnaris). As a variant, on the left side of our infant Foyo the epitrochleoanconeus was seemingly not present as a separate muscle (we could not check this feature on the other side of this specimen, nor in our other nonfetal specimens).
- Usual innervation: Ulnar nerve (Miller 1952: *P. paniscus*) [same innervation in *P. troglodytes*: Howell and Straus 1932, Ziegler 1964].
- Comparison with common chimpanzees: In *P. troglodytes* the epitrochleanconeus is usually present as a distinct muscle, running from the medial epicondyle of the humerus to the olecranon process of the ulna. There are some cases in which the muscle is not described, but it is not clear if this is because it is absent. Some authors (e.g., Macalister 1871 – 1 specimen; Sonntag 1923 – 1 specimen) comment when a muscle is absent, but it is more common for authors to state that the muscle was present (e.g., in the specimen described by Gratiolet and Alix 1866), on one side of the body of the specimen described by Ziegler 1964 (he could not discern if the muscle was present or not on the other side), in the specimen illustrated by Swindler and Wood 1973 (in their p. 327 they state that this muscle is absent in *P. troglodytes*, but in p. 146 they state that it is present and they clearly show it in the specimen illustrated in a figure in their p. 147), and in 2 of the 5 specimens dissected by us in which we could discern if the muscle was present or not (in the other 3 infant specimens the muscle seemed to be undifferentiated). It should be noted that the epitrochleoanconeus should be, by definition, as shown in plate 71 of Swindler and Wood (1973) (i.e., it should be a small muscle that derives from the anlage that gives rise to the flexor carpi ulnaris and it should run mainly from the medial epicondyle of the humerus to the olecranon process of the ulna. It is not clear if the muscle shown in the common chimpanzee of plate 1 of Howell and Straus (1932) is an epitrochleoanconeus, because this muscle can be seen in a ventral view of the upper limb (and usually it should only be seen from a dorsal view), because it is rather well-developed and not small, and also because the muscle does not seem to insert onto the olecranon process of the ulna.
- Synonymy: Anconeus quartus, anconeus medialis, anconeus sextus, anconeus parvus or tensor fasciae antebrachii (Jouffroy 1971).

***Flexor carpi radialis*** (infant Foyo 2.41 g; fetus 0.36 g; Figs. 3.12, 3.13, 3.23, 3.26 and 3.37)

- Attachments in bonobos: From medial epicondyle of humerus and sometimes also from radius (adult bonobo of Miller 1952), to base of MII (Miller's 1952 adult bonobo and our fetus and infant Foyo), and sometimes also

MIII (our fetus). Variations were found in the adult bonobo of Miller (1952), where the flexor carpi radialis was fused with the pronator teres, and on the right side in our adult Kidogo, where it was fused with the tendon of the flexor digitorum superficialis to digit 4.

- Usual innervation: Median nerve (Miller 1952: *P. paniscus*).
- Comparison with common chimpanzees: In *P. troglodytes* the muscle usually has bony origins from at least the humerus and the radius. This condition was found in the *P. troglodytes* specimens described by Gratiolet and Alix (1866), Hartmann (1886), Dwight (1895), Sonntag (1923), Ziegler (1964) and Swindler and Wood (1973) and dissected by us; there are very few descriptions of common chimpanzees that refer to a bony origin from the humerus only (e.g., Hepburn 1892 and Beddard 1893). Also, in *P. troglodytes* the muscle often inserts onto both MII and MIII. In the survey of Gibbs (1999) this author stated that the muscle is inserted onto both MII and MIII in 5 out of 10 common chimpanzees, but Jouffroy (1971) argue that an insertion onto both these metacarpals is the usual condition for *Pan*. An exclusive insertion onto MII was described in *P. troglodytes* by Gratiolet and Alix (1866), Beddard (1893), Dwight (1895), but in the *P. troglodytes* specimens dissected by Hepburn (1892), Sonntag (1923), Ziegler (1964) and Swindler and Wood (1973), as well as in the three *P. troglodytes* specimens dissected by us in which we could discern this feature in detail, the insertion was onto both MII and MIII.
- Synonymy: Métacarpien palmaire du deuxiéme doigt or radial antérieur ou grand palmaire (Gratiolet and Alix 1866; Jouffroy and Lessertisseur 1959, 1960).

***Pronator teres*** (infant Foyo 2.60 g; fetus 0.29 g; Figs. 3.23 and 3.37)

- Attachments in bonobos: From humerus and often ulna to radius, extending to midshaft of radius in our fetus and adult Kidogo and adolescent Jasiri, and attaching onto middle 1/3 of radius in Miller's (1952) adult bonobo. As a variation, the pronator teres has to separate distal tendons on the left side of the adult Kidogo.
- Usual innervation: Median nerve, which usually passes between the ulnar and humeral heads of the muscle (Miller 1952: *P. paniscus*) [same innervation in *P. troglodytes*: Champneys 1872, Chapman 1879, Dwight 1895, Hepburn 1892, Sonntag 1923, Swindler and Wood 1973].
- Comparison with common chimpanzees: Gratiolet and Alix (1866) described a humeral head only in *P. troglodytes*, and Lewis (1989) stated that a distinct ulnar head is not present in apes. However, the common chimpanzees reported by Tyson (1699), Macalister (1871), Champneys (1872), Chapman (1879), Hepburn (1892), Beddard (1893), Dwight (1895), Sonntag (1923, 1924), Swindler and Wood (1973), Stern and Larson (2001) and dissected by us have an ulnar head and a humeral head separated by the median nerve, and Oishi et al. (2009) found two heads in two of the four *P. troglodytes* specimens dissected by them, but they could not find an ulnar head in the other two. According to Parsons (1898b), Loth (1931) and Jouffroy (1971) two heads are present in 90%, 91% and 90% of common chimpanzees, respectively, but Gibbs (1999) stated that two heads are found in 5 out of 9 *Pan* specimens.
- Synonymy: Grand pronateur or rond pronateur (Gratiolet and Alix 1866); pronator radii teres (Champneys 1872; Hepburn 1892; Beddard 1893; Sonntag 1923).

***Palmaris brevis***

- Attachments in bonobos: From the flexor retinaculum to the skin of the medial border of the palm (Miller's 1932 adult bonobo). We could not find this muscle in any of the specimens dissected by us, but this does not mean the muscle was necessary missing in all of them, as this is a very small, thin muscle that is strongly attached to skin.

- Usual innervation: Superficial branch of ulnar nerve (Miller 1952: *P. paniscus*) [same innervation in *P. troglodytes*: Swindler and Wood 1973].
- Comparison with common chimpanzees: Within *P. troglodytes* the palmaris brevis was not present in the specimen dissected by Hepburn (1892), or in the two specimens of Sarmiento (1994) and in the specimen of Wilder (1862), but it was present in the specimen of Vrolik (1841), the specimen of Gratiolet and Alix (1866), the two specimens of Humphry (1867), the specimen of Champneys (1872), the specimen of Sonntag (1923), the specimen of Miller (1952), the specimen illustrated by Dylevsky (1967), the specimen of Swindler and Wood (1973), and in 4 of the 5 dissected specimens in which we could discern this feature. According to the reviews of the literature by Sarmiento (1994) and Gibbs (1999), this muscle is present in 7/9 and 4/5 *P. troglodytes*, respectively, and according to our own review of the literature and the data obtained from our dissections of common chimpanzees, it is present in 14/19 common chimpanzees (we are not including data from our dissections of bonobos, as we could not find this muscle in any of them, but we are not sure if it was always truly absent; see just above). The *flexor digitorum brevis manus* and *palmaris superficialis* (e.g., Diogo et al. 2009) are usually not present as distinct muscles in chimpanzees.
- Synonymy: Palmaire cutané (Gratiolet and Alix 1866).

***Lumbricales*** (fetus lumbricalis 1 = 0.04 g, lumbricalis 2 = 0.01 g, lumbricalis 3 = 0.01 g, LSB lumbricalis 4 = 0.01 g; Figs. 3.13, 3.26 and 3.45)

- Attachments in bonobos: From the dorsal surfaces of the tendons of the flexor digitorum profundus to digits 2, 3, 4 and 5, to the radial side of the proximal phalanx and the extensor expansion of digits 2 (lumbricalis I), 3 (lumbricalis II), 4 (lumbricalis III) and 5 (lumbricalis IV). Two variations were found on the right side of our adult Kidogo: the lumbricalis IV (normally going to digit 5) was absent, and the lumbricalis II (normally going to digit 3) inserted instead onto digit 2.
- Usual innervation: First and second lumbricales by the median nerve and the third and fourth lumbricales by the deep branch of the ulnar nerve (Miller 1952: *P. paniscus*) [in *P. troglodytes* is same innervation according to Swindler and Wood 1973; however it is first, second and third lumbricales by median nerve, and the fourth by the deep branch of ulnar nerve according to Hepburn 1892].
- Function: See flexores breves profundi, below.
- Comparison with common chimpanzees: In *P. troglodytes* the lumbricales usually run from the dorsal surfaces of the tendons of the flexor digitorum profundus to digits 2, 3, 4 and 5, to the radial side of the proximal phalanx and the extensor expansion of digits 2 (lumbricalis I), 3 (lumbricalis II), 4 (lumbricalis III) and 5 (lumbricalis IV). The *intercapitulares* (see, e.g., Jouffroy 1971) are usually not present in common chimpanzees.

***Contrahentes digitorum*** (fetus contrahentes to digits 4 and 5 = 0.01 g; Figs. 3.14 and 3.47)

- Attachments in bonobos: Mainly from the contrahens fascia to the base of proximal phalanx of the ulnar side of digits 4 and 5 (Miller's 1952 adult bonobo and our infant Etje and fetus). In the infant Foyo and seemingly also in the adolescent Jasiri (Fig. 3.14), the contrahentes to digits 4 and 5 were reduced to a very thin layer of connective tissue, apparently without muscle fibres, i.e., they did not seem to be present as distinct, separate, fleshy muscles.
- Usual innervation: Deep branch of the ulnar nerve (Miller 1952: *P. paniscus*) [same innervation in *P. troglodytes*: Hepburn 1892].
- Comparison with common chimpanzees: In the *P. troglodytes* specimen dissected by Gratiolet and Alix (1866), the two specimens dissected by Jouffroy and Lessertisseur (1959), the two specimens reported by Day and Napier (1963), and four of the five specimens dissected by us in which we could

discern this feature, the contrahentes were not present as fleshy, well-defined, distinct structures (the exception being PFA 1077). In the *P. troglodytes* specimen dissected by Brooks (1886), there were three contrahentes to digits 2, 4 and 5. However, according to Lewis (1989) in *P. troglodytes*, there are usually contrahentes to digits 4 and 5; this was the case in a specimen described by Hartmann (1886), in the specimen dissected by Hepburn (1892), the specimen dissected by Sarmiento (1994), and the specimen reported by Swindler and Wood (1973). In the review of the literature by Sarmiento (1994), the contrahentes were said to be present in 7 out of 7 *Pan*, seemingly including the *P. paniscus* specimen described by Miller (1952).

- Synonymy: Part of the adductors of digits (Brooks 1886); contrahentes manus (Swindler and Wood 1973).

***Adductor pollicis* + *Adductor pollicis accessorius*** (infant Foyo adductor pollicis 0.45 g; fetus adductor pollicis 0.07 g, with oblique head 0.04 g and transverse head 0.03 g; Figs. 3.13, 3.14, 3.26, 3.45, 3.46, 3.47 and 3.48)

- Attachments in bonobos: In our fetus, the oblique head originates from MI, trapezium and contrahens raphe (was from MII and MIII in Miller's 1952 adult bonobo and on the right side of our adolescent Jasiri, but only from MIII in both sides of our adult Kidogo), while the transverse head originates mainly from MIII (but MIII + MIV on both sides of our adult Kidogo, and MII + MIII on the right side of our adolescent Jasiri) and contrahens raphe, both heads inserting onto the proximal phalanx and/or adjacent structures such as the medial (ulnar) sesamoid bone (as was also reported in Miller's 1952 adult bonobo). In our fetus there is a relatively well-developed *adductor pollicis accessorius* (which was not described by Miller 1952, but this might be because most authors had no knowledge about this muscle at that time) from the base of MI to the ulnar side of the base of the proximal phalanx (Fig. 3.48). The adductor pollicis accessorius (see just below) was also present in our infant Etje, running from MI and II and contrahens raphe to the proximal phalanx of digiti 1 and associated to the lateral sesamoid bone, together (fused distally with) the two heads of the adductor pollicis. We could not discern if the adductor pollicis accessorius was also present, or not, in our infant Foyo nor in our adolescent Jasisi and adult Kidogo.
- Usual innervation: Deep branch of the ulnar nerve (Miller 1952: *P. paniscus*) [*P. troglodytes*: deep branch of the ulnar nerve (Brooks 1887; Hepburn 1892); ulnar nerve (Swindler and Wood 1973).
- Comparison with common chimpanzees: In *P. troglodytes* the caput obliquum and caput transversum are usually well-differentiated, connecting MII and/or MIII, the contrahens fascia, and often at least some carpal bones and/or ligaments, to the metacarpophalangeal joint, as well as to the proximal phalanx and sometimes also to the distal phalanx of the thumb (directly by means of a thin tendon and/or indirectly by means of a ligament) and occasionally to MI (e.g., Gratiolet and Alix 1866; Jouffroy and Lessertisseur 1960). When present, the adductor pollicis accessorius (see below) often connects the proximo-medial portion of the MI and/or adjacent carpal structures to the metacarpophalangeal joint and/or the proximal portion of the proximal phalanx of digit 1.There has been much controversy regarding the homologies of the thenar muscles of primate and nonprimate mammals. This subject has been discussed in detail in the recent studies of Diogo et al. (2009) and particularly of Diogo and Abdala (2010) and Diogo and Wood (2011), and here we summarize the main conclusions of those studies. The *interosseous volaris primus of Henle* of modern human anatomy corresponds to a *thin, deep additional slip of the adductor pollicis* (*TDAS-AD sensu* Diogo and Abdala 2010 and Diogo and Wood 2011; *adductor pollicis accessorius sensu* Diogo et al. in press and sensu the present study), and not to the *flexor brevis profundus 2* of 'lower' mammals, as

suggested by some authors and as suggested by the use of the name 'interosseous volaris primus of Henle'). Within common chimpanzees, the adductor pollicis accessorius is present in some, but not in most, cases (i.e., in <50% of the cases). It was present in the specimen dissected by Champneys (1872) and was probably present in the specimen dissected by Gratiolet and Alix (1866) and the specimen reported by Brooks (1886), and in 25% of common chimpanzees according to Tuttle (1970), in 2 out of 7 common chimpanzees according to the review of the literature by Sarmiento (1994), and in at least one of the hands of the 4 specimens dissected by us in which we could discern this feature in detail. It was not found in the two specimens dissected by Susman et al. (1999) and in the two specimens dissected by Sarmiento (1994). Abramowitz (1955) also supports the idea that the adductor pollicis accessorius is present in at least some common chimpanzees. The adductor pollicis accessorius is also present in some, but not in most (i.e., in <50% of the cases) gorillas, and it is present in more than 50% of modern humans (Diogo and Wood 2011, 2012). Regarding the *deep head of the flexor pollicis brevis* of modern human anatomy, this corresponds to the flexor brevis profundus 2 of 'lower' mammals. This structure is clearly found in four of the five common chimpanzees dissected by us in which we could discern this feature in detail (including the right hand of one of the infants, PFA 1077; we could not find it in the left hand of this infant nor in the single – left – dissected hand of the infant PFA UNC) and in most of the *P. troglodytes* specimens dissected by other authors, who often refer to this structure as the 'deep head of the flexor pollicis brevis'. Regarding the *superficial head of the flexor pollicis brevis* of modern human anatomy, this thus seems to correspond to a true *flexor pollicis brevis*. The true flexor pollicis brevis and the *opponens pollicis* likely derive from the *flexor brevis profundus 1* of 'lower' mammals, and they are present as distinct muscles in the vast majority of common chimpanzees (see below).

- Synonymy: Adductor pollicis and probably the 'ulnar division of flexor pollicis brevis' (Humphry 1867); adducteur du pouce (Gratiolet and Alix 1866); adductor obliquus pollicis plus adductor transversus pollicis (Sonntag 1923); contrahentes I (Jouffroy 1971).

***Interossei palmares*** (fetus flexor brevis profundus 4, or 'interosseous palmaris 1' = 0.04 g; fetus flexor brevis profundus 7, or 'interosseous palmaris 2' = 0.03 g; fetus flexor brevis profundus 9, or 'interosseous palmaris 3' = 0.01 g; infant Foyo all three palmar interossei together =1.32 g; Fig. 3.48)

- Attachments in bonobos: In most bonobo upper limbs dissected by us, the intermetacarpales are mainly fused with four flexores breves profundi to form the dorsal interossei (see Fig. 3.49), with exception to the left upper limb of the adult Kidogo (Figs. 3.15 and 3.16) and to Miller's 1952 adult bonobo, in which only the 1st dorsal interosseus is a complex structure formed by the fused 1st intermetacarpalis and 3rd flexor brevis profundus, as in the other bonobos, but the other three intermetacarpales are mainly not fused with the flexores breves profundi to form dorsal interossei, as they do in the other bonobos. In all specimens, the flexores breves profundi 3 (or the part of the 1st dorsal interossei to which it corresponds) and 4 (designated 1st palmar interosseus when the dorsal interossei complexes are present) run mainly from MII (sometimes also from MI, e.g., left and of adult Kidogo) to the radial and ulnar sides of the proximal phalanx of digit 2, respectively; flexores breves profundi 5 and 6 (or the part of the 2st and 3rd dorsal interossei to which they correspond) run mainly from MIII to the radial and ulnar sides of the proximal phalanx of digit 3, respectively; flexores breves profundi 7 (designated 2st palmar interosseus when the dorsal interossei complexes are present) and 8 (or the part of the 4st dorsal interossei to which it corresponds) run mainly from MIV to the radial and ulnar sides of the proximal phalanx of digit 4, respectively; flexor brevis profundus 9 (designated 3st palmar interosseus when the dorsal interossei complexes are present) runs mainly from MV to the ulnar side of the proximal phalanx of digit 5. The intermetacarpales 1, 2, 3 and 4 (or the part of the 1st, 2st, 3rd and 4th dorsal interossei to which they correspond, respectively) run mainly from MI-MII, MII-MIII, MIII-MIV and MIV-MV, respectively, to the radial side of digit 2, radial side of digit 3, ulnar side of digit 3, and ulnar side of digit 4, respectively. A major difference between common chimpanzees and bonobos is therefore that in the former the intermetacarpales are usually present as distinct muscles, while in bonobos the usual condition seems to be that in which the intermetacarpales are fused with four flexores breves profundi to form the dorsal interossei, because this happens in 9 of the 12 upper limbs dissected by us and by Miller 1952 (the exception being the two limbs of Miller's 1952 adult bonobo, and one side of our adult Kidogo: see notes below for *P. troglodytes*).

- Usual innervation: Deep branch of the ulnar nerve (Miller 1952: *P. paniscus*) [same innervation in *P. troglodytes*: Hepburn 1892].
- Function: EMG study of the lumbricales and of the 'palmar and dorsal interossei' (flexores breves profundi plus intermetacarpales sensu the present study) of common chimpanzees indicates that the function of these muscles is essentially similar to that of modern humans, the 'interossei' being mainly associated with flexion at the metacarpophalangeal joints and extension at the interphalangeal joints, as in modern humans (could not be accessed if the 'interossei' of common chimpanzees were also associated with abduction/adduction as they are in humans and/or rotation of the digits) and the lumbricales with extension at the interphalangeal joints, as in modern humans (Susman and Stern 1980).
- Comparison with common chimpanzees: As explained above (see adductor pollicis and interossei palmares), the *flexor brevis profundus 2* of 'lower' mammals corresponds very likely to the *deep head of the flexor pollicis brevis* of modern human anatomy; the *superficial head of the flexor pollicis brevis* of modern human anatomy, as well as the *opponens pollicis*, derive very likely from the *flexor brevis profundus 1* of 'lower' mammals, while the *flexor digiti minimi brevis* and the *opponens digiti minimi* derive very likely from the *flexor brevis profundus 10* of 'lower' mammals (for recent reviews, see Diogo et al. 2009, Diogo and Abdala 2010 and Diogo and Wood 2012). In extant hominoids except *P. troglodytes*, the *flexores breves profundi* 3, 5, 6 and 8 are fused with the *intermetacarpales* 1, 2, 3 and 4, forming the interossei dorsales 1, 2, 3 and 4, respectively; the *interossei palmares* 1, 2 and 3 thus directly correspond respectively to the flexores breves profundi 4, 7 and 9 of 'lower' mammals (e.g., Lewis 1989, Diogo et al. 2009 and Diogo and Abdala 2010). The *interossei accessorii* are not present as distinct muscles in common chimpanzees.
- Synonymy: Palmar or volar interossei and part of dorsal interossei (Huxley 1864; Champneys 1872; Dwight 1895; Miller 1952; Tuttle 1969, 1970; Swindler and Wood 1973; Gibbs 1999; Ogihara et al. 2005; Tocheri et al. 2008); palmar interossei plus interosseous volaris primus (Brooks 1887; Hepburn 1892); palmar interossei and possibly part of flexor brevis pollicis (Sonntag 1923); flexores breves (Lewis 1989).

***Interossei dorsales*** (fetus intermetacarpalis 1 + flexor brevis profundus 3, or 'interosseous dorsalis' 1 = 0.06 g; LSB intermetacarpalis 2 + flexor brevis profundus 5, or 'interosseous dorsalis' 2 = 0.10 g; LSB intermetacarpalis 3 + flexor brevis profundus 6, or 'interosseous dorsalis' 3 = 0.06 g; LSB intermetacarpalis 4 + flexor brevis profundus 8, or 'interosseous dorsalis' 4 = 0.04 g; infant Foyo all four dorsal interossei together =0.94 g; Figs. 3.46, 3.48 and 3.49)

- Attachments in bonobos: See interossei palmares above.
- Usual innervation: Deep branch of the ulnar nerve (Miller 1952: *P. paniscus*) [*P. troglodytes*: deep branch of the ulnar nerve (Hepburn 1892); ulnar nerve (Swindler and Wood 1973)].

- Comparison with common chimpanzees: The attachments of the intermetacarpales are similar in *P. troglodytes* and *P. paniscus*, the main difference being that these muscles are usually present as distinct muscles in common chimpanzees, and fused with four flexores breves profundi to form the dorsal interossei in most bonobos, as they are in most modern humans (see above). The *interdigitales*, present in primates such as lorisoids, are usually not present in common chimpanzees.
- Synonymy: Part or totality of dorsal interossei (Gratiolet and Alix 1866; Champneys 1872; Brooks 1886; Hepburn 1892; Sonntag 1923, 1924; Miller 1952; Tuttle 1969, 1970; Swindler and Wood 1973; Susman and Stern 1980; Lewis 1989; Gibbs 1999; Ogihara et al. 2005; Tocheri et al. 2008).

***Flexor pollicis brevis*** and ***flexor brevis profundus 2*** (infant Foyo flexor pollicis brevis + flexor brevis profundus 2, or 'superficial + deep heads of flexor pollicis brevis' = 0.28 g; fetus flexor pollicis brevis, or 'superficial head of flexor pollicis brevis' = 0.03 g; fetus flexor brevis profundus 2, or 'deep head of flexor pollicis brevis' = 0.01 g; Figs. 3.14, 3.26, 3.45 and 3.46)

- Attachments in bonobos: As explained above (see adductor pollicis), the flexor brevis profundus 2 of 'lower' mammals corresponds to the *deep head of the flexor pollicis brevis* of modern human anatomy; the *superficial head of the flexor pollicis brevis* of modern human anatomy thus corresponds to a true flexor pollicis brevis, being likely derived, together with the *opponens pollicis*, from the *flexor brevis profundus 1* of 'lower' mammals (see, e.g., Diogo et al. 2009, Diogo and Abdala 2010 and Diogo and Wood 2012). The flexor brevis profundus 2 of bonobos usually runs mainly from the carpal region (trapezium and/or flexor retinaculum in most our specimens; scaphoid and MII in Miller's 1952 adult bonobo) to the ulnar side of the proximal phalanx of digit 1 and/or the small ulnar sesamoid bone lying between MI and the proximal phalanx of the thumb, while the true flexor pollicis brevis (part of the flexor brevis profundus 1) runs mainly from the carpal region (trapezium and/or flexor retinaculum in most our specimens; scaphoid and flexor retinaculum in Miller's 1952 adult bonobo) to the radial side of the proximal phalanx of digit 1 and/or the small radial sesamoid bone lying between MI and the proximal phalanx of the thumb, lying superficially to the flexor brevis profundus 2. In Miller's 1952 adult bonobo and on the left hand of our adult Kidogo, both these structures were fused with the abductor pollicis brevis, while on the right hand of our adult Kidogo and the right hand of our adolescent Jasiri the flexor pollicis brevis distally joins the tendon of the abductor pollicis brevis to the radial sesamoid bone. In the single adult bonobo dissected by Payne (2001), the origin of the 'flexor pollicis brevis' was stated to be directly from the palmar aponeurosis.
- Usual innervation of the true flexor pollicis brevis and of the flexor brevis profundus 2: 'flexor pollicis brevis' (i.e., flexor pollicis brevis plus flexor brevis profundus 2 sensu the present study) is innervated by the ulnar and the median nerves (Miller 1952: *P. paniscus*) [In *P. troglodytes*: 'flexor pollicis brevis' (which might include the flexor pollicis brevis plus flexor brevis profundus 2 sensu the present study) is innervated by the median nerve (Champneys 1872: *P. troglodytes*); 'superficial head of the flexor pollicis brevis' of modern human anatomy (i.e., flexor pollicis brevis sensu the present study) is innervated by the median nerve (Brooks 1887, Hepburn 1892: *P. troglodytes*); median nerve ('superficial head', because the 'deep head' of modern human anatomy is, according to these authors, absent in *Pan*; Swindler and Wood 1973: *P. troglodytes*); the true flexor pollicis brevis is innervated by the median nerve, while the flexor brevis profundus 2 innervated by the ulnar nerve (our *P. troglodytes* specimens GWUANT PT1 and GWUANT PT2)].
- Comparison with common chimpanzees: As there is much confusion in the literature about the flexor brevis profundus 2 and the true flexor pollicis brevis *sensu* the present study, we will

describe in some detail the configuration found in some of the common chimpanzees dissected by us. For more details about the descriptions of other authors and for information about the synonymy, homologies and attachments of these muscles, see Diogo and Wood (2012) and see also the sections above about adductor pollicis, interossei palmares and intermetacarpales. In *P. troglodytes* the flexor brevis profundus 2 usually runs mainly from the carpal region to the ulnar side of the proximal phalanx of digit 1, while the true flexor pollicis brevis runs mainly from the carpal region to the radial side of the proximal phalanx of digit 1, lying superficially to the flexor brevis profundus 2. In our *P. troglodytes* adult specimens GWUANT PT1 and GWUANT PT2, the true flexor pollicis brevis originates from the flexor retinaculum and trapezium, while the flexor brevis profundus 2 originates from the capitate and trapezoid. The true flexor pollicis brevis inserts onto the lateral side of the palmar surface of the base of the proximal phalanx of digit 1 and/or on the adjacent sesamoid bone, while the flexor brevis profundus 2 inserts onto the medial side of the palmar surface of the proximal phalanx of digit 1 and/or on the adjacent sesamoid bone. On the left side of our *P. troglodytes* infant specimen PFA 1077, only one 'head' of the so-called flexor pollicis brevis seems to be present (i.e., the flexor brevis profundus 2 does not seem to be present as a separate muscle). The only 'head' that is present is the 'superficial head' (i.e., the true flexor pollicis brevis sensu the present study, which runs from the flexor retinaculum, the trapezium and the ligaments of the carpal region to the central and lateral portions of the ventral region of the metacarpophalangeal joint of the thumb). However, in the right hand of that infant specimen, there is what clearly seems to be a 'deep head of the flexor pollicis brevis'; the flexor brevis profundus 2 does seem to be present as a distinct muscle, being somewhat blended with the opponens pollicis. On the left dissected hand of our *P. troglodytes* infant specimen PFA UNC, the 'deep head of the flexor pollicis brevis' (i.e., the flexor brevis profundus 2) is not present as a distinct muscle, for the single 'head of the flexor pollicis brevis' (i.e., the true flexor pollicis brevis *sensu* the present study) is blended dorsally to the opponens pollicis.

- Synonymy of the true flexor pollicis brevis: Part or totality of court fléchisseur du pouce (Gratiolet and Alix 1866); part or totality of flexor brevis pollicis, and probably of the 'interosseous volaris primus" (Champneys 1872); part or totality of flexor brevis pollicis and possibly part of the abductor pollicis brevis (Sonntag 1923); 'superficial head of the flexor pollicis brevis' of human anatomy (Terminologia Anatomica 1998).

***Opponens pollicis*** (infant Foyo 0.07 g; fetus 0.01 g; Figs. 3.14 and 3.46)

- Attachments in bonobos: From flexor retinaculum and scaphoid (Miller's 1952 adult bonobo) and/or trapezium and/or adjacent sesamoid bone, to whole length of the MI shaft.
- Usual innervation: Median nerve (Miller 1952: *P. paniscus*) [same innervation in *P. troglodytes*: Brooks 1887, Hepburn 1892, Swindler and Wood 1973].
- Comparison with common chimpanzees: In *P. troglodytes* it often runs from the flexor retinaculum, trapezium and/or adjacent sesamoid bone to the whole length of the MI shaft.
- Synonymy: Opposant du pouce (Gratiolet and Alix 1866).

***Flexor digiti minimi brevis*** (infant Foyo 0.18 g; fetus 0.05 g; Figs. 3.14 and 3.45)

- Attachments in bonobos: From hamate and/or flexor retinaculum, to the ulnar side of the proximal phalanx of digit 5 and sometimes also the metacarpophalangeal joint of this digit. Miller (1952) described an origin from the flexor retinaculum only, in her adult bonobo, while we found an origin from the flexor retinaculum and hamate in our specimens. In the left hand of our adult Kidogo the flexor digiti minimi brevis was apparently completely fused with the abductor digiti minimi.

- Usual innervation: Ulnar nerve (Miller 1952: *P. paniscus*) [same innervation in *P. troglodytes*: Hepburn 1892, Swindler and Wood 1973].
- Comparison with common chimpanzees: In *P. troglodytes* an origin from the hamate and flexor retinaculum was found by Gratiolet and Alix (1866), Sonntag (1923), and Swindler and Wood (1973), the insertion being similar to that of bonobos (see above).
- Synonymy: Court fléchisseur du cinquième doigt (Gratiolet and Alix 1866); flexor brevis minimi digiti (Champneys 1872; Sonntag 1923); flexor digiti quinti (Miller 1952).

***Opponens digiti minimi*** (infant Foyo 0.15 g; fetus 0.01 g; Fig. 3.48)

- Attachments in bonobos: From the hamate and, often, also from the flexor retinaculum to the whole length of MV.
- Usual innervation: Ulnar nerve (Miller 1952: *P. paniscus*) [same innervation in *P. troglodytes*: Hepburn 1892, Swindler and Wood 1973].
- Comparison with common chimpanzees: In *P. troglodytes* it runs from the hamate and, often, also from the flexor retinaculum to the whole length of the MV shaft, being often divided into a superficial and a deep head, as noted below; occasionally the muscle may also originate from structures such as the contrahens aponeurosis (e.g., Gratiolet and Alix 1866). As noted by Brooks (1886), Lewis (1989) and Diogo et al. (2009) and Diogo and Wood (2011, 2012) and corroborated by our dissections, in hominoids the opponens digiti minimi is usually slightly differentiated into superficial and deep bundles. Lewis (1989) stated that the opponens digiti minimi is more markedly divided in hominoids such as *P. troglodytes* and modern humans than in hominoids such as hylobatids, and Brooks (1886) stated that, contrary to *P. troglodytes* and modern humans, in hominoids such as *Pongo* there are no superficial and deep bundles of the muscle separated by the deep branch of the ulnar nerve. Regarding our dissections, in hylobatids, *Gorilla* and *Pongo*, the deep branch of the ulnar nerve runs mainly radial to both these bundles, and not mainly superficially (palmar) to the deep bundle and deep (dorsal) to the superficial bundle as is usually the case in *P. troglodytes* and particularly in modern humans.
- Synonymy: Opposant du cinquième doigt (Gratiolet and Alix 1866); opponens minimi digiti (Champneys 1872; Sonntag 1923); opponens digiti quinti (Miller 1952).

***Abductor pollicis brevis*** (infant Foyo 0.38 g; fetus 0.04 g; Figs. 3.14, 3.26 and 3.45)

- Attachments in bonobos: From the flexor retinaculum (only origin in Miller's 1952 adult bonobo) and structures such as the trapezium and scaphoid (our fetus) to the base of the proximal phalanx of digit 1 and/or the radial sesamoid associated with it.
- Usual innervation: Median nerve (Miller 1952: *P. paniscus*) [same innervation in *P. troglodytes*: Brooks 1887, Hepburn 1892, Swindler and Wood 1973].
- Comparison with common chimpanzees: In common chimpanzees this muscle usually runs from the flexor retinaculum and structures such as the trapezium, the adjacent sesamoid bone, the transverse carpal ligament and/or the scaphoid, to the radial side of the metacarpophalangeal joint, its sesamoid bone and/or the base of the proximal phalanx of digit 1; occasionally the insertion of the muscle extends to the distal phalanx of this digit (e.g., Hepburn 1892; Dwight 1895) and/or to the distal portion of MI (e.g., Hepburn 1892; Aziz and Dunlap 1986; Landsmeer 1986). In common chimpanzees, the abductor pollicis brevis may also be divided into slips (e.g., Aziz and Dunlap 1986) and may be reinforced by slips from the 'flexor pollicis brevis' (true flexor pollicis brevis plus flexor brevis profundus 2 *sensu* the present study) according to Hepburn (1892).
- Synonymy: Court abducteur du pouce (Gratiolet and Alix 1866); abductor pollicis (Champneys 1872); part or totality of abductor pollicis brevis (Sonntag 1923).

***Abductor digiti minimi*** (infant Foyo 0.23 g; fetus 0.04 g; Figs. 3.13, 3.14, 3.26 and 3.45)

- Attachments in bonobos: From pisiform and sometimes from flexor retinaculum to ulnar side of proximal phalanx of digit 5 (our fetus and infants Etje and Foyo), and in some cases also to the metacarpophalangeal joint of this digit (adult Kidogo and adolescent Jasiri), or to the extensor expansion of this digit (Miller's 1952 adult bonobo). The abductor digit minimi has a single head in all bonobos dissected by us and by Miller (1952).
- Usual innervation: Radial nerve (Miller 1952; *P. paniscus*) [same innervation in *P. troglodytes*: Hepburn 1892, Sonntag 1923, Swindler and Wood 1973].
- Comparison with common chimpanzees: In *P. troglodytes* it runs from the pisiform and sometimes from the flexor retinaculum (e.g., Swindler and Wood 1973; some of our specimens) and/or hamate (e.g., Swindler and Wood 1973; some of our specimens), to the ulnar side of the proximal phalanx of digit 5 and sometimes to the adjacent metacarpophalangeal joint (e.g., some of our specimens), and/or to the adjacent ulnar sesamoid bone (e.g., some of our specimens).
- Synonymy: Métacarpien du index or métacarpien du deuxième doigt or deuxiéme métacarpien dorsal or prémier radial externe (Gratiolet and Alix 1866).

***Extensor carpi radialis longus*** (infant Foyo 1.85 g; fetus 0.30 g; Figs. 3.42 and 3.44)

- Attachments in bonobos: From lateral supracondylar ridge (Miller's 1952 adult bonobo, our adult Kidogo and adolescent Jasiri) and often also from lateral epicondyle of the humerus (our fetus), to base of MII and sometimes also of MIII (Payne's 2001 adult bonobo) or MI (left side of adult Kidogo).
- Usual innervation: Radial nerve (Miller 1952: *P. paniscus*) [same innervation in *P. troglodytes*: Hepburn 1892, Sonntag 1923, Swindler and Wood 1973].
- Comparison with common chimpanzees: In *P. troglodytes* it usually runs from the lateral supracondylar ridge and often also from the lateral epicondyle of the humerus, to the base of MII and often also to base of MI (e.g., Bojsen-Møller 1978; our *P. troglodytes* specimen HU PT1 and one side of specimen PFA 1077); in a few cases it also attaches to the ligament connecting the MII to the trapezium (e.g., Gratiolet and Alix 1866).
- Synonymy: Premier radial (Deniker 1885); extensor carpi radialis longior (Hartmann 1886, Hepburn 1892; Champneys 1872; Beddard 1893; Sonntag 1923).

***Extensor carpi radialis brevis*** (infant Foyo 1.34 g; fetus 0.29 g; Figs. 3.17 and 3.42)

- Attachments in bonobos: From lateral epicondyle of humerus to base of MIII, and often also of MII (our fetus).
- Usual innervation: Radial nerve (Miller 1952: *P. paniscus*) [*P. troglodytes*: posterior interosseous nerve (Hepburn 1892); radial nerve (Swindler and Wood 1973)].
- Comparison with common chimpanzees: In *P. troglodytes* it runs from the lateral epicondyle of the humerus and occasionally also from structures such as the radial collateral ligament (e.g., Sonntag 1923), to the base of MIII and often also to the base of MII (e.g., Ribbing and Hermansson 1912; our *P. troglodytes* specimens HU PT1, PFA 1077 and PFA UNC).
- Synonymy: Métacarpien du médius or métacarpien du troisième doigt or troisième métacarpein dorsal or deuxième radial externe (Gratiolet and Alix 1866); extensor carpi radialis brevior (Champneys 1872; Hepburn 1892; Beddard 1893; Sonntag 1923).

***Brachioradialis*** (infant Foyo 3.59 g; fetus 0.53 g; Figs. 3.9, 3.23, 3.26, 3.37, 3.38, 3.39, 3.41 and 3.42)

- Attachments in bonobos: From humerus (its origin extending more proximally than in modern humans, e.g., origin from distal 1/3 of humerus in our fetus and from distal 1/2 of humerus in Miller's 1952 adult bonobo) to shaft of radius (not reaching the styloid process in our fetus, and seemingly also not in Miller's 1952 adult bonobo but seemingly reaching it in our adult Kidogo and adolescent Jasiri; it was only bound to the radial shaft for 1/2 of an inch in Payne's 2001 adult bonobo), being usually fused with the brachialis (Miller's 1952 adult bonobo).

- Usual innervation: Radial nerve (Miller 1952: *P. paniscus*) (same innervation in *P. troglodytes*: Hepburn 1892, Sonntag 1923, Straus 1941a, b, Swindler and Wood 1973).
- Comparison with common chimpanzees: In *P. troglodytes* it usually also runs from the shaft and distal end of the humerus to the shaft of the radius (usually not reaching the styloid process; exceptions were reported by, e.g., Hepburn 1892, Ziegler 1964 and Swindler and Wood 1973 and found in our *P. troglodytes* specimens PFA 1077 and PFA UNC). The origin of the brachioradialis also extends more proximally in *P. troglodytes* than in modern humans (e.g., often lying at the level of the deltoid insertion: Sonntag 1923 and Swindler and Wood 1973 and our *P. troglodytes* specimens GWUANT PP1, GWUANT PP2, PFA 1016, PFA 1051 and HU PT1).
- Synonymy: Supinator longus (Wyman 1855; Wilder 1862; Gratiolet and Alix 1866; Macalister 1871; Barnard 1875; Duckworth 1904; Sonntag 1923); supinator radii longus (Champneys 1872; Hepburn 1892; Beddard 1893).

***Supinator*** (infant Foyo 1.93 g; fetus 0.37 g; Figs. 3.17, 3.23, 3.38, 3.39, 3.41 and 3.44)

- Attachments in bonobos: From lateral epicondyle of humerus (caput humerale, or superficiale) and proximal ulna (caput ulnare, or profundum), to proximal radius, seemingly with the exception only of the left upper limb of Kidogo, in which we could not find an origin from the humerus. The muscle was seemingly not perforated by the radial nerve in the two specimens in which we analysed this link between nerve and muscle in detail (i.e., in our fetus and infant Foyo).
- Usual innervation: Posterior interosseous (deep radial) nerve (Miller 1952: *P. paniscus*) [*P. troglodytes*: posterior interosseous (deep radial) nerve (Champneys 1872, Hepburn 1892, Sonntag 1923, Straus 1941a, b); radial nerve (Swindler and Wood 1973)].
- Comparison with common chimpanzees: In *P. troglodytes* it usually runs from the lateral epicondyle of the humerus (caput humerale, or superficiale) and usually also from the proximal part of the ulna (caput ulnare, or profundum; e.g., Gratiolet and Alix 1866, Champneys 1872, Beddard 1893, Straus 1941a, Swindler and Wood 1973 and our dissections), to the proximal radius. The supinator is usually perforated by the radial nerve in common chimpanzees (e.g., Macalister 1871, Beddard 1893, Straus 1941a, Lewis 1989, and our *P. troglodytes* specimens PFA 1077 and PFA UNC), but was not perforated by this nerve in our bonobo fetus and infant Foyo, as noted above.
- Synonymy: Court supinateur (Gratiolet and Alix 1866); supinator radii brevis (Macalister 1871; Champneys 1872; Hepburn 1892; Beddard 1893); supinator brevis (Sonntag 1923).

***Extensor carpi ulnaris*** (infant Foyo 1.03 g; fetus 0.21 g; Fig. 3.43)

- Attachments in bonobos: From lateral epicondyle of humerus and ulna (caput humerale; Miller's 1952 adult bonobo) or only from ulna (our fetus and infant Foyo and adult Kidogo and adolescent Jasiri) to base of MV. A difference between bonobos and common chimpanzees might thus be that in most bonobos there is seemingly no direct origin from humerus, while an origin from the humerus is present in most common chimpanzees (see just below).
- Usual innervation: Posterior interosseous (deep radial) nerve (Miller 1952: *P. paniscus*) [*P. troglodytes*: posterior interosseous (deep radial) nerve (Hepburn 1892); radial nerve (Swindler and Wood 1973)].

- Comparison with common chimpanzees: In *P. troglodytes* it runs from the lateral epicondyle of the humerus (caput humerale) and usually also the ulna (caput ulnare; e.g., Gratiolet and Alix 1866, Champneys 1872, Sonntag 1923, Straus 1941a, Swindler and Wood 1973 and our dissections) to the base of MV and occasionally to the proximal phalanx of digit 5 (e.g., Macalister 1871, who stated that in the *P. troglodytes* specimen dissected by him the extensor carpi ulnaris had a distinct *ulnaris-quinti* tendon to the proximal phalanx of digit 5).
- Synonymy: Métacarpien du cinquième doigt or cubital postérieur or cubital dorsal or cinquième métacarpien dorsal (Gratiolet and Alix 1866).

***Anconeus*** (infant Foyo 0.31 g; fetus 0.01 g; Fig. 3.43)

- Attachments in bonobos: From lateral epicondyle of humerus to olecranon process of ulna, and often also to adjacent areas of ulna (Miller's 1952 adult bonobo and Payne's 2001 adult bonobo).
- Usual innervation: Radial nerve (Swindler and Wood 1973: *P. troglodytes*; Miller 1952: *P. paniscus*); ulnar nerve (Ziegler 1964: *P. troglodytes*).
- Comparison with common chimpanzees: The anconeus is usually present as a distinct muscle in *P. troglodytes* according to Gratiolet and Alix (1866), Macalister (1871), Champneys (1872), Sonntag (1923), Swindler and Wood (1973), and to our dissections, running from the lateral epicondyle of the humerus to the olecranon process of the ulna and in some cases also to adjacent areas of the ulna (e.g., Swindler and Wood 1973, and our dissections).

***Extensor digitorum*** (infant Foyo 2.58 g; fetus 0.57 g; Figs. 3.17 and 3.43)

- Attachments in bonobos: From lateral epicondyle of humerus, to middle phalanges and (via the extensor expansions) to distal phalanges of digits 2, 3, 4 and 5 (our fetus and infant Etje and Miller's 1952 adult bonobo). Variations include insertion only onto digits 2–4 in our infant Foyo, and the presence of an extra tendon to digit 4 originating from the tendon to digit 5 in the right upper limb of our adult Kidogo.
- Usual innervation: Posterior interosseous (deep radial) nerve (Hepburn 1892: *P. troglodytes*; Miller 1952: *P. paniscus*); radial nerve (Swindler and Wood 1973: *P. troglodytes*).
- Comparison with common chimpanzees: In *P. troglodytes* it usually runs from the lateral epicondyle of the humerus and occasionally also from the radius (e.g., MacDowell 1910) and/or the ulna (e.g., MacDowell 1910 and our *P. troglodytes* specimens PFA 1077 and PFA UNC), to the middle phalanges and (via the extensor expansions) to the distal phalanges of digits 2, 3, 4 and 5; occasionally the muscle has an insertion onto the interphalangeal joints (e.g., Sonntag 1923) or has no insertion onto digit 5 (e.g., Vrolik 1841, Wilder 1862 and Straus 1941a).
- Synonymy: Extensor communis digitorum (Wilder 1862; Champneys 1872; Hepburn 1892; Beddard 1893; Sonntag 1923); extenseurs directes des doigts (Gratiolet and Alix 1866); extensor digitorum longus (Macalister 1871); extensor digitorum communis (Barnard 1875; Dwight 1895; MacDowell 1910; Straus 1941a, b; Miller 1952).

***Extensor digiti minimi*** (infant Foyo 0. 54 g; fetus 0.11 g; Fig. 3.43)

- Attachments in bonobos: From lateral epicondyle of humerus and/or common extensor tendon, to middle phalanx and (via the extensor expansion) to middle and distal phalanges of digit 5 (e.g., our infant Etje and Miller's 1952 adult bonobo and our adult Kidogo and adolescent Jasiri), and often also of digit 4 (our fetus and infant Foyo).
- Usual innervation: Posterior interosseous (deep radial) nerve (Miller 1952: *P. paniscus*) [*P. troglodytes*: posterior interosseous (deep radial) nerve (Hepburn 1892); radial nerve (Swindler and Wood 1973)].

- Comparison with common chimpanzees: In *P. troglodytes* it usually runs from the lateral epicondyle of the humerus and/or the common extensor tendon, and/or occasionally from the ulna (e.g., MacDowell 1910), to the middle phalanx and (via the extensor expansion) to the distal phalanx of digit 5. In the *P. troglodytes* specimens described by Wilder (1862), Gratiolet and Alix (1866), Macalister (1871), Champneys (1872), Chapman (1879), Hepburn (1892), MacDowell (1910), Sonntag (1923), Swindler and Wood (1973), Aziz and Dunlap (1986), Landsmeer (1986) and Oishi et al. (2009), as well as in 3 of the 6 *P. troglodytes* specimens dissected by us in which we could discern this feature in detail, as well as in three specimens dissected by Straus (1941a) and in one specimen dissected by Kaneff (1980a), the muscle goes to digit 5 only. In the 3 other *P. troglodytes* specimens dissected by us in which we could discern this feature in detail, in one specimen dissected by Huxley (1864), in one specimen dissected by Dwight (1895), in one specimen dissected by Straus (1941a), in 4 specimens dissected by Kaneff (1980a), and in one specimen dissected by Lewis (1989), the muscle goes to digits 4 and 5. According to the review of the literature by Gibbs (1999), an exclusive insertion onto digit 5 occurs in 26 out of 31 (84%) of *Pan*; it occurs in 73% *P. troglodytes* according to the review of the literature done by Straus (1941a). The *extensor digiti quarti* is usually not present as a distinct muscle in chimpanzees.
- Synonymy: Extensor minimi digiti (Wilder 1862; Macalister 1871; Champneys 1872; Hepburn 1892; Beddard 1893; Dwight 1895; Sonntag 1923); extenseur latéral du cinquième doigt (Gratiolet and Alix 1866); extensor digiti quinti proprius (MacDowell 1910); extensor digiti – quarti et – quinti proprius (Straus 1941a, b); extensor digiti quinti proprius (Miller 1952); extensor digitorum lateralis bidigitalis (Kaneff 1979, 1980a); extensor digitorum proprius or profundus 4 and 5 (Lewis 1989).

***Extensor indicis*** (infant Foyo 0.37 g; fetus 0.04 g; Fig. 3.44)

- Attachments in bonobos: From ulna to distal phalanx of digit 2 through the extensor expansion.
- Usual innervation: Posterior interosseous (deep radial) nerve (Miller 1952: *P. paniscus*) [*P. troglodytes*: posterior interosseous (deep radial) nerve (Hepburn 1892); radial nerve (Swindler and Wood 1973)].
- Comparison with common chimpanzees: In *P. troglodytes* it usually runs from the ulna and often also from the interosseous membrane and occasionally also from the radius (e.g., Sonntag 1923) and/or the humerus/common extensor tendon (e.g., Gratiolet and Alix 1866), to digit 2; occasionally there is also an insertion onto digit 4 and/or digit 3 and/or no insertion onto digit 2. An insertion onto digit 2 only is described in *P. troglodytes* by most authors, including Wilder (1862), Gratiolet and Alix (1866), Broca (1869), Champneys (1872), Chapman (1879), Beddard (1893), Dwight (1895), Sonntag (1923), Fick (1925), Tuttle (1969), Swindler and Wood (1973), Dunlap et al. (1985), Landsmeer (1986), Lewis (1989) and Oishi et al. (2009) and found in the vast majority of the specimens dissected by us. However, Hepburn (1892) found an insertion onto digits 2 and 4, Straus (1941a) and Kaneff (1980a, b) onto digit 2 or onto digits 2 and 3, and Humphry (1867), Macalister (1871), Hartmann (1886), MacDowell (1910), Ribbing and Hermansson (1912), and Jouffroy and Lessertisseur (1959) onto digits 2 and 3. According to the review of the literature by Straus (1941a, b), in *P. troglodytes* an insertion onto digits 2, 3 and 4 occurs in about 4% of the cases, onto digits 2 and 3 occurs in about 21% of the cases, onto digits 2 and 4 occurs in about 4% of the cases, only onto digit 3 occurs in about 4% of the cases, and only onto digit 2 occurs in about 68% of the cases. In our *P. troglodytes* infant specimen PFA 1077 the extensor indicis has tendons to digits 2 and 3, but the tendon to digit 3 is very 'degenerated' distally; this suggests that the

tendon to digit 3 could become completely degenerate/lost during ontogeny, thus leading to an adult configuration in which there is only a tendon to digit 2, as is usually the case in adult common chimpanzees. However, in the bonobo infants and the bonobo fetus dissected by us, the extensor indicis attached to digit 2 only. The *extensor digiti III proprius* and the *extensor brevis digitorum manus* (e.g., Diogo et al. 2009) are usually not present as distinct muscles in common chimpanzees.
- Synonymy: Extenseur latéral de l'index (Gratiolet and Alix 1866); extenseur propre de l'index (Broca 1869); indicator (Macalister 1871); part or totality of extensor profundus digitorum or extensor digitorum profundus (Barnard 1875; Hepburn 1892; Straus 1941a, b); extensor indicis proprius (Dwight 1895; MacDowell 1910; Miller 1952); extensor digiti indicis (Swindler and Wood 1973); extensor digitorum profundus proprius or extensor indicis-et-medii digiti (Aziz and Dunlap 1986); extensor digitorum proprius or profundus 2 (Lewis 1989).

***Extensor pollicis longus*** (infant Foyo extensor pollicis longus + abductor pollicis longus =1.87 g; fetus extensor pollicis longus 0.05 g; Figs. 3.17 and 3.44)

- Attachments in bonobos: From ulna and interosseous membrane to distal phalanx of thumb and occasionally – seemingly a variant – also to radial side of base of MI (in our fetus). Another variant was found in the right upper limb of Jasiri, where the extensor pollicis longus seemingly attached only onto the proximal phalanx of the thumb.
- Usual innervation: Posterior interosseous (deep radial) nerve (Miller 1952: *P. paniscus*) [same innervation in *P. troglodytes*: Hepburn 1892].
- Comparison with common chimpanzees: In *P. troglodytes* it usually runs from the ulna and interosseous membrane and occasionally from the humerus/common extensor tendon (e.g., Gratiolet and alix 1866) and/or the radius (e.g., some of our *P. troglodytes* specimens), to the distal phalanx of the thumb and sometimes also to its proximal phalanx (e.g., Champneys 1872, Sutton 1883, Ziegler 1964 and our *P. troglodytes* specimen PFA UNC). The structure designated as *extensor communis pollicis et indicis* in one of the five common chimpanzees dissected by Kaneff (1980a) does not correspond to the extensor communis pollicis et indicis *sensu* the present study. Instead, it corresponds to the extensor pollicis longus *sensu* the present study, which in this specimen goes to both digits 1 and 2. In the four other common chimpanzees dissected by Kaneff 1980a, the extensor pollicis longus goes to digit 1 only, as is usually the case in most common chimpanzees.
- Synonymy: Extenseur latéral du pouce (Gratiolet and Alix 1866); extensor secundi internodii pollicis (Champneys 1872; Sutton 1883; Hepburn 1892; Beddard 1893; Sonntag 1923); part or totality of extensor profundus digitorum or extensor digitorum profundus (Straus 1941a, b); extensor digitorum proprius or profundus 1 (Lewis 1989).

***Abductor pollicis longus*** (infant Foyo extensor pollicis longus + abductor pollicis longus =1.87 g; fetus extensor pollicis longus 0.27 g; Figs. 3.17 and 3.44).

- Attachments in bonobos: From interosseous membrane and ulna and radius, dividing into two well-separated tendons to insert onto MI and onto trapezium, in all our bonobos and in the adult bonobo of Miller (1952) (see, e.g., Fig. 3.17).
- Usual innervation: Posterior interosseous (deep radial) nerve (Miller 1952: *P. paniscus*) [*P. troglodytes*: posterior interosseous (deep radial) nerve (Hepburn 1892); radial nerve (Swindler and Wood 1973)].
- Comparison with common chimpanzees: In *P. troglodytes* it runs from the interosseous membrane and/or ulna and/or radius, to MI and trapezium; sometimes the muscle inserts also/instead onto the scaphoid and/or sesamoid bone adjacent to the trapezium, and occasionally the muscle extends to the proximal

phalanx of digit 1. Apart from modern humans, in all of the primate specimens dissected by us only in hylobatids is there a distinct *extensor pollicis brevis* that is only partially blended, proximally, with the belly of the abductor pollicis longus (e.g., Diogo and Wood 2011, 2012). It should be noted that some authors have described an 'extensor pollicis brevis' and an 'abductor pollicis longus' in primate taxa other than hylobatids and modern humans, including common chimpanzees (see Synonymy below). For example, in gorillas the name 'extensor pollicis brevis' has been used (e.g., Hepburn 1892; Straus 1941a, b) to refer to a tendon of the abductor pollicis longus (sensu the present study) that inserts onto the proximal phalanx of the thumb (i.e., to the typical insertion point of the extensor pollicis brevis of modern humans). However, as stressed by authors such as Kaneff (1979, 1980a, b) and Aziz and Dunlap (1986) and corroborated by our dissections in gorillas, there is usually a single fleshy belly of the abductor pollicis longus that then gives rise to the so-called tendons of the extensor pollicis brevis and of the abductor pollicis longus; this configuration is usually also found in *Pongo* and *Pan*. Thus, contrary to the condition in *Homo* and hylobatids, in *Pongo*, *Pan* and *Gorilla* the extensor pollicis brevis is usually *not* present as a separate muscle. However, these two tendons present consistently in great apes clearly seem to be the precursors of the tendons of the abductor pollicis longus and of the extensor pollicis brevis of modern humans. In *P. troglodytes* the insertion of the abductor pollicis longus is onto the MI, the adjacent sesamoid bone and trapezium according to Hepburn (1892), Dwight (1895), and Sonntag (1923), to the scaphoid and MI according to Humphry (1867), to the sesamoid, scaphoid and MI according to Ziegler (1964), and to MI and trapezium according to Vrolik (1841), Wyman (1855), Wilder (1862), Huxley (1864), Champneys (1872), Hartmann (1886), Beddard (1893), Sutton (1883), Gratiolet and Alix (1866), Macalister (1871), MacDowell (1910), Ogihara et al. (2005), and according to our dissections. An insertion onto MI and trapezium is the common condition in *Pan* according to Gibbs (1999), and in the review of the literature done by Keith (1899) an extension to the proximal phalanx of the thumb was only found in 1 out of 20 *Pan*, while Sarmiento (1994) did not find such an extension in any of the two *Pan* specimens dissected by him, and in the review of the literature done by Sarmiento (1994) there was no such extension in 10 out of 10 *Pan*.

- Synonymy: Extensor ossis metacarpi plus short extensor of the thumb (Wyman 1855); extensor ossis metacarpi pollicis plus extensor primi internodii pollicis (Wilder 1862; Beddard 1893; Sonntag 1923); extensor ossis metacarpi pollicis (Huxley 1864; Champneys 1872; Sutton 1883; Hepburn 1892; MacDowell 1910; Wood Jones 1920); carpo-métacarpiens du pouce (Gratiolet and Alix 1866); adductor pollicis longus plus extensor pollicis brevis (Ziegler 1964).

## 3.2 Brachial Plexus

### 3.2.1 Description of the *plexus brachialis* of Adolescent Female Jasiri (Fig. 3.50)

The left *plexus brachialis* of Jasiri converges in the interscalene triangle, between the *m. scalenus anterior* and *m. scalenus medius*. The *plexus brachialis* comprised of nerve roots C5, C6, C7, C8 and T1, and is broadly similar in morphology to the plexus brachialis in *P. troglodytes* and *Homo sapiens*, although some variations in the segmentation exist that we believe are idiosyncratic to this specimen, and not necessarily representative of the most common condition likely to be seen in bonobos. These roots form three trunks: C5 + C6 form the *truncus superior plexus brachialis*, C7 forms the *truncus intermedius plexus brachialis*, and C8 + T1 form the *truncus inferioris plexus brachialis*. The *truncus superior plexus brachialis* forms shortly after the nerve roots emerge from between the *scaleni* muscles, persists only for a short distance, and then splits into an anterior and posterior division which contribute to the *fasciculus lateralis plexus brachialis* and the *fasciculus posterior plexus brachialis*. The *truncus intermedius plexus brachialis* arises solely from C7, persists for a small distance, and then gives off segment of the *n. musculocutaneous* from its ventral surface, an anterior division which contributes to the *fasciculus intermedius plexus brachialis*, a smaller branch that arises directly after a contribution from the *truncus inferioris plexus brachialis*, and a posterior division which contributes to the *fasciculus posterior plexus brachialis*. The small contribution from the *truncus inferioris plexus brachialis* combines with the *truncus intermedius plexus brachialis* near where it begins to branch into divisions. The *truncus inferioris plexus brachialis* forms from the roots of C8 + T1, and almost immediately gives off a branch to the *truncus intermedius plexus brachialis* and the *n. subscapularis inferior*. It contributes an anterior division to the *fasciculus intermedius plexus brachialis* and a posterior division to the *fasciculus posterior plexus brachialis*. The *n. phrenicus* arises from the anterior portion of the root of C4 and with a contribution from C5 before descending superficial to the *m. scalenus anterior* to pass into the thoracic cavity. As no histological studies of nerve fibre contributions were conducted for this atlas, all assignations of root contributions to nerves, designated by the nerve number in brackets – e.g., (C5-6) – are hypotheses that remain to be tested based on what nerve roots could have possibly contributed to the distal segment. The discrete nerves found and followed to their terminal points in the arm of Jasiri are as follows.

The major nerves of the ventral division of the *plexus brachialis* are: *n. suprascapularis*, *n. pectoralis medialis*, *n. pectoralis lateralis*, *n. musculocutaneous*, *n. cutaneus antebrachii lateralis*, n. *medianus* and *n. ulnaris*. The *n. suprascapularis* (C5-C6) is the first branch off the *truncus superior plexus brachialis*. A single unit branches off from the posterior portion of the *truncus superior plexus brachialis* and travels through the *incisura scapulae* to innervate the *m. supraspinatus,* and through the *incisura spinoglenoid* to innervate the *m. infraspinatus* muscle, both via their ventral surfaces. The *n. pectoralis medialis* arises from the *fasciculus medialis plexus brachialis* (C8-T1), and passes through the subclavicular space to innervate *m. pectoralis minor* via its deep surface, but does not pierce the muscle, unlike in humans where the *n. pectoralis medialis* frequently pierces *m. pectoralis minor* on its way to also innervate *m. pectoralis major*. Instead, it passes inferior to the *ligamentum costocoracoideum*, and into the deep aspect of *m. pectoralis minor*. The *n. pectoralis medialis* divides into three segments distally, and forms an *ansa pectoralis* with the *n. pectoralis lateralis* via its most lateral segment where it continues on to innervate the *m. pectoralis major*. This pattern is consistent in all bonobo specimens dissected. This nerve also innervated the *m. chondroepitrochlearis*, which was present on the lateral aspect of the deep surface of *m. pectoralis major* in Jasiri. The *n. pectoralis lateralis* arises from *fasciculus lateralis plexus brachialis* (C5-C6), passes beneath the *ligamentum costocoracoideum*, medial and deep to *m. pectoralis minor* (where it does not give a direct contribution or pierce the muscle, unlike in most *Homo sapiens*). It continues on to innervate the *m. pectoralis major* through its deep surface. The *n. pectoralis lateralis* receives a contribution from the *n. pectoralis medialis* to form an *ansa pectoralis*. The *n. musculocutaneous* forms in two segments, first a large branch from the anterior portion of the middle trunk (C7) that directly pierces the *m. coracobrachialis*, and a second branch more distally splitting from the combined *n. medianus* an *n. ulnaris* bundle (C5-C7), just proximal to the junction with the ventral branches of the inferior trunk. This smaller portion pierces the *m.* coracobrachialis and continues distally between the deep belly of the *m. biceps brachii* and the *m. brachialis* to become the *n. cutaneus antebrachii lateralis*. The *n. medianus* arises from the junction of the ventral branches of all roots (C5-T1) and runs with the *n. ulnaris* for a significant distance before splitting off. The *n. ulnaris* branches off of the *n. medianus*, with which it shares a common set of roots (C5-T1). This distal branching is atypical in modern humans, but has been observed more frequently in nonhuman primates (Shearer, unpublished data).

Regarding the dorsal division, the major non-cutaneous nerves of this division of the plexus brachialis are: *n. thoracicus longus*, *thoracodorsalis*, *n. subscapularis superior* and *n. subscapularis inferior*, *n. axillaris* and *n. radialis*. The *n. thoracicus longus* originates on the posterior portion of the root of C7, and descends caudally to receive contributions from the combined *truncus inferioris plexus brachialis* (C8 + T1). This nerve inserted directly into the superficial surface of *m. serratus anterior*. The *n. thoracodorsalis* branches from the caudal portion of the *fasciculus posterior plexus brachialis (*posterior cord, C5-T1) proximal to the split of the *n. axillaris* and the *n. radialis*. It proceeds caudally to innervate the *m. latissimus dorsi* on its superficial surface. The *n. subscapulari* branch from two portions of the *fasciculus posterior plexus brachialis* (posterior cord, C5-C6) and insert into the superior, middle and inferior ventral aspects of the *M. subscapularis* (we here do not refer to the *n. thoracodorsalis* as the *n. subscapularis medius*, or middle subscapular nerve, but address its distribution above in a separate section). The *n. subscapularis inferior* continues on to innervate the *m. teres major* via its ventral surface. The *n. subscapularis superior* bifurcates and passes superficially to *n. axillaris* and *n. radialis* before inserting into the superior and middle aspects of the ventral *m. subscapularis*. The *n. axillaris* branches from the *fascicularis posterior plexus brachialis*, where it splits from the *n. radialis*, and runs through the *foramen humerotricipitale* (quadrangular space). Upon passing through this space, it divides into three segments, the superior and inferior portions of which insert into the deep surface of the *m. deltoideus*, and the remaining portion that inserts into the dorsal (superficial) surface of *m. teres minor*. The *n. radialis* arises from the *fasciculus posterior plexus brachialis*, where it receives contributions from all nerve roots of the *plexus brachialis* (C5-T1), contrary to a previous report stating that the root of T1 did not contribute to this structure. As in modern humans we find that this nerve innervates the entire posterior compartment of the arm, including the *m. dorsoepitrochlearis*, contrary to some reports that suggested this muscle is innervated by a branch of the *n. thoracodorsalis*. This pattern of innervation is consistent in many genera of extant primates (Shearer, unpublished data), and supports the idea that the dorsoepitrochlearis is in fact developmentally/evolutionarily derived from the triceps brachii, and not from the latissimus dorsi (Diogo and Abdala 2010).

### 3.2.2 Intraspecific Variations or Polymorphisms of the Brachial Plexus

The following are descriptions of the intraspecific variation in our study sample. For consistency, we used the left side on all specimens as a template, and here expand upon these notes with observations from the right side where dissected.

Figure 3.51, adolescent female Jasiri: The right *plexus brachialis* of Jasiri presented a set of variations, mainly in the distal segments of the *fascicula plexus brachialis*. The root contributions were identical to the left side (C5-T1), but rather than branching into a *fasciculus medialis plexus brachialis* and a separate *fasciculus lateralis plexus brachialis*, the ventral aspects conjoin to form a distinct combined f*asciculus ventralis plexus brachialis*. This pattern is similar to that seen in the left *plexus brachialis* of Jasiri, but it is more defined and does not give off an additional branch of *n. musculocutaneous,* nor does it receive any additional contributions from inferior portions of the plexus as seen on the left side. The combined *fasciculus ventralis* receives nerve contributions from C5-T1, and branches into the *n. musculocutaneous*, the *n. medianus* and the *n. ulnaris* just distal to the convergence of the nerve bundle divisions. These ventral branches follow the normal pathways for their distribution after the split. The proximal segments of the *fasciculus posterior brachialis plexus* form normally.

Figures 3.52, 3.53, 3.54 and 3.55, adult male Kidogo: The left *plexus brachialis* of Kidogo begins to converge in the interscalene space between the *m. scalenus anterior* and *m. scalenus medius*, and is largely similar in structure to the left *plexus brachialis* of Jasiri in root contributions to nerves, distribution pathways, and individual neuromuscular units. The *plexus brachialis* is comprised of nerve roots C5, C6, C7, C8 and T1. These roots form three trunks: C5 + C6 form the *truncus superior plexus brachialis*, C7 forms the *truncus intermedius plexus brachialis*, and C8 + T1 form the *truncus inferioris plexus brachialis*. The *n. phrenicus* arises from the anterior portion of the root of C4 and with a contribution from C5 before descending superficial to the *m. scalenus anterior* to pass into the thoracic cavity. The *plexis cervicalis* does not appear to provide any significant contribution to the *plexus brachialis*, contrary to some reports. The *truncus intermedius plexus brachialis* contributes a branch to the pectoral nerves, which combines with the *n. pectoralis lateralis* distal to its origin point.

Figure 3.56, infant male Foyo, left side: The left *plexus brachialis* of Foyo presents with a number of variations in the proximal and distal segments. C5-7 combine to form the *truncus superior plexus brachialis,* while C8 and T1 form the *truncus inferioris plexus brachialis*; no true *truncus intermedius plexus brachialis* is formed. The *truncus superior* gives rise to the *n. musculocutaneous*, contributes to the lateral root of the *n. medianis* and provides dorsal contributions to the *fasciculus posterior plexus brachialis*. The ventral aspect of the *truncus inferioris* contributes to the lateral root of the *n. medianis* and provides much of the dorsal contribution for the *fasciculus posterior plexus brachialis*. The *n. medianis* branches off from a common stalk with the *n. radialis* and *n. ulnaris* (which run together in this specimen). The other posterior nerves (e.g., *n. axillaris, n. subscapularis*, etc.) arise from a common cord with the *n. radialis/ulnaris*, but arise more dorsally. The lack of diversification in the *fasciculi* is not commonly seen among human or nonhuman primates.

Figures 3.57, infant male Foyo, right side: The right plexus brachialis of Foyo viewed from a dorsal view demonstrates the branching patterns contributing to the dorsal nerves. C4 provides a very small contribution to the *truncus superior plexus brachialis*, while C5-6 provide the major nerve segments. The *n. dorsalis scapulae* can be viewed emerging from C4-5 deep to the *m. scalenus posterior*, and the *n. suprascapularis* can be seen arising from the truncus superior plexus brachialis shortly before it splits into anterior and posterior divisions. C7 gives rise to the *truncus intermedius plexus brachialis*, which continues for a short distance before being joined by the dorsal divisions of C8-T1 to form the *fasiculus posterior plexus brachialis*. The *n. axillaris* branches very proximally, arising from the trunks of C5-7.

Figures 3.58, infant male Etje: Etje presents a series of atypical configurations in the proximal segments of the right *plexus brachialis*. The roots form a *truncus superior plexus brachialis* (C5-6) and a *truncus inferior plexus brachialis* (C7-T1), with no true *intermedius* being formed. Similar to the right side of Jasiri, the trunks then merge to form a combined *fasciculus medialis/posterior/lateralis plexus brachialis*. The *fasciculus posterior plexus brachialis* is formed somewhat separately from the other two, but still is more closely adherent than is usually seen. This bundle makes it impossible to hypothesize the potential nerve contributions to each terminal nerve, although each appears to derive from or near the usual position (e.g., superior, inferior, posterior).

Figures 3.59, female fetus Ano: Ano presents a mostly typical formation of the right *plexus brachialis,* despite its relatively early developmental stage. Root contributions are C5-T1, and the typical *truncus superioris plexus brachialis* (C5-6)*, truncus medialis plexus brachialis* (C7) and *truncus inferioris plexus brachialis* (C8-T1) are formed. All distal nerve formations are as seen in the typical modern human *plexus brachialis*.

### 3.2.3 Summary

Overall, the left *plexus brachialis* of our primary study specimen, Jasiri, is somewhat atypical compared to the other bonobos in our sample, although it is not outside of the range of variation seen within modern humans and nonhuman primates. Kidogo, Ano and the right *plexus brachialis* of Foyo exhibit a pattern very similar in morphology to that described as most common for modern humans and published literature on common chimpanzees. Some notable variations seen in the left *plexus brachialis* in our primary study specimen (Jasiri) include: additional contributions from the *truncus inferioris plexus brachialis* to the *truncus intermedius plexus brachialis*; the *n. musculocutaneous* arising from the *truncus intermedius plexus brachialis;* two contributions from the *truncus intermedius plexus brachialis* to the *fasciculus medius plexus brachialis* instead of one; a combined *n. medianus* and *n. ulnaris* traveling together and splitting distally in brachium; the a small branch of the *n. musculocutaneous* arising from the combined *m. medianus* and *m. ulnaris* nerve bundle; the *n. subscapularis inferior* arising from the *truncus inferior plexus brachialis* instead of its normal point of departure from the *fasiculus posterior plexus brachialis* distal to the contributions from the dorsal divisions of the other *fasiculii*.

Based on the total overview of the different specimens dissected for this atlas, we can surmise that the following description summarizes the morphology most common to *Pan paniscus*:

- Root contributions: C5-T1, small (potential) contribution from C4; no observed contributions from T2. This is similar to the most common condition found in modern humans and *P. troglodytes*.
- Three separate trunks form off of the roots: *truncus superioris plexus brachialis* formed by C5-C6 (small contribution from C4 is possible); *truncus intermedius plexus brachialis* is formed exclusively by C7; *truncus inferior plexus brachialis* is formed by C8-T1, with no visible contributions from T2 in any of our specimens. Variations exist here, the most common being a combination of two of the trunks (most commonly the superior and inferior branches).
- Dorsal and ventral segments strongly separate, even where more proximal segments are merged together (e.g., combined *fasciculus medialis/lateralis plexus brachialis* in the right *plexus brachialis* of Jasiri).
- Variation exists in the origin of the pectoral nerves, though the *n. pectoralis medialis* consistently did not pierce the *m. pectoralis minor* to innervate the *m. pectoralis major*, contrary to the normal condition in modern humans and *P. troglodytes*.
- The *n. ulnaris* occasionally runs with the *n. medianus* well into the axillary sheath, where it branches distally to continue into the antebrachium.
- Distal contributions to particular neuromuscular units are similar to those seen in modern humans and *P. troglodytes*, although the actual nerve root contributions for *P. paniscus* to each muscle and the overall distribution patterns for nerves remain to be investigated.

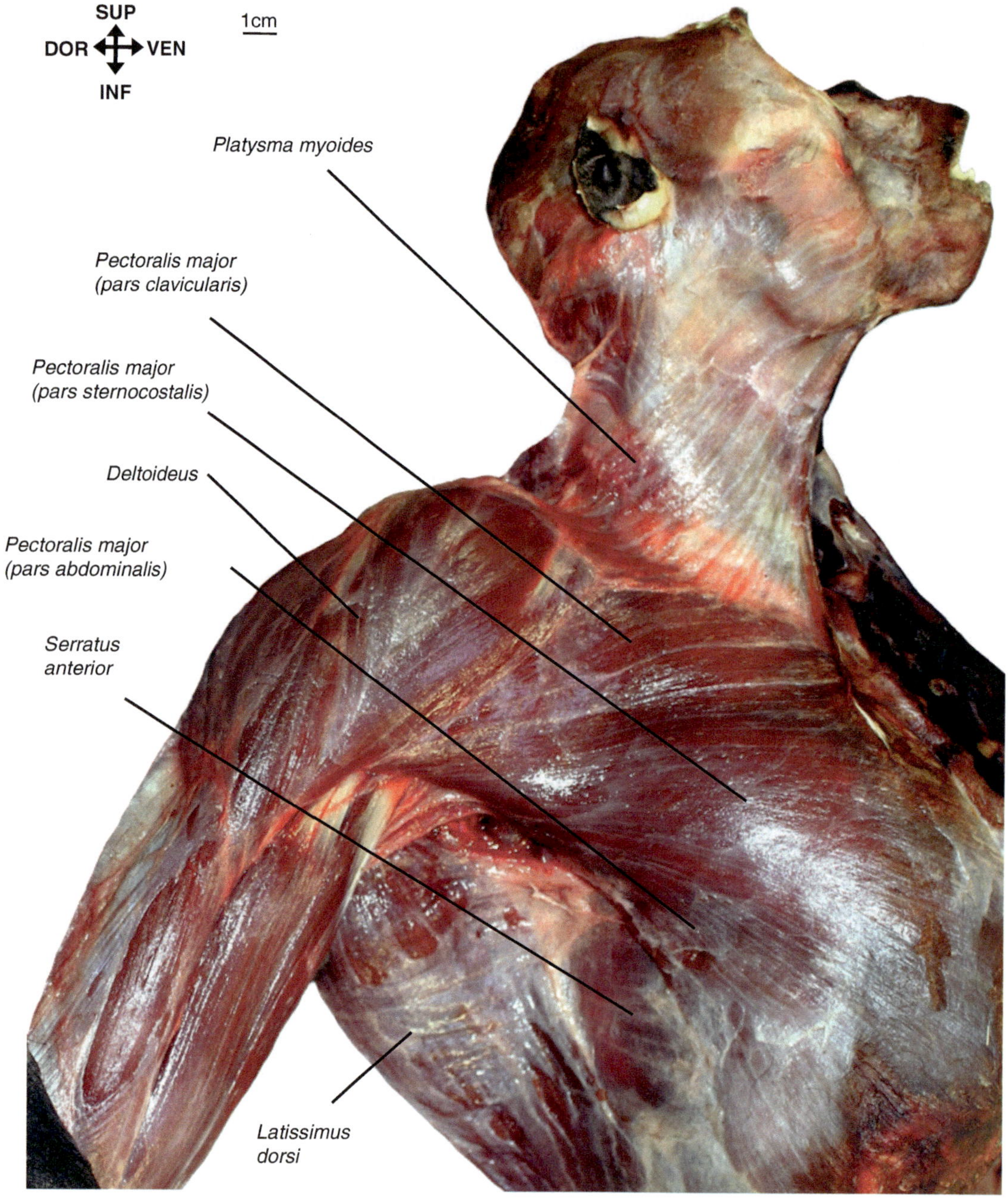


**Fig. 3.1** *Pan paniscus* (ZIMS 164031, Kidogo, adult male): ventro-lateral view of the shoulder and trunk muscles

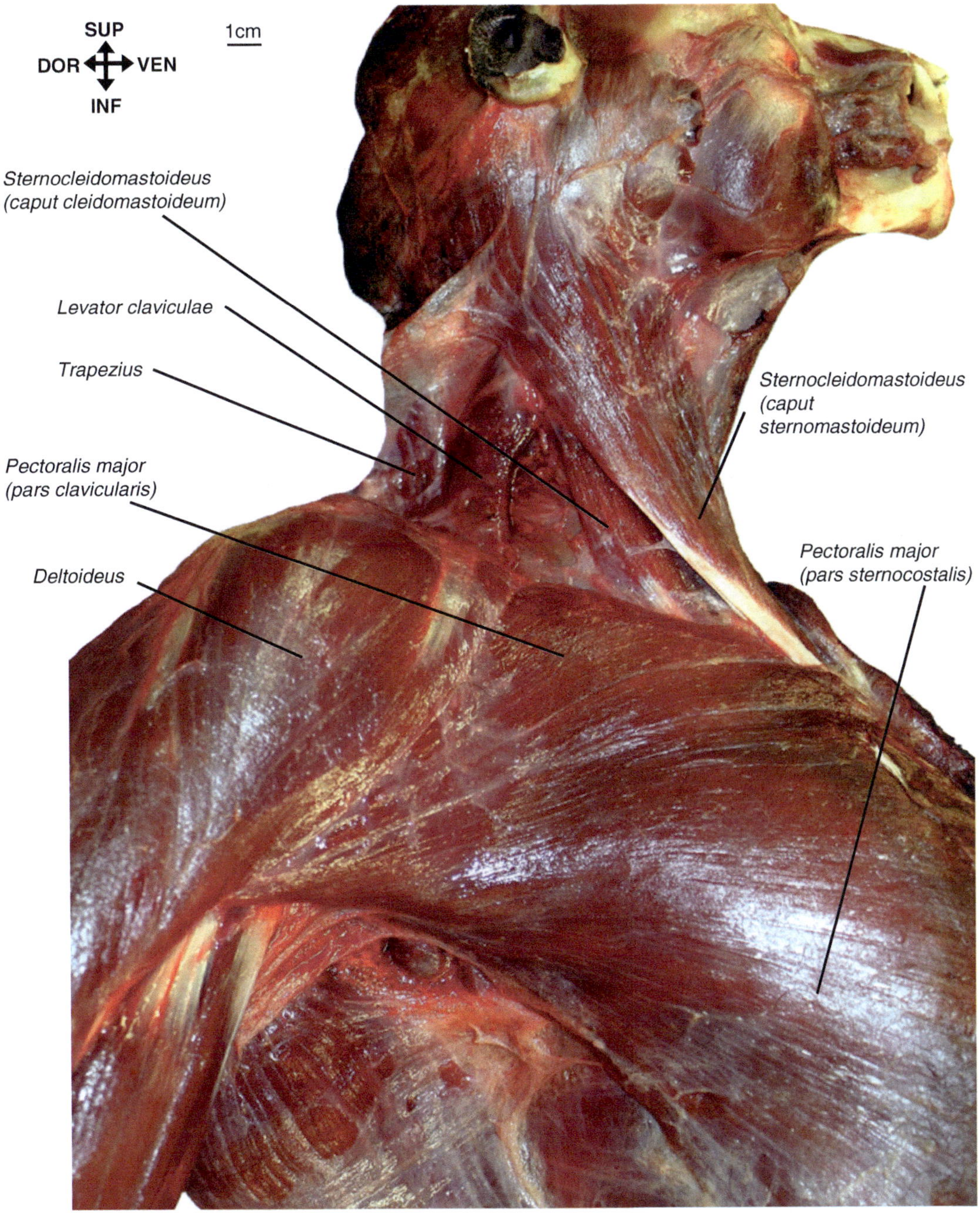


**Fig. 3.2** *Pan paniscus* (ZIMS 164031, Kidogo, adult male): ventro-lateral view of the shoulder and neck muscles after removal of platysma myoides

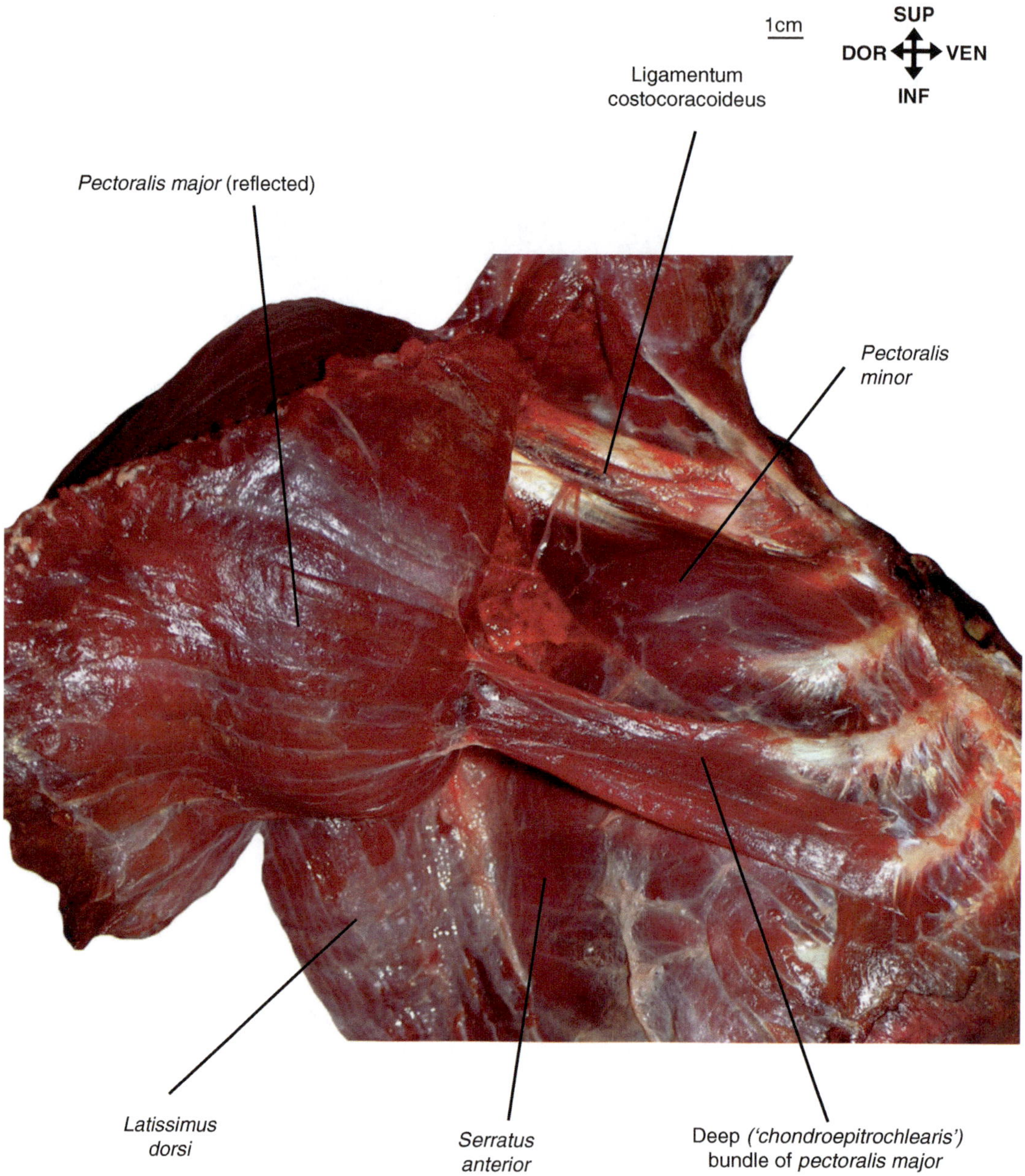


**Fig. 3.3** *Pan paniscus* (ZIMS 164031, Kidogo, adult male): ventro-lateral view of the shoulder muscles after reflection of the main body of the pectoralis major, showing the deep ('chondroepitrochlearis') bundle of this muscle

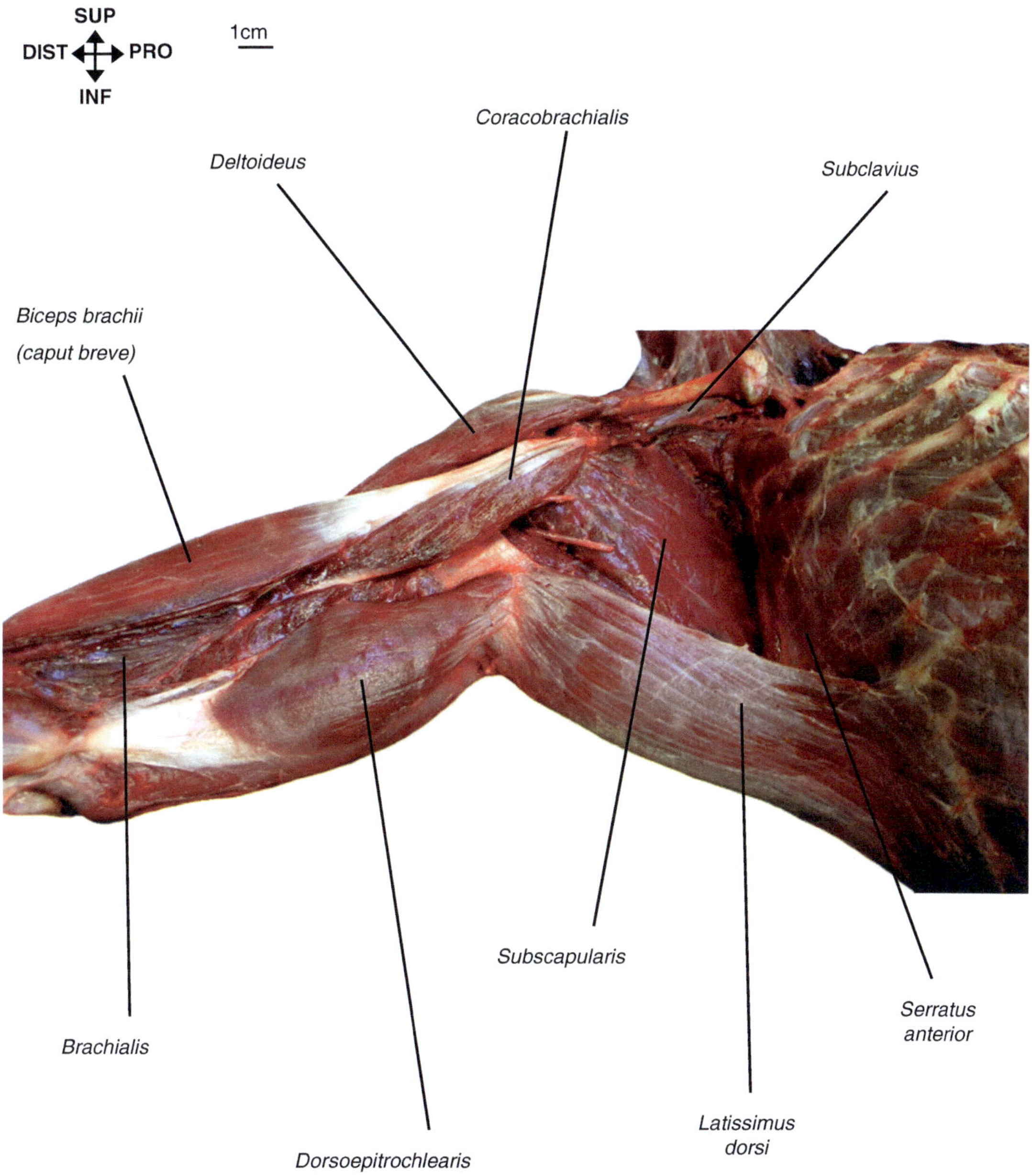


**Fig. 3.4** *Pan paniscus* (ZIMS 164031, Kidogo, adult male): ventro-inferior view of the shoulder, arm and trunk muscles after removal of pectoralis major and pectoralis minor

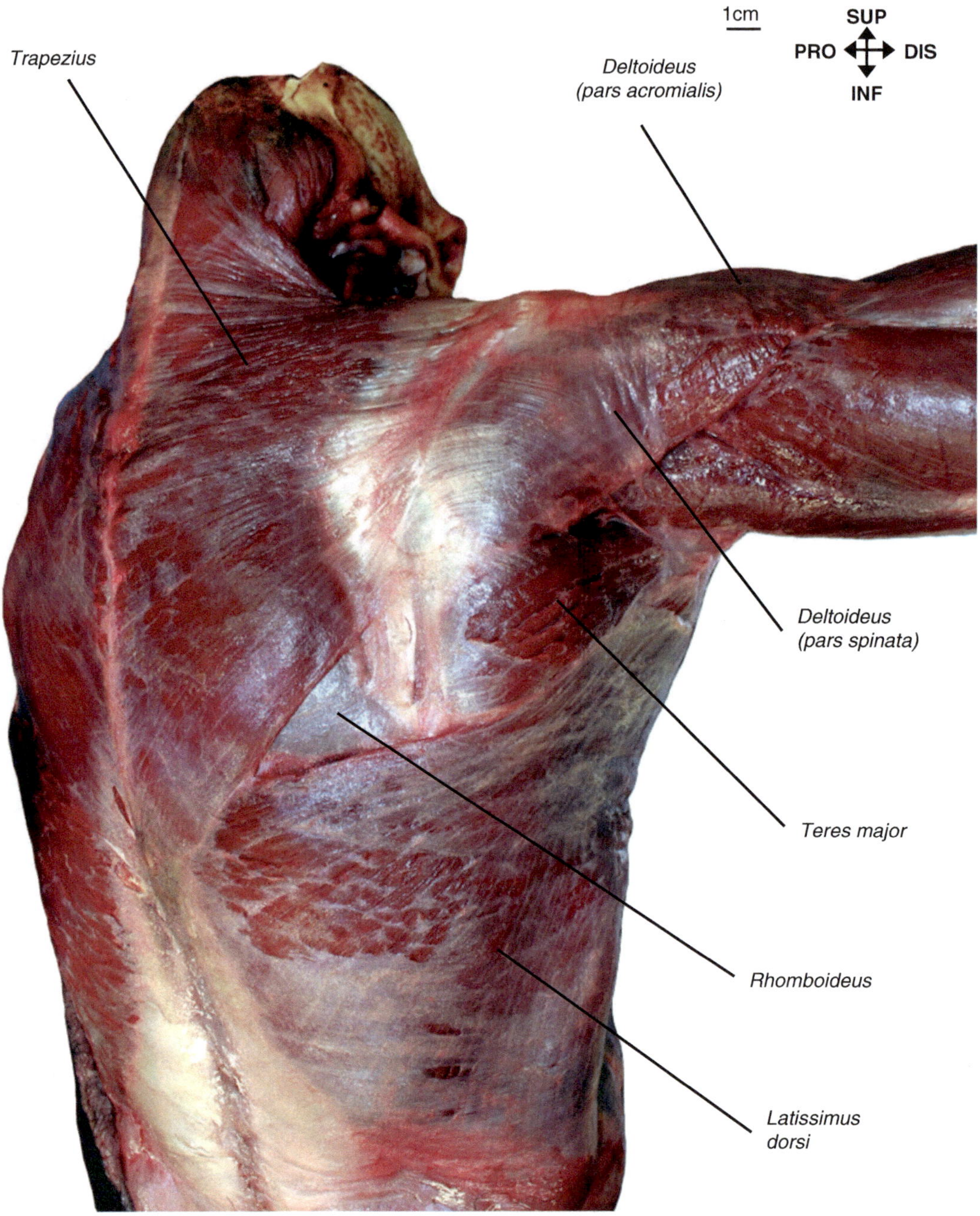


**Fig. 3.5** *Pan paniscus* (ZIMS 164031,Kidogo, adult male): dorso-lateral view of the neck, shoulder and trunk muscles

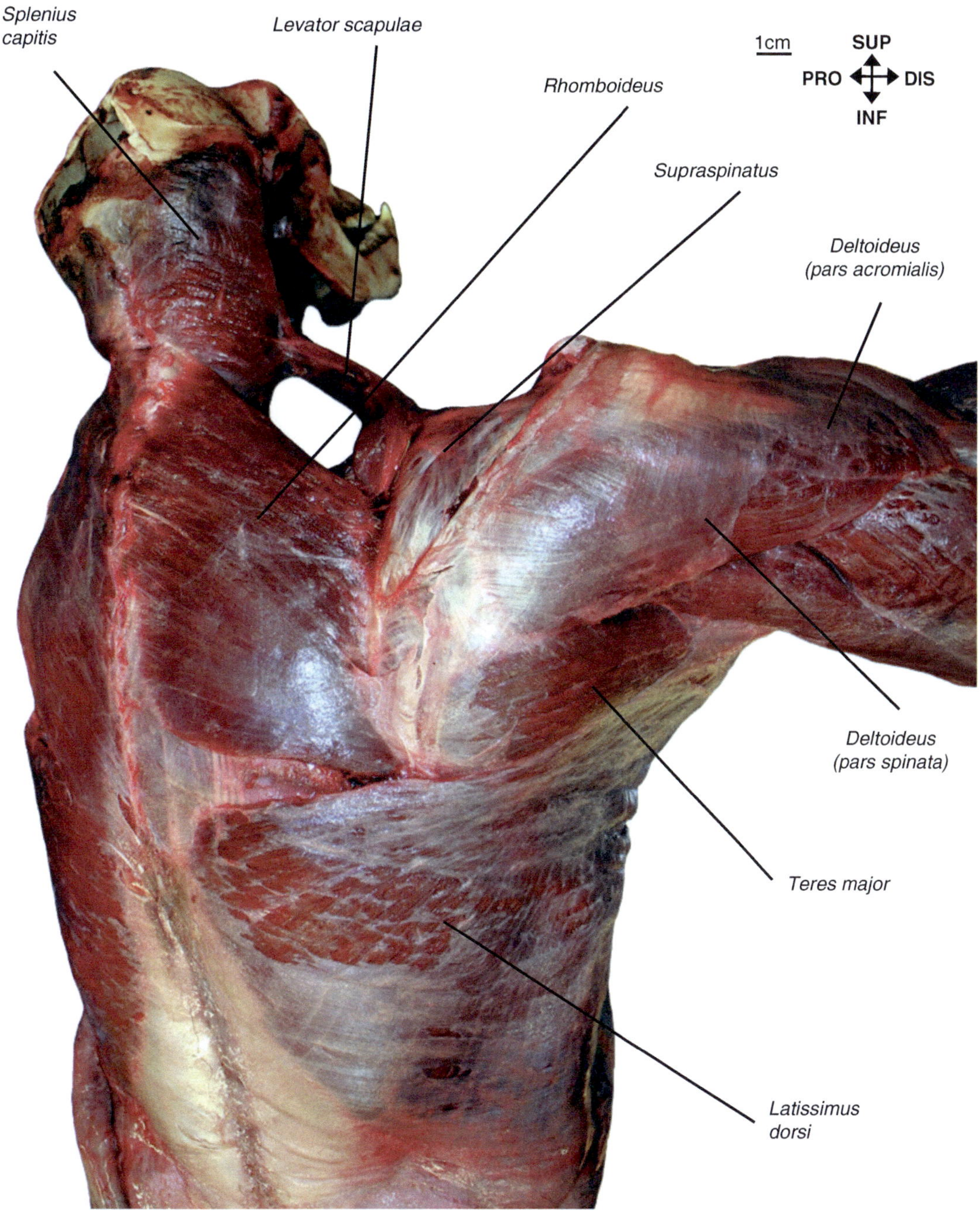


**Fig. 3.6** *Pan paniscus* (ZIMS 164031, Kidogo, adult male): dorso-lateral view of the neck, shoulder and trunk muscles after removal of trapezius

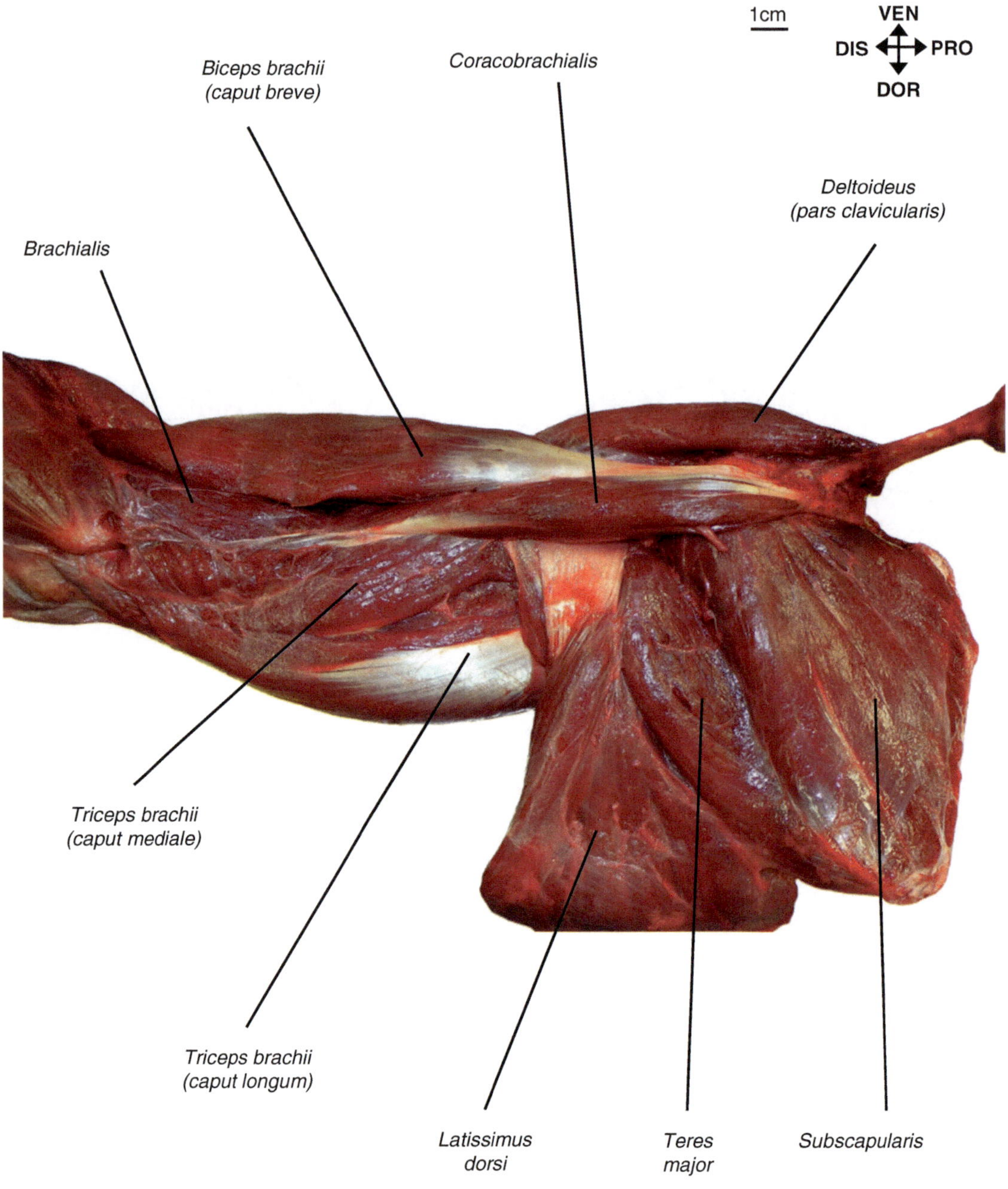


**Fig. 3.7** *Pan paniscus* (ZIMS 164031, Kidogo, adult male): medial view of the shoulder and arm muscles after removal of dorsoepitrochlearis

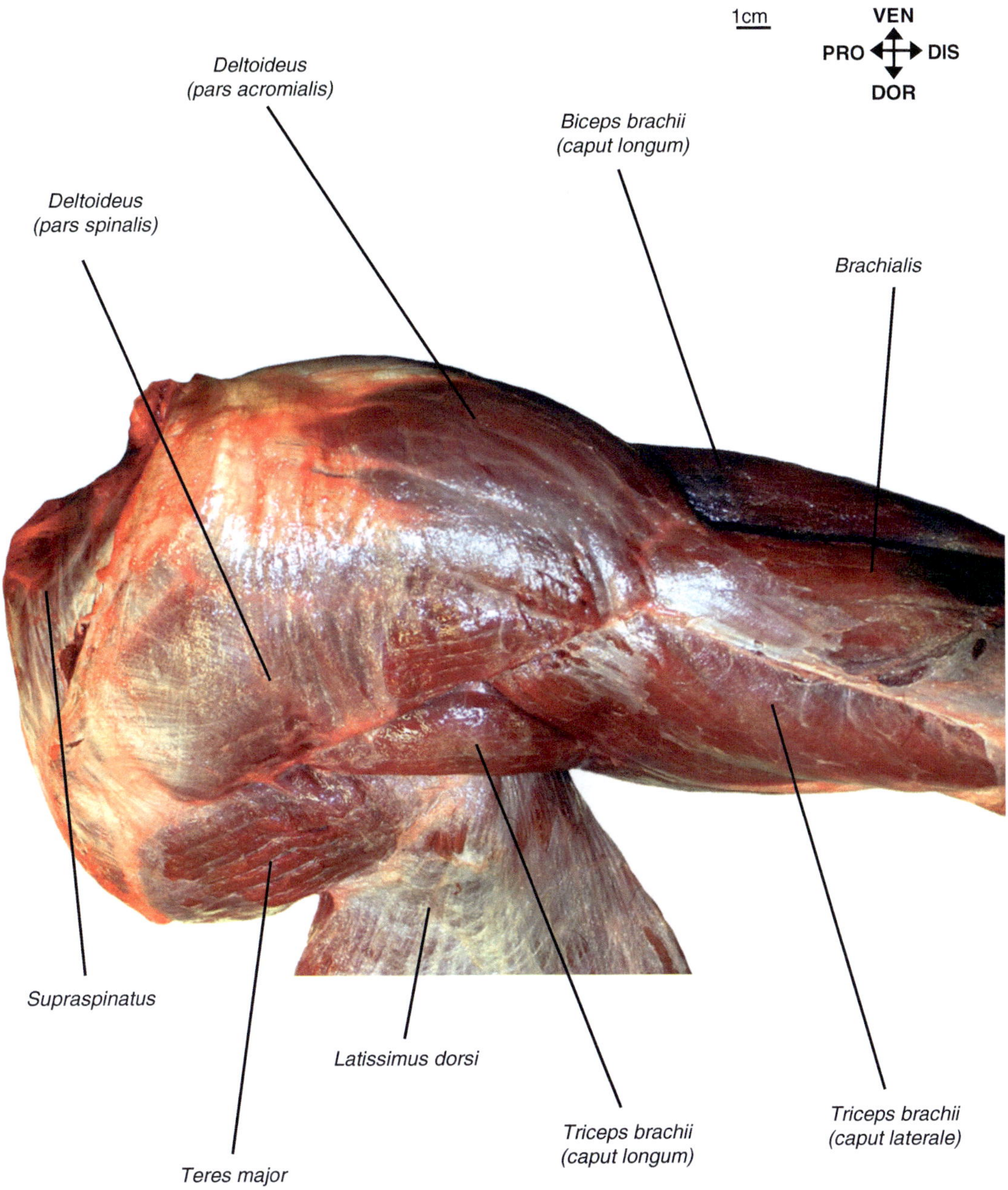


**Fig. 3.8** *Pan paniscus* (ZIMS 164031, Kidogo, adult male): dorso-lateral view of the shoulder and arm muscles

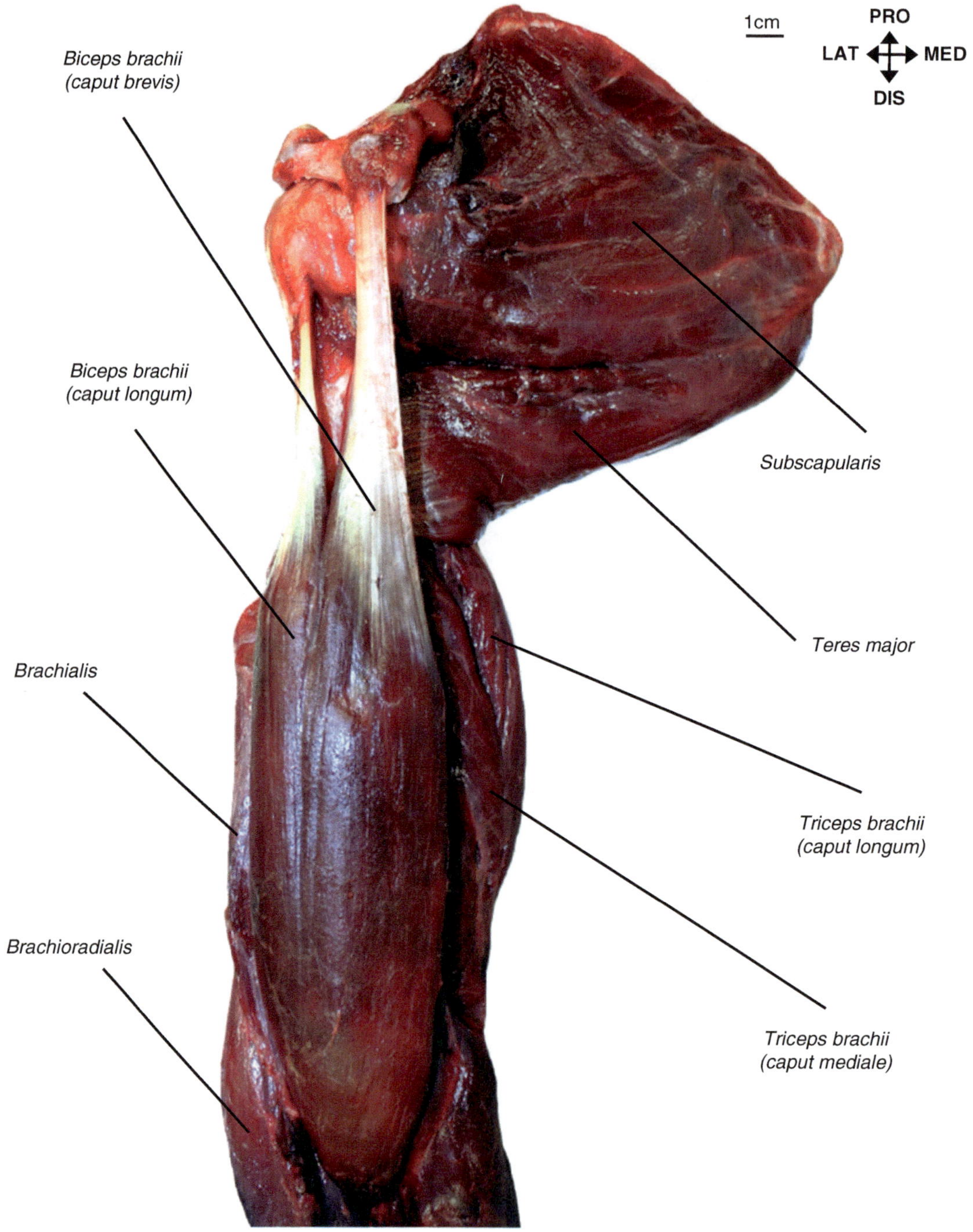


**Fig. 3.9** *Pan paniscus* (ZIMS 164031, Kidogo, adult male): ventro-medial view of the shoulder and arm muscles after removal of deltoideus and latissimus dorsi

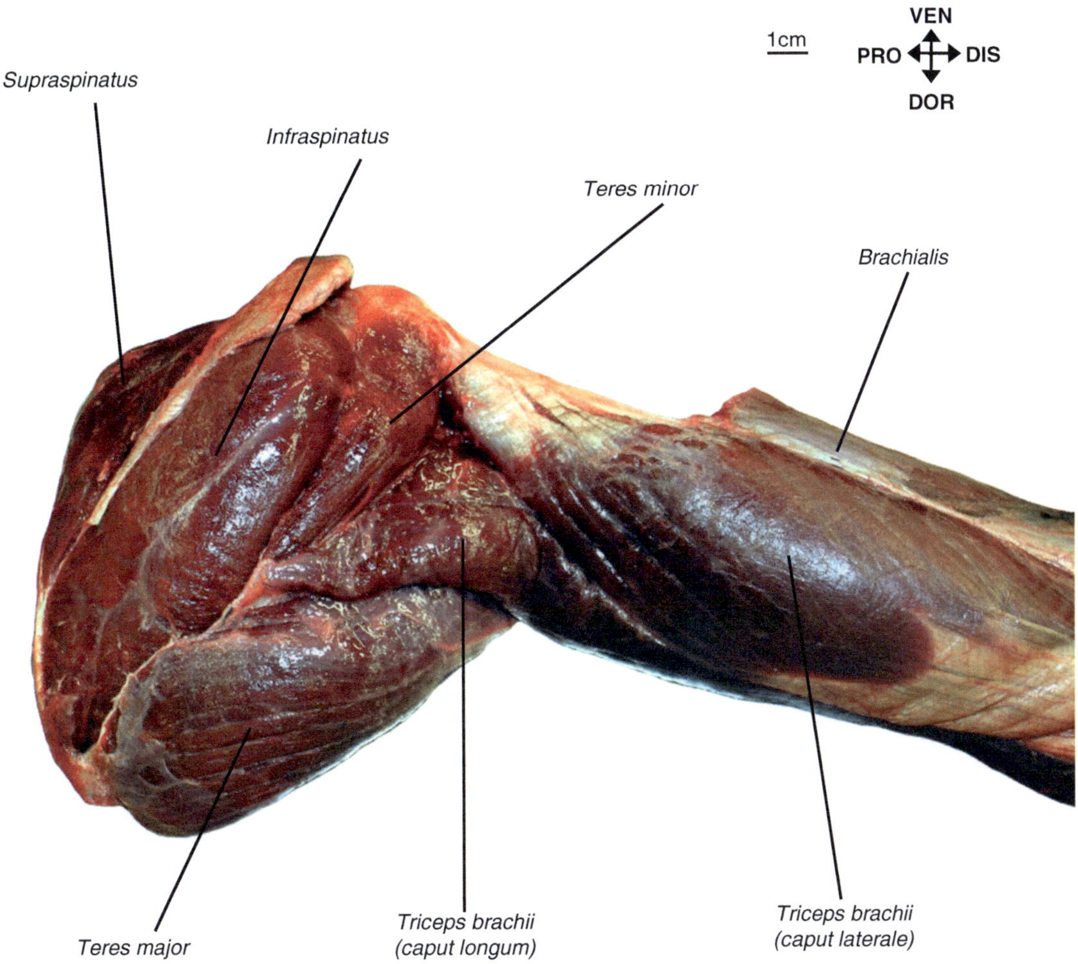


**Fig. 3.10** *Pan paniscus* (ZIMS 164031, Kidogo, adult male): dorso-lateral view of the shoulder and arm muscles after removal of deltoideus and latissimus dorsi

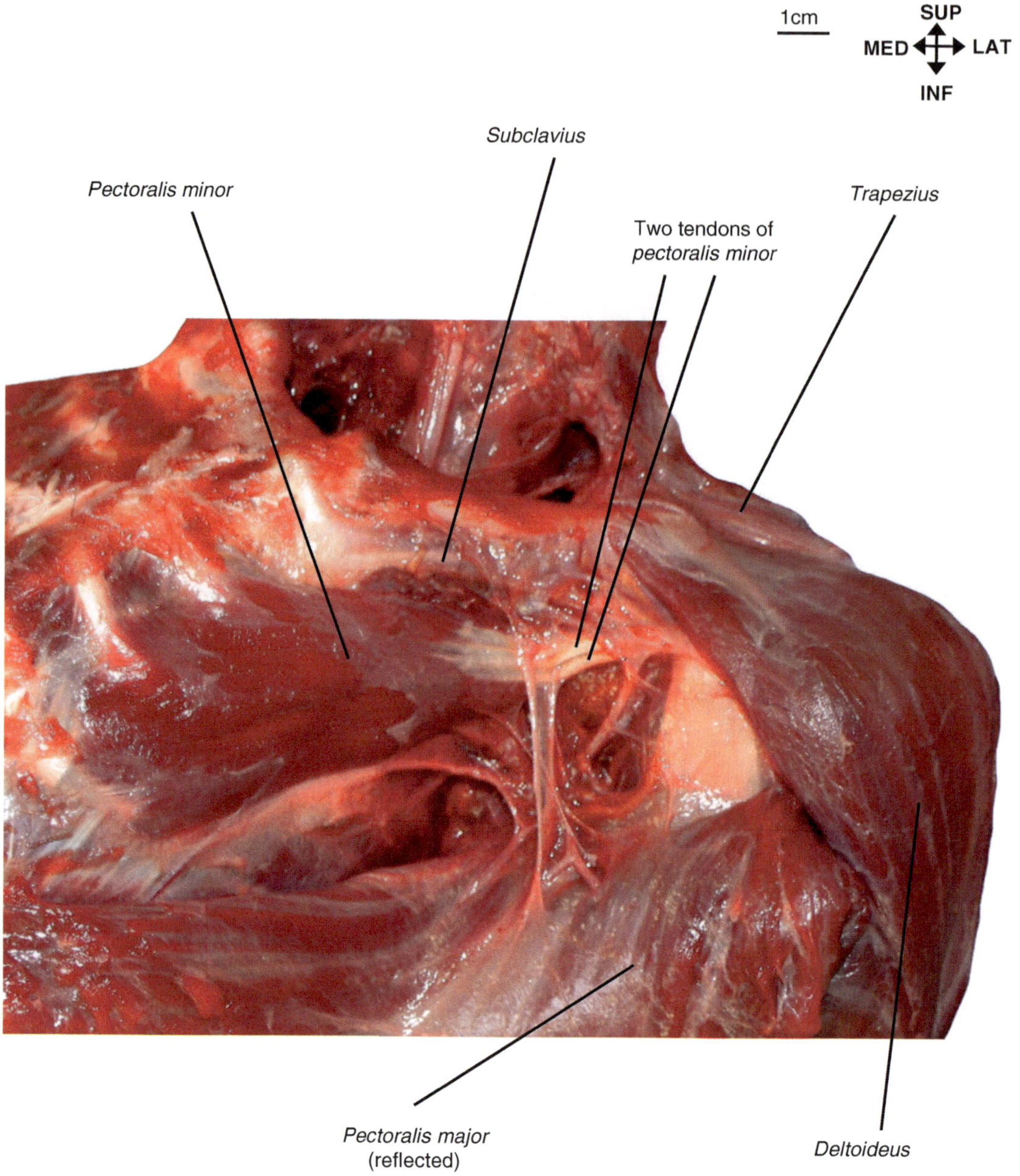


**Fig. 3.11** *Pan paniscus* (ZIMS 164047, Jasiri, adolescent female): ventro-lateral view of the shoulder muscles after reflection of pectoralis major, showing a variation: the presence of two tendons of the pectoralis minor (inserting onto the coracoid process of scapula)

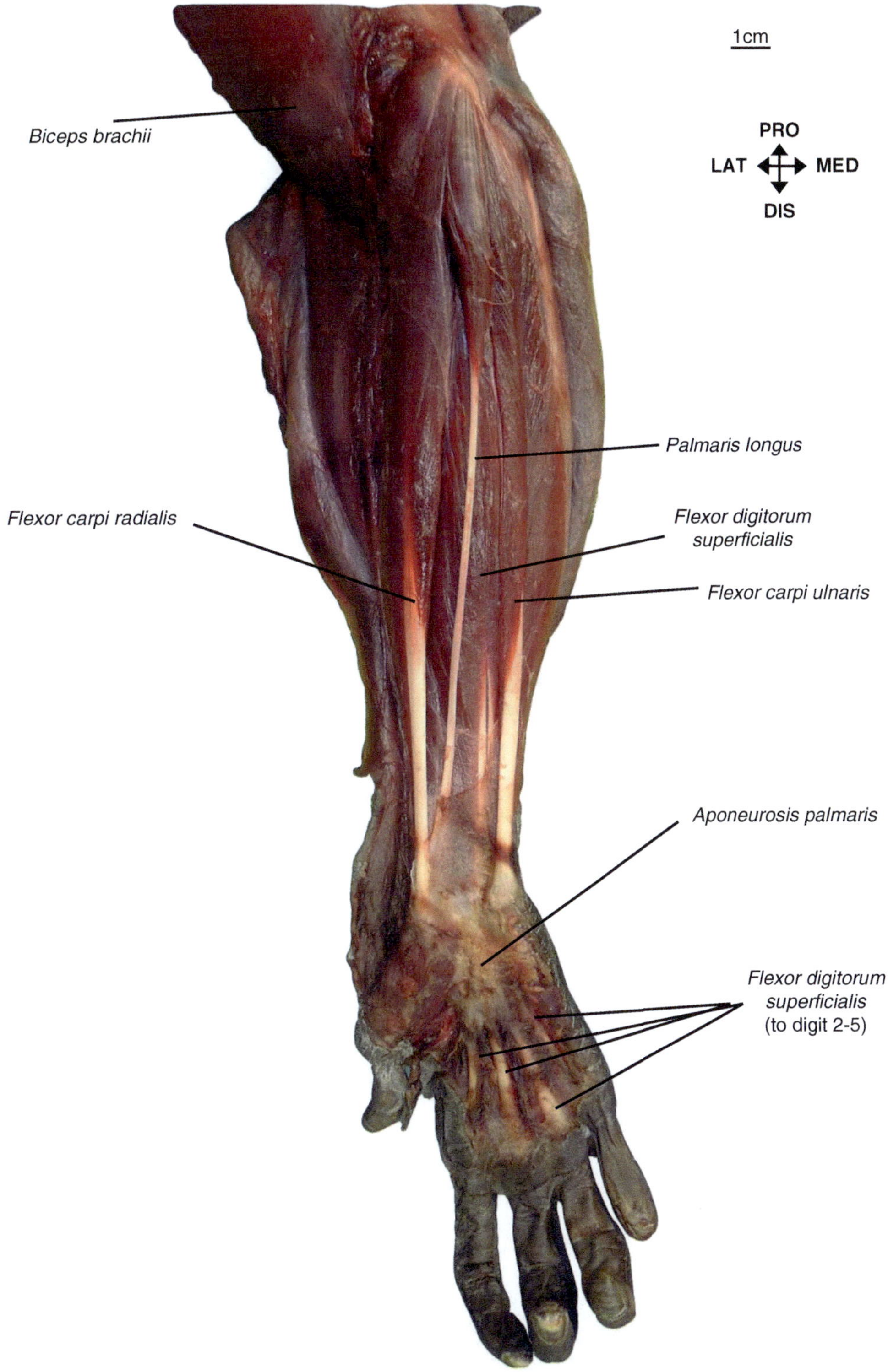


**Fig. 3.12** *Pan paniscus* (ZIMS 164031, Kidogo, adult male): ventral view of the superficial muscles of the right forearm

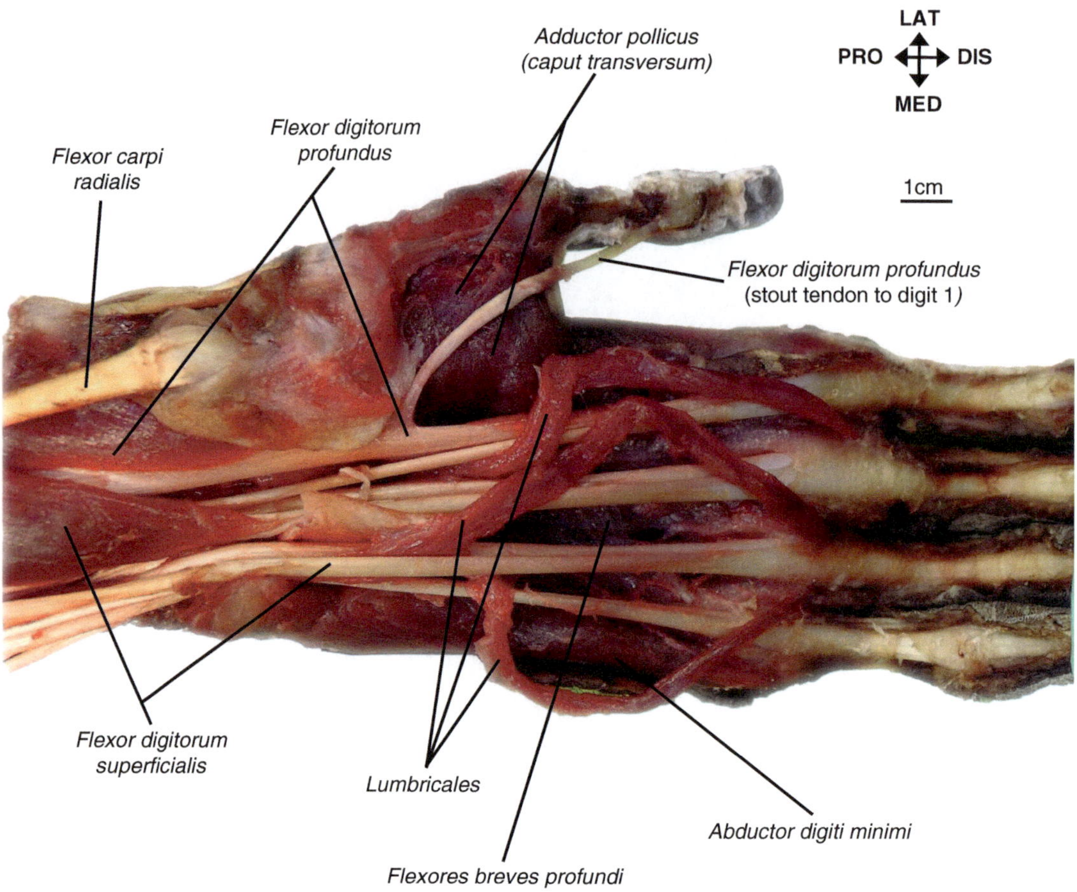


**Fig. 3.13** *Pan paniscus* (ZIMS 164031, Kidogo, adult male): ventral view of the superficial muscles of the left hand

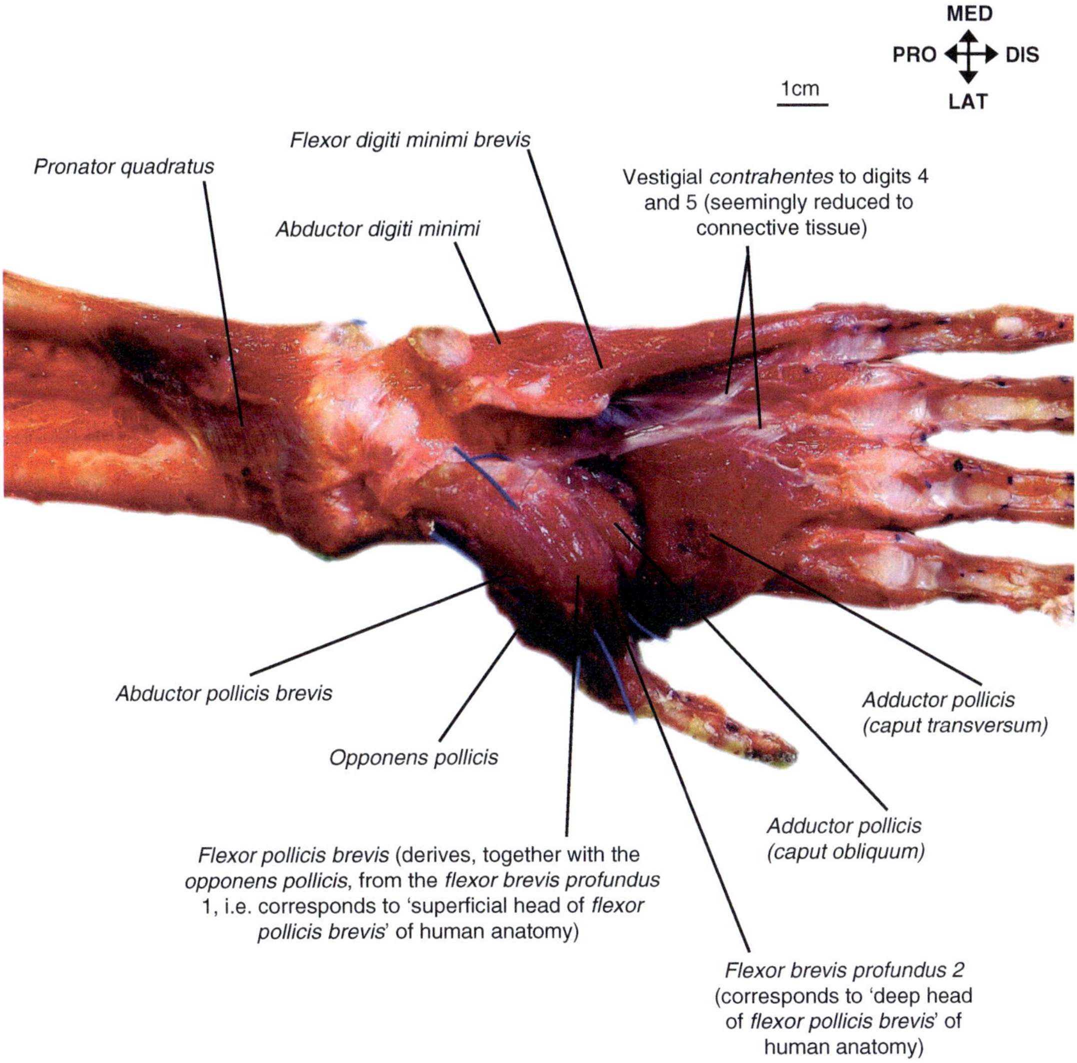


**Fig. 3.14** *Pan paniscus* (ZIMS 164047, Jasiri, adolescent female): ventral view of intrinsic muscles of the right hand

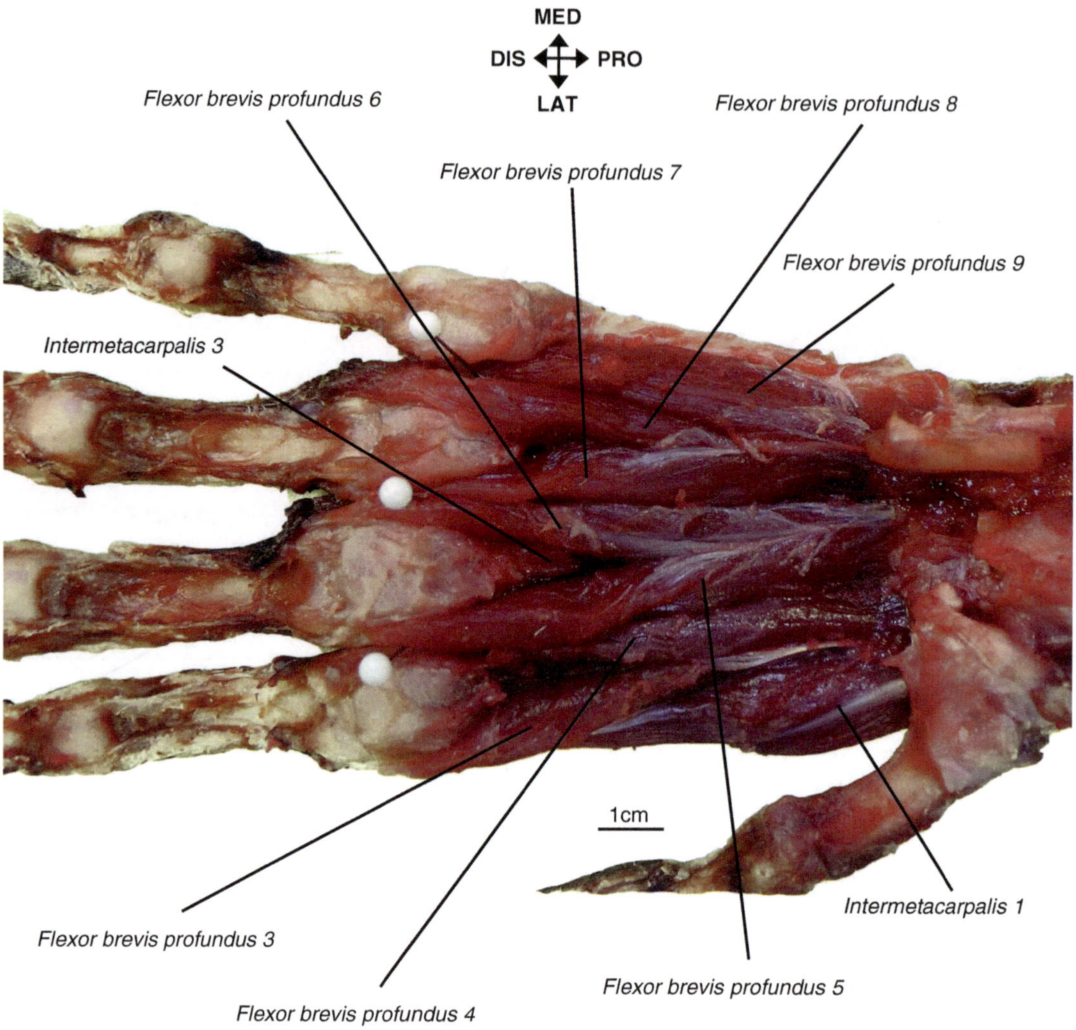


**Fig. 3.15** *Pan paniscus* (ZIMS 164031, Kidogo, adult male): ventral view of the flexores breves profundi and intermetacarpales of the left hand; note that the flexores breves profundi 3, 5, 6 and 8 are not fused to the intermetacarpales 1, 2, 3 to form the dorsal interossei 1, 2, 3 and 4, respectively, as they usually do in modern human adults (the flexores breves profundi 4, 7 and 9 correspond to the modern human palmar interossei 1, 2 and 3, respectively)

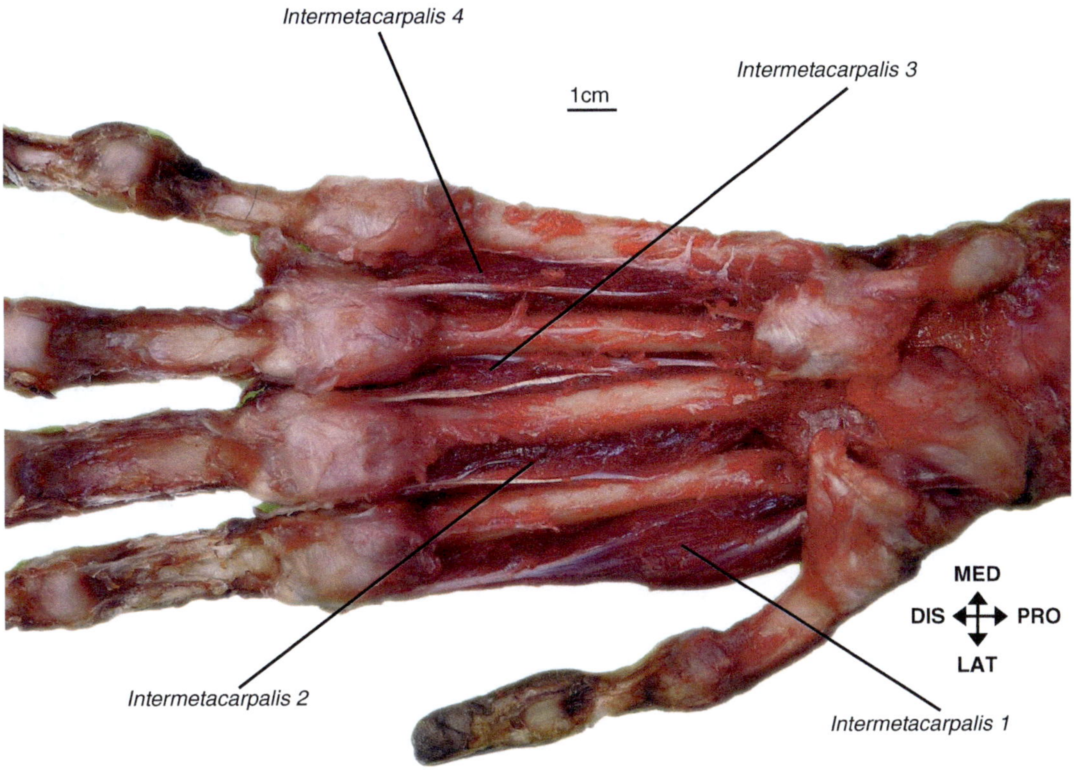


**Fig. 3.16** *Pan paniscus* (ZIMS 164031, Kidogo, adult male): ventral view of the *intermetacarpales* of the left hand

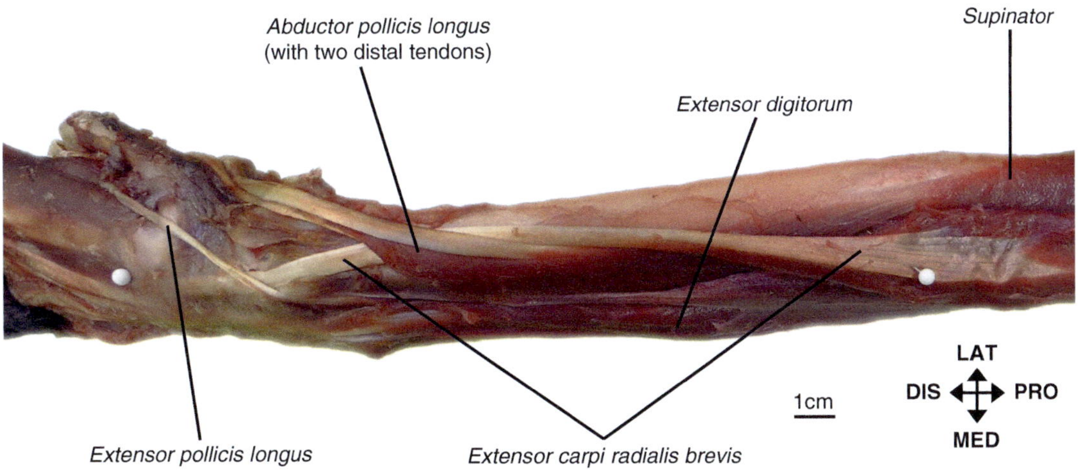


**Fig. 3.17** *Pan paniscus* (ZIMS 164031, Kidogo, adult male): dorso-lateral view of the left forearm muscles after removal of the some superficial muscles, such as the brachioradialis

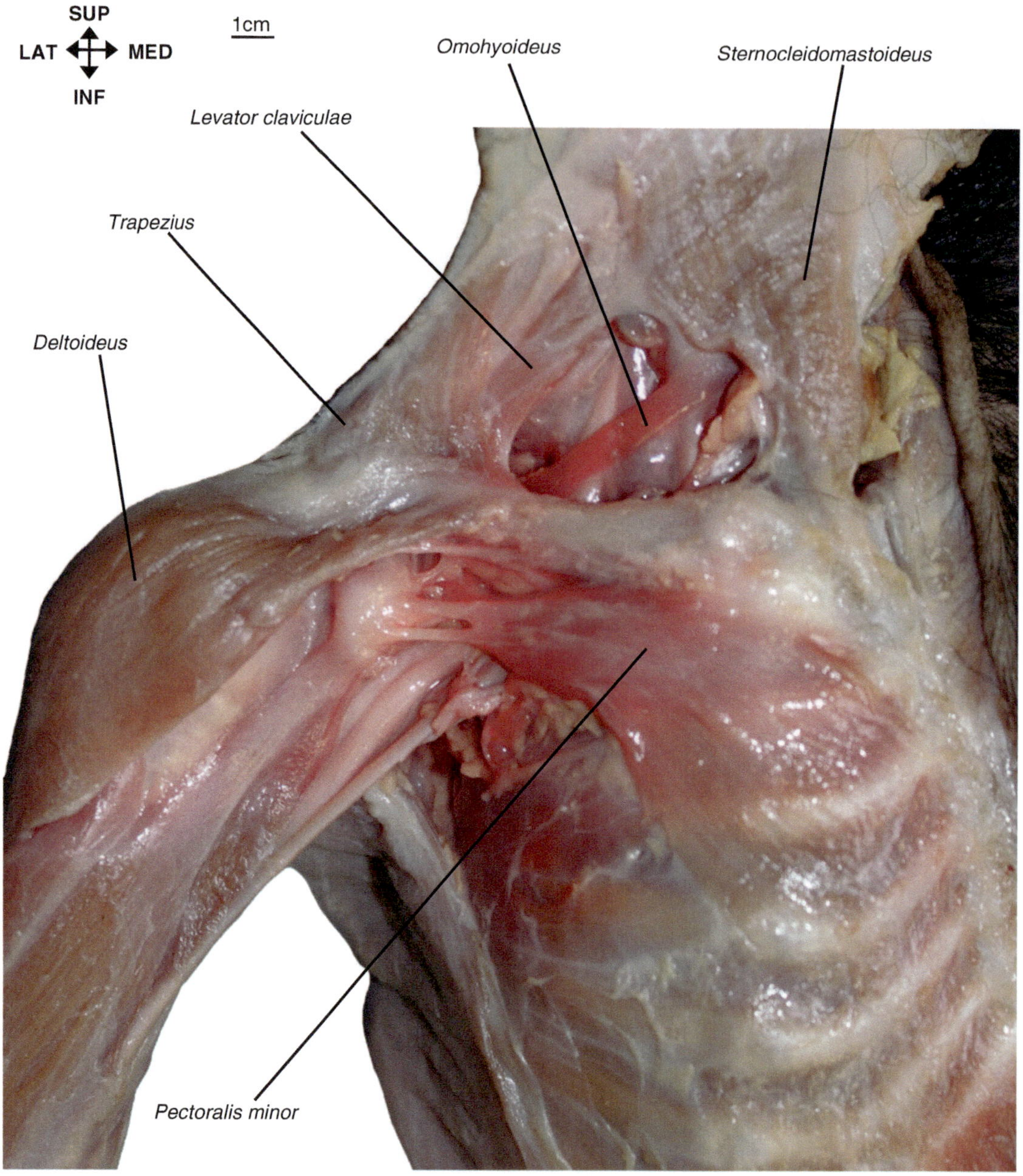


**Fig. 3.18** *Pan paniscus* (ZIMS 164041, Foyo, infant male): ventro-lateral view of the shoulder and neck muscles, including the single belly of the omohyoideus, after removal of pectoralis major, showing two variations: the presence of two tendons of the pectoralis minor inserting onto the coracoid process of scapula, and the absence of a clavicular insertion of the sternocleidomastoideus

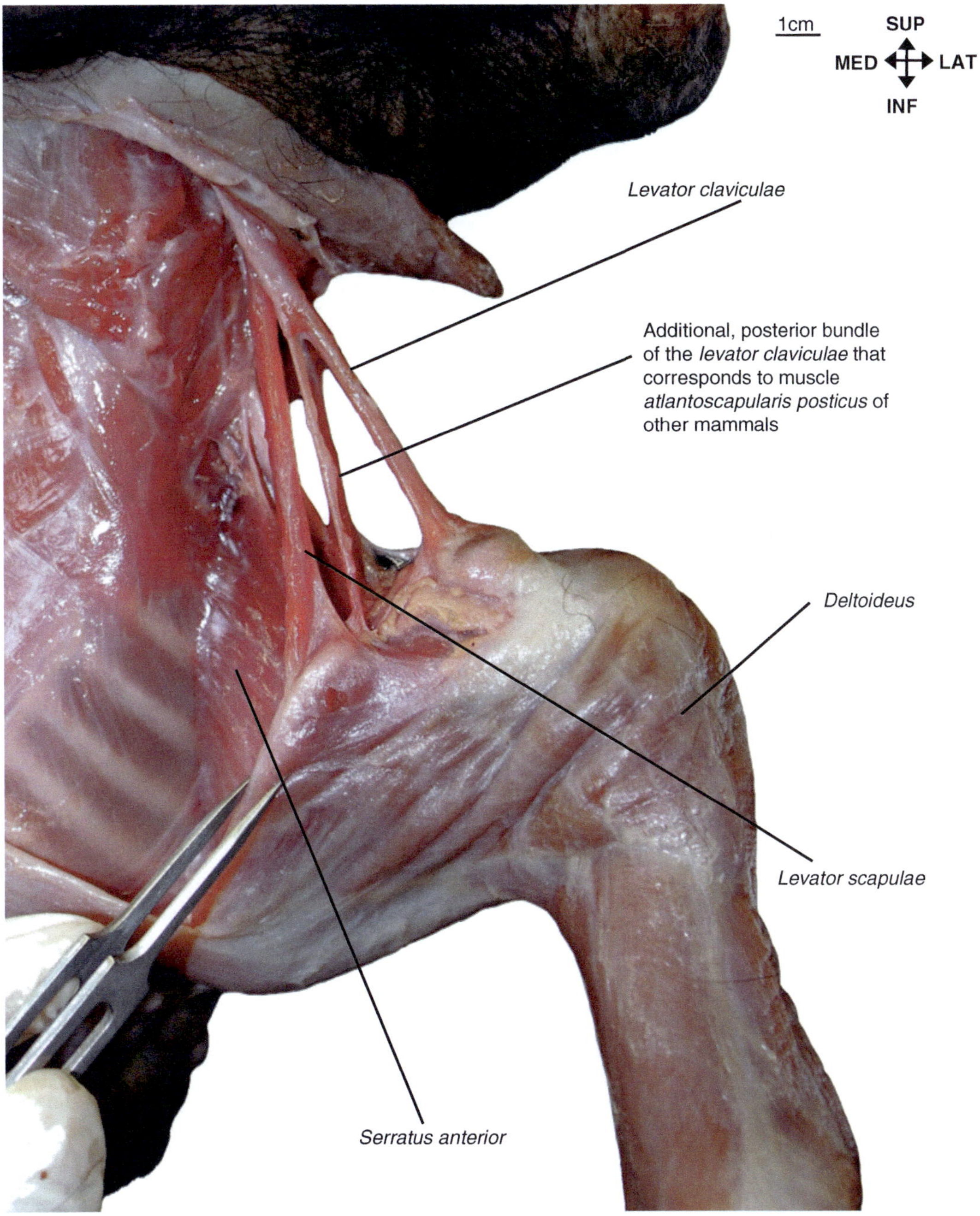


**Fig. 3.19** *Pan paniscus* (ZIMS 164041, Foyo, infant male): dorso-lateral view of the neck, shoulder and trunk muscles after removal of trapezius and rhomboideus, showing a variation: the presence of a posterior bundle of the levator claviculae that corresponds to the muscle atlantoscapularis posticus found in many other mammals

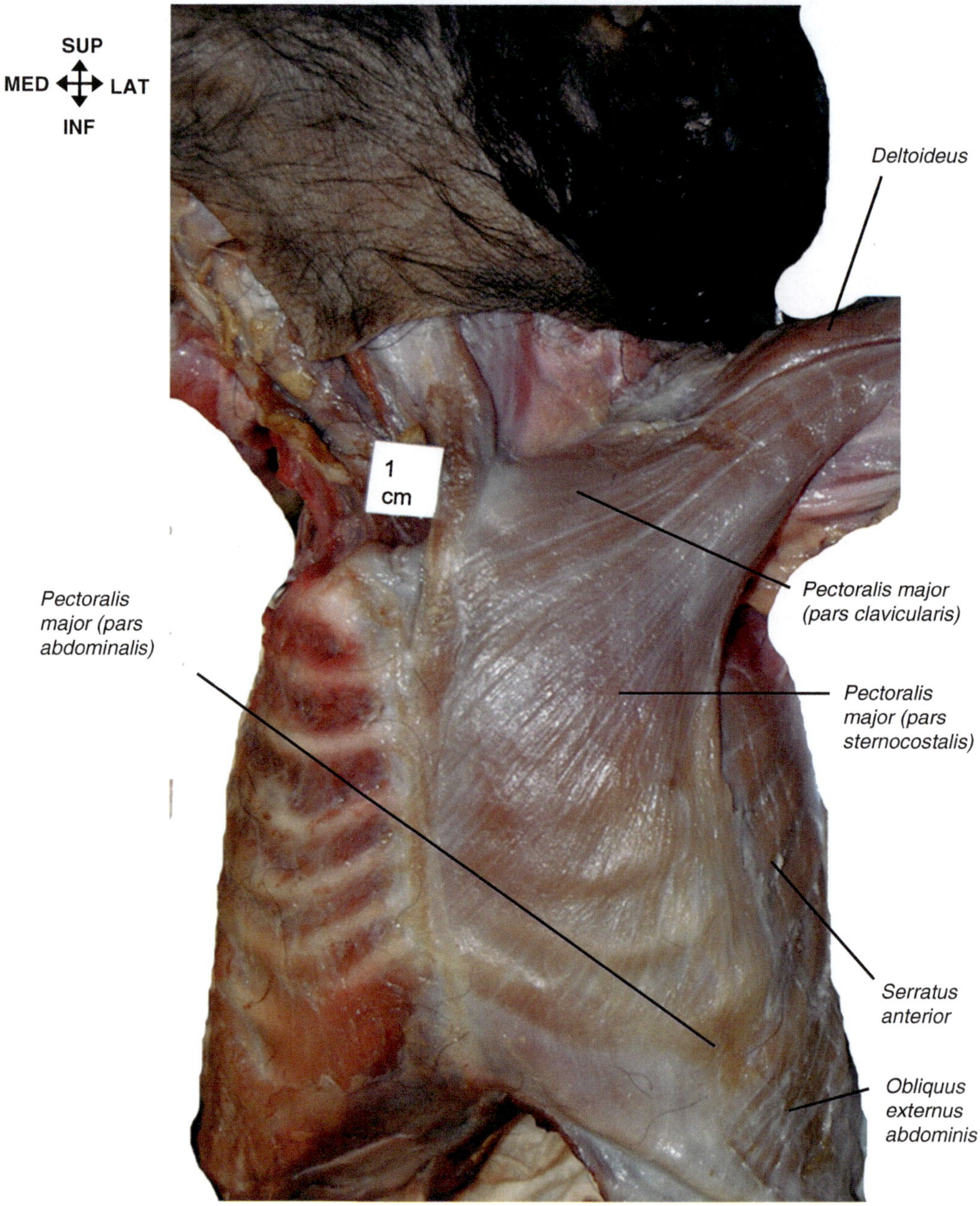


**Fig. 3.20** *Pan paniscus* (AU PP3, Foyo, infant male): ventro-lateral view of the shoulder and trunk muscles

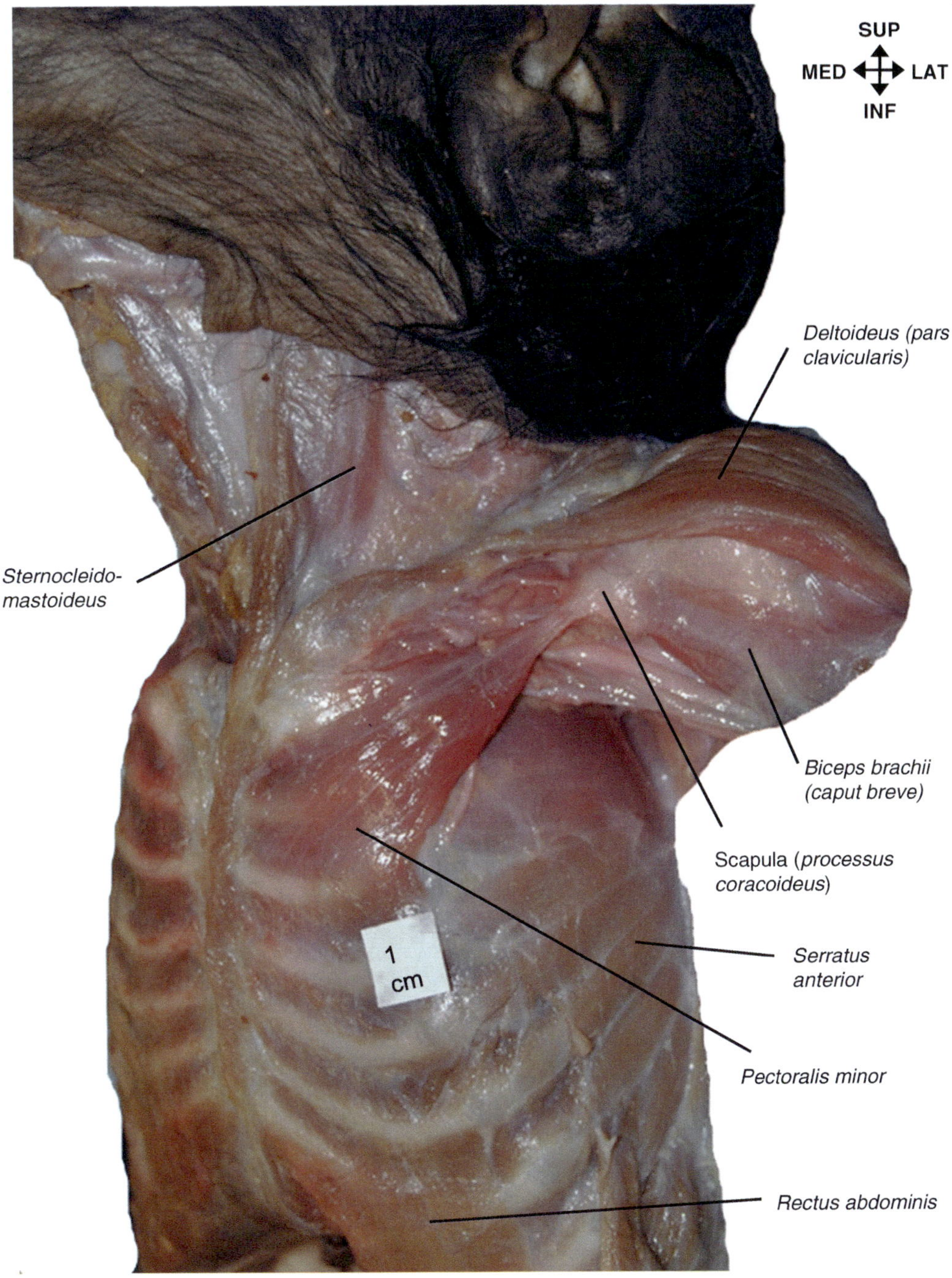


**Fig. 3.21** *Pan paniscus* (AU PP3, Foyo, infant male): ventro-lateral view of the left shoulder and trunk muscles after removal of pectoralis major

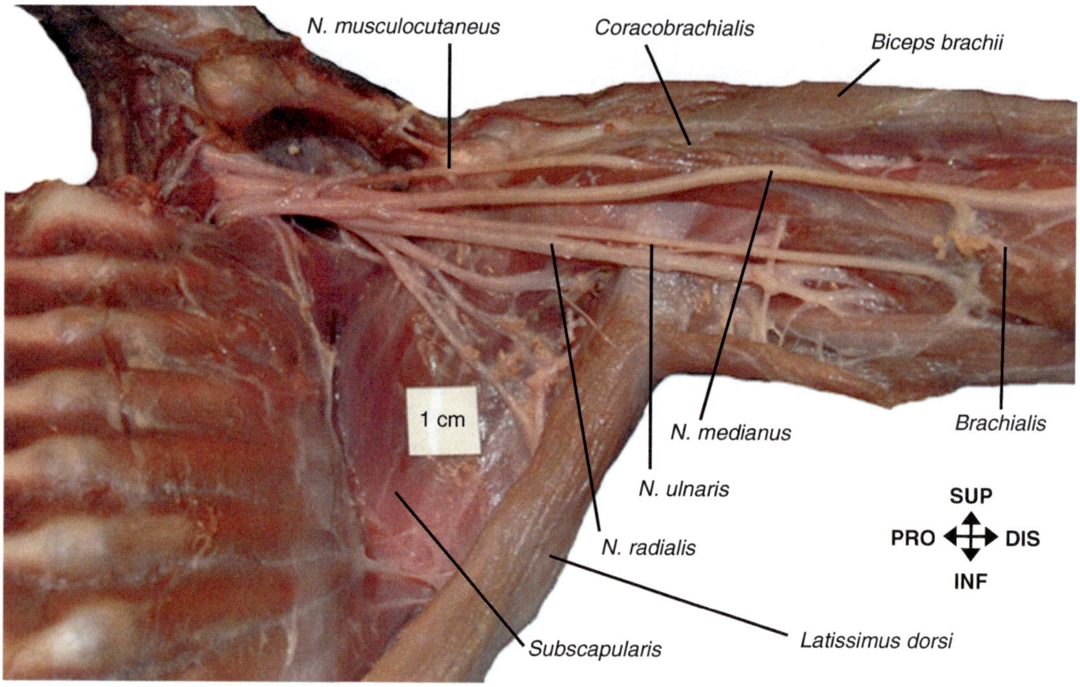


**Fig. 3.22** *Pan paniscus* (AU PP3, Foyo, infant male): ventral view of the plexus brachialis and the left pectoral and arm muscles after removal of pectoralis major and minor

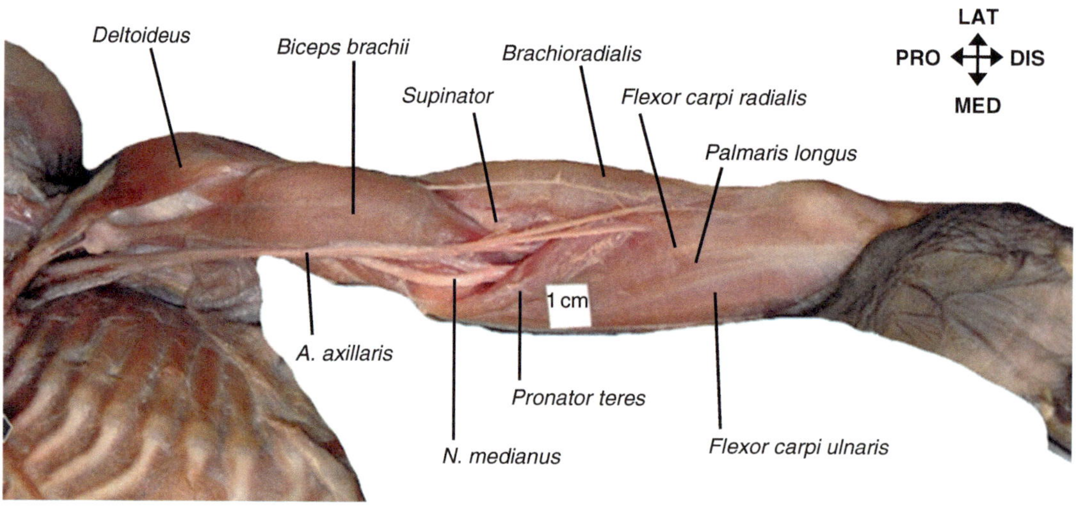


**Fig. 3.23** *Pan paniscus* (AU PP3, Foyo, infant male): ventral view of the left arm and forearm muscles

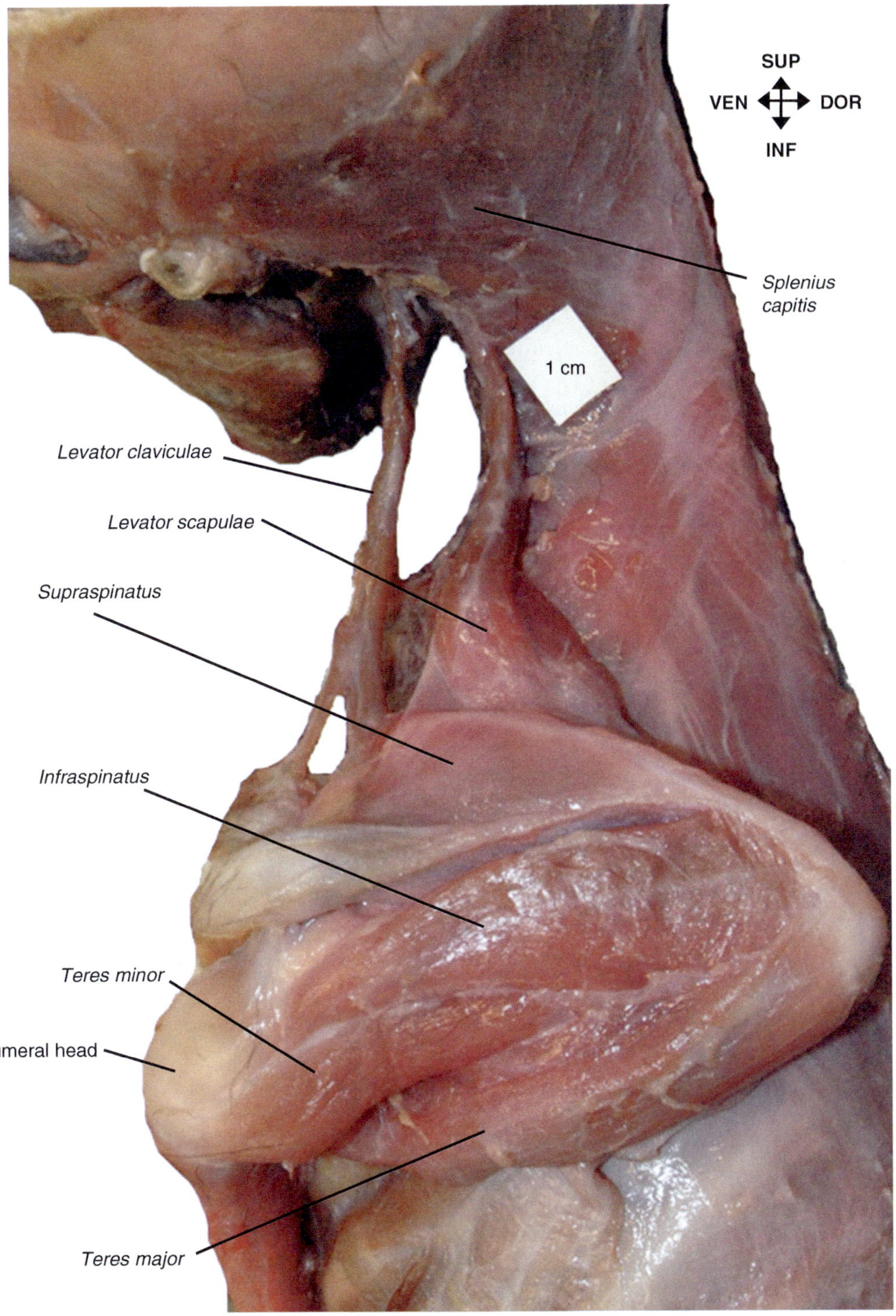


**Fig. 3.24** *Pan paniscus* (AU PP3, Foyo, infant male): dorso-lateral view of the shoulder and trunk muscles after removal of deltoideus and trapezius

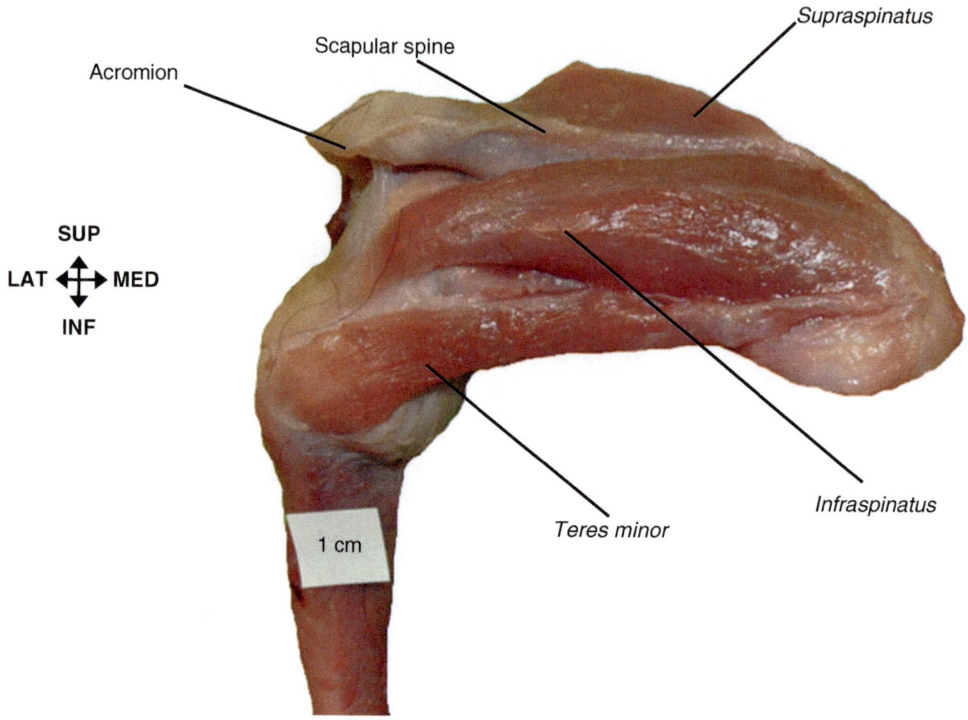


**Fig. 3.25** *Pan paniscus* (AU PP3, Foyo, infant male): dorsal view of shoulder muscles

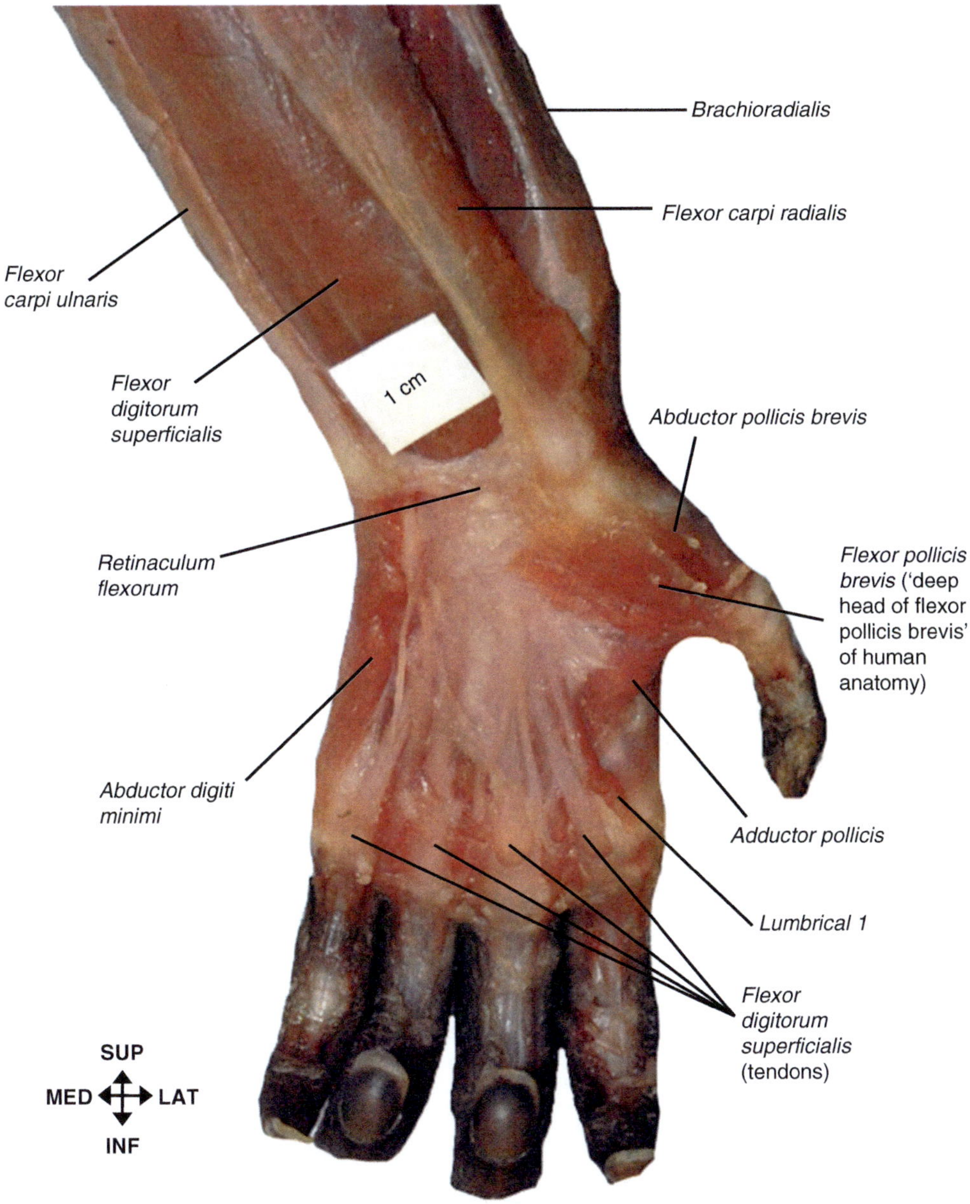


**Fig. 3.26** *Pan paniscus* (AU PP3, Foyo, infant male): ventral view of the left forearm and hand muscles

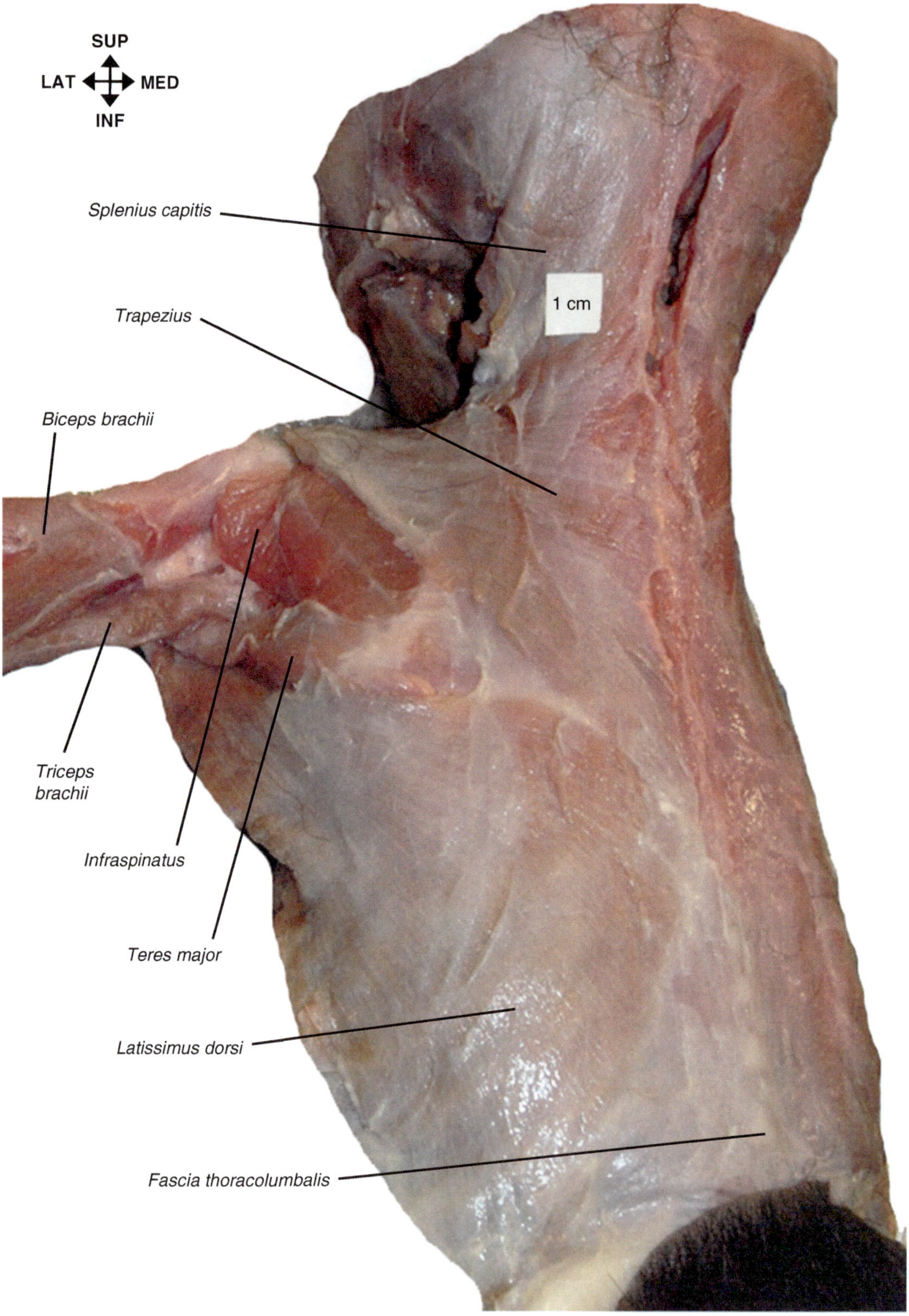


**Fig. 3.27** *Pan paniscus* (AU PP3, Foyo, infant male): dorso-lateral view of the left shoulder and trunk muscles

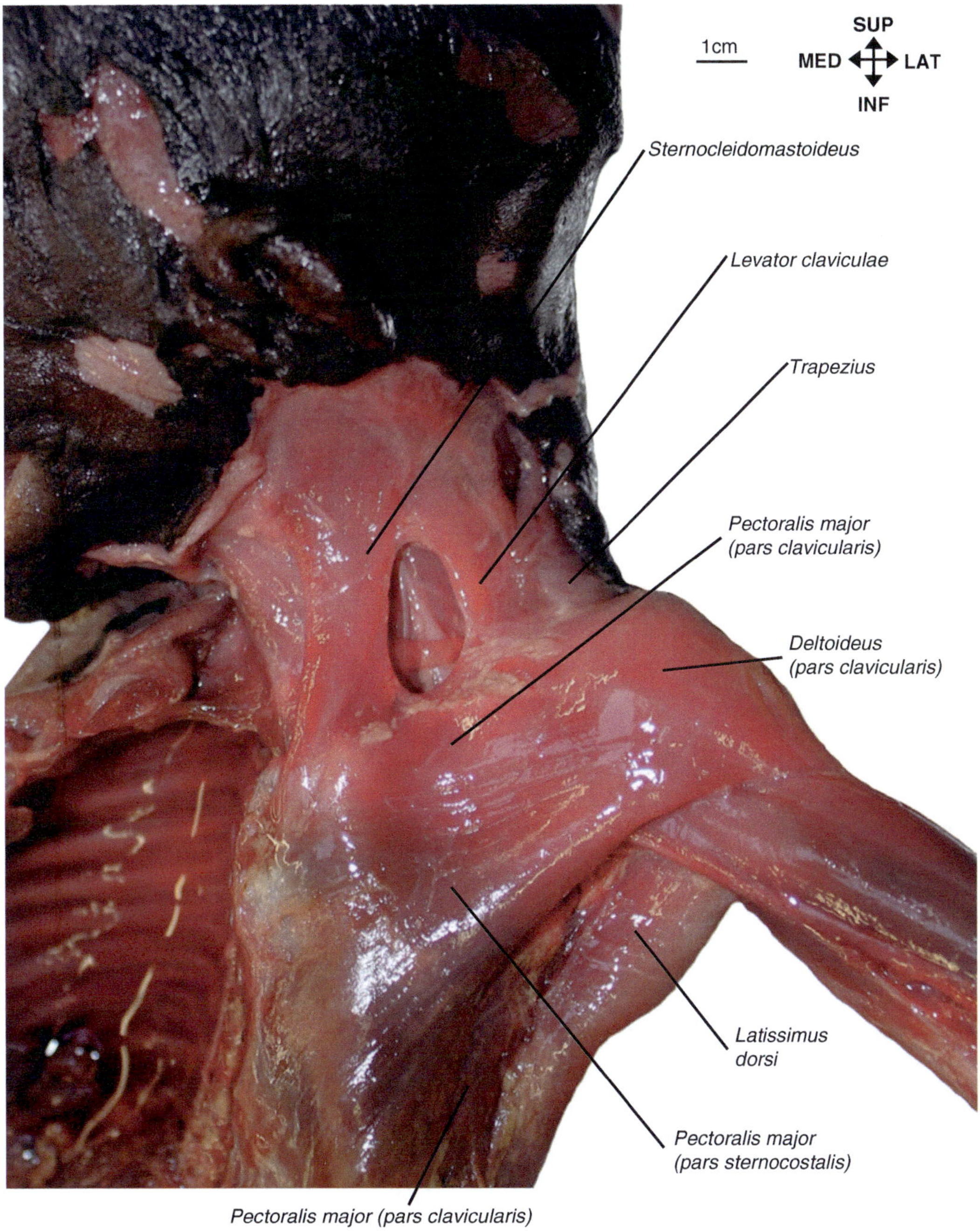


**Fig. 3.28** *Pan paniscus* (ZIMS 164052, Ano, fetus female): ventro-lateral view of the neck, shoulder and trunk muscles

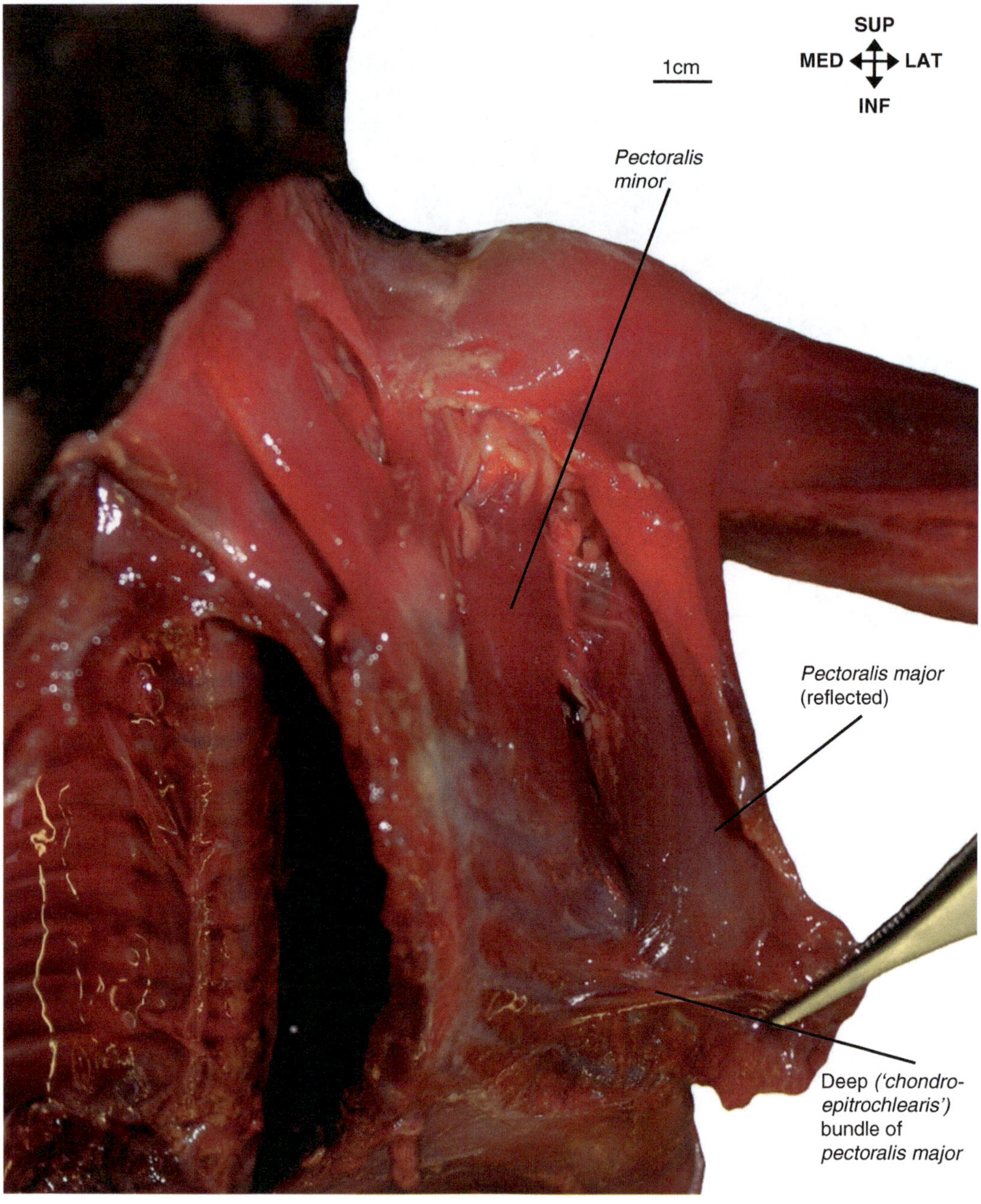


**Fig. 3.29** *Pan paniscus* (ZIMS 164052, Ano, fetus female): ventro-lateral view of the shoulder muscles after reflection of the main body of the pectoralis major, showing the deep ('*chondroepitrochlearis*') bundle of this muscle

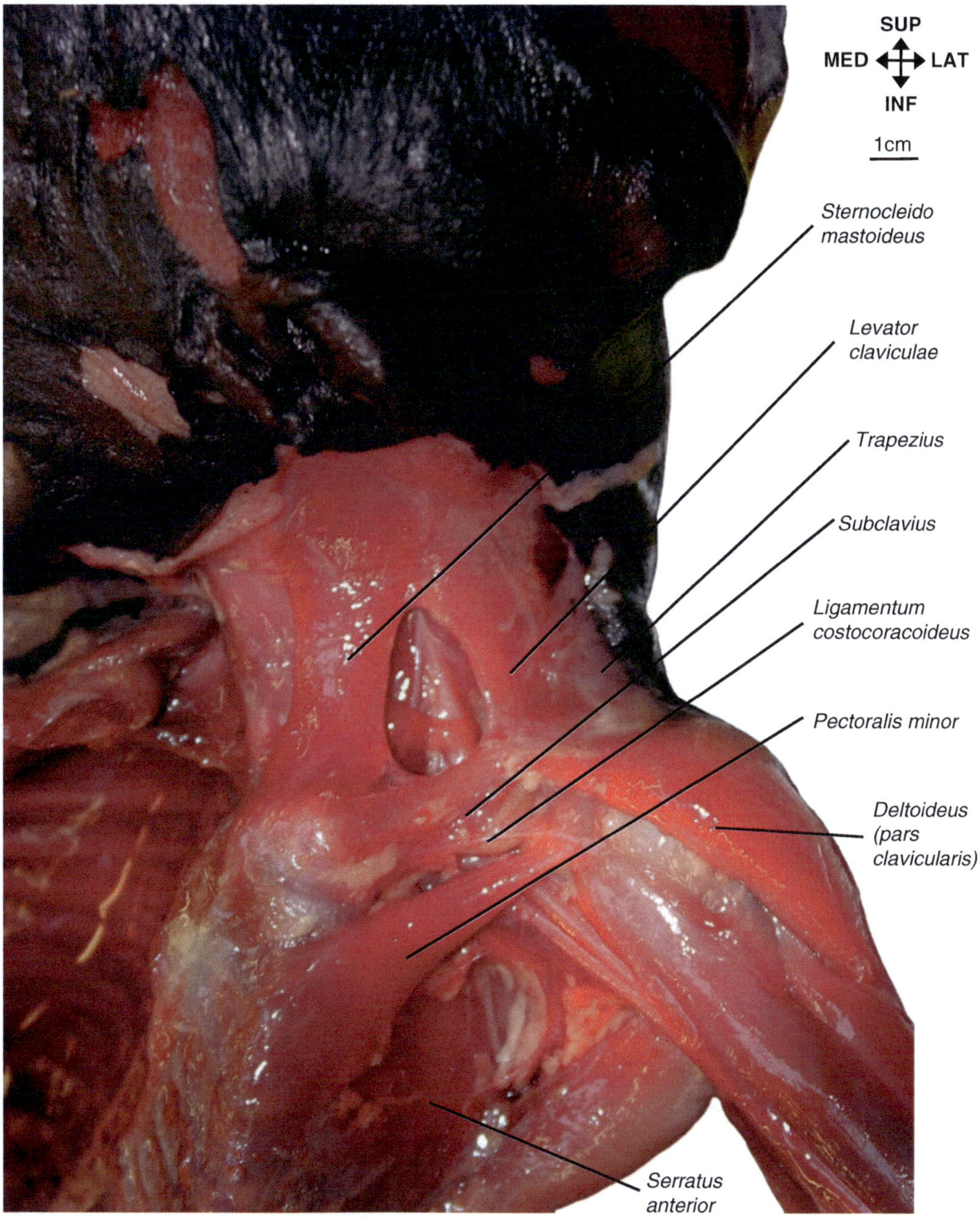


**Fig. 3.30** *Pan paniscus* (ZIMS 164052, Ano, fetus female): ventro-lateral view of the neck, shoulder and trunk muscles after removal of pectoralis major

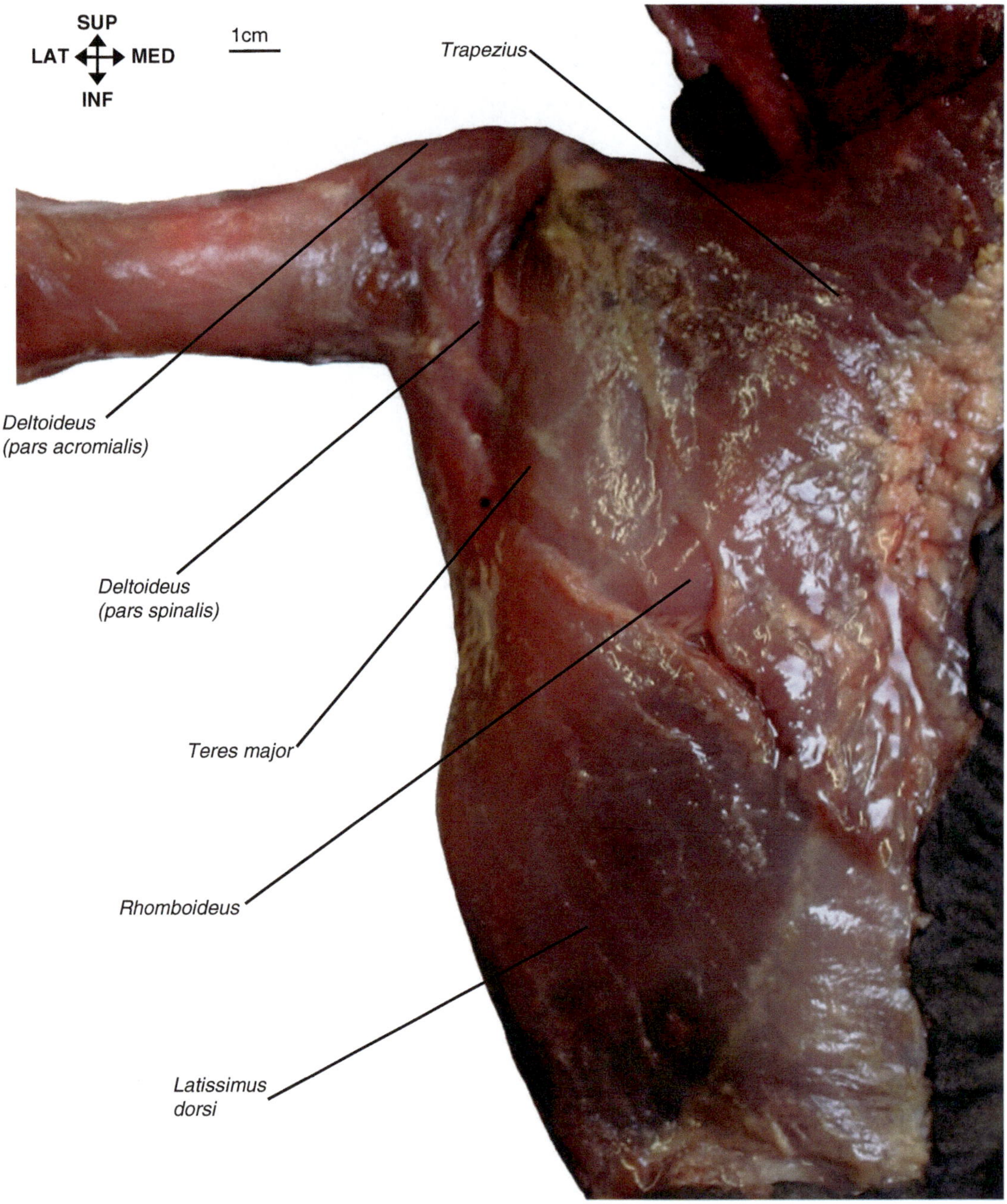


**Fig. 3.31** *Pan paniscus* (ZIMS 164052, Ano, fetus female): dorsal view of the neck, shoulder and trunk muscles

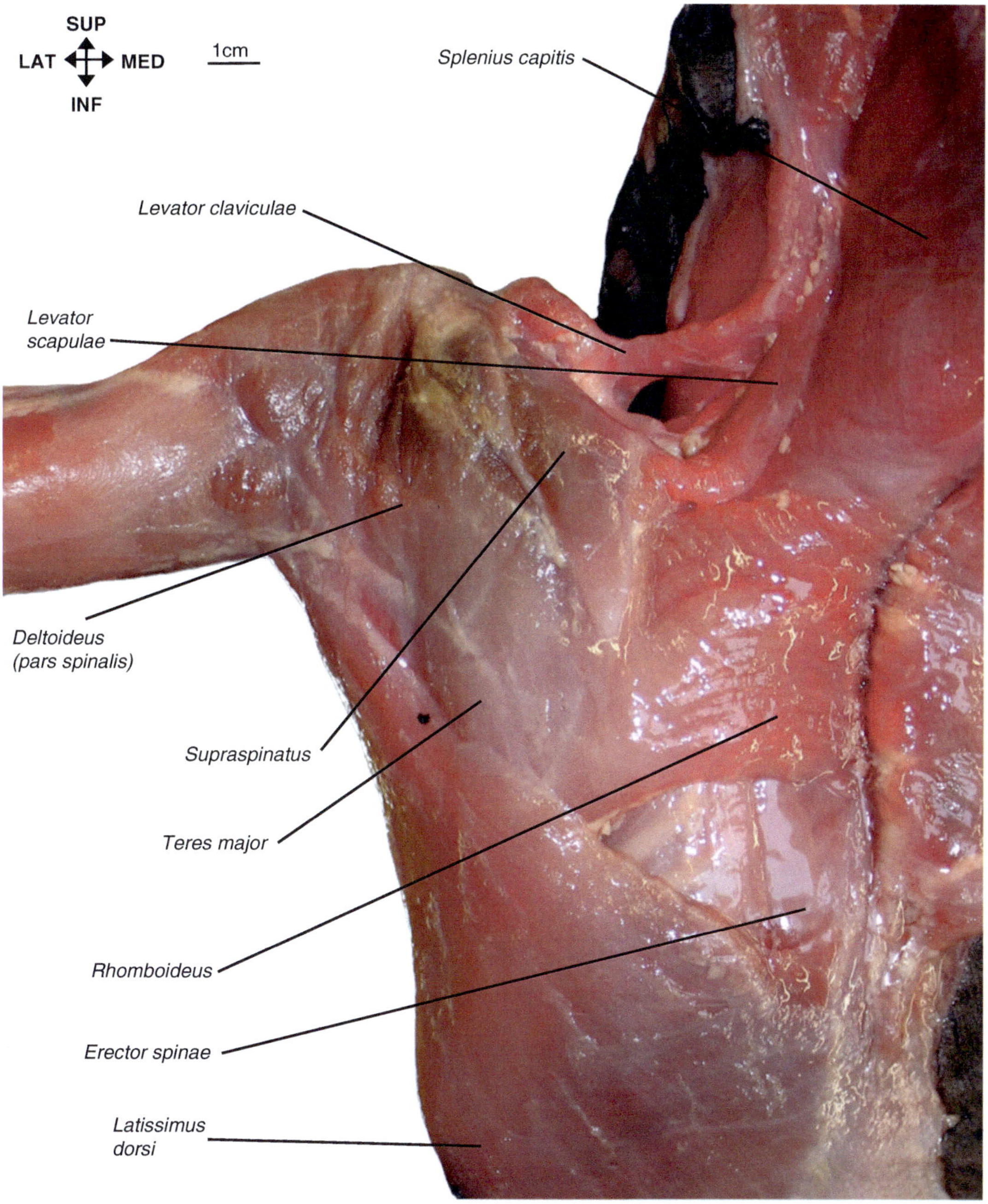


**Fig. 3.32** *Pan paniscus* (ZIMS 164052, Ano, fetus female): dorsal view of the neck, shoulder and trunk muscles after removal of trapezius

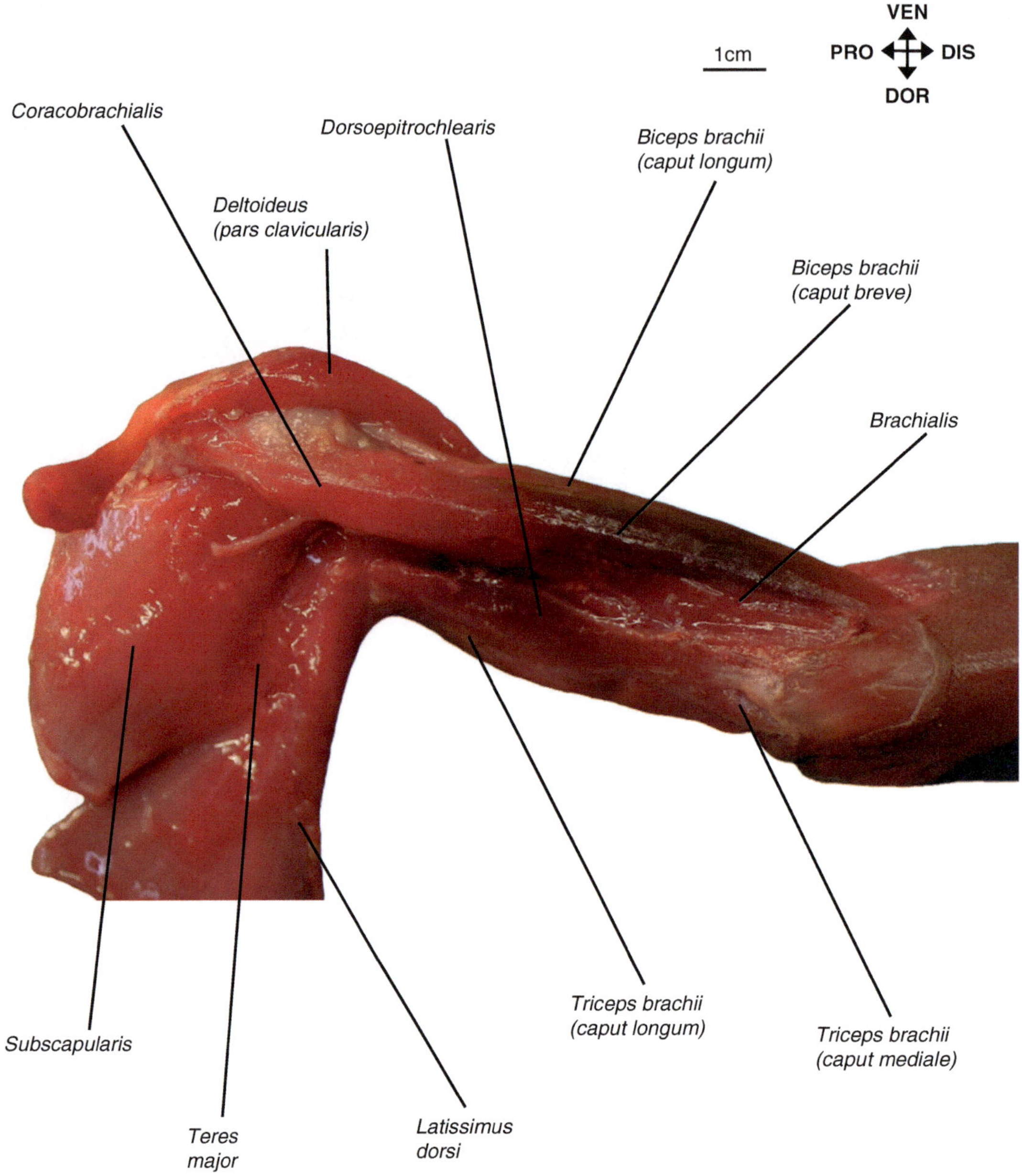


**Fig. 3.33** *Pan paniscus* (ZIMS 164052, Ano, fetus female): medial view of the shoulder and arm muscles

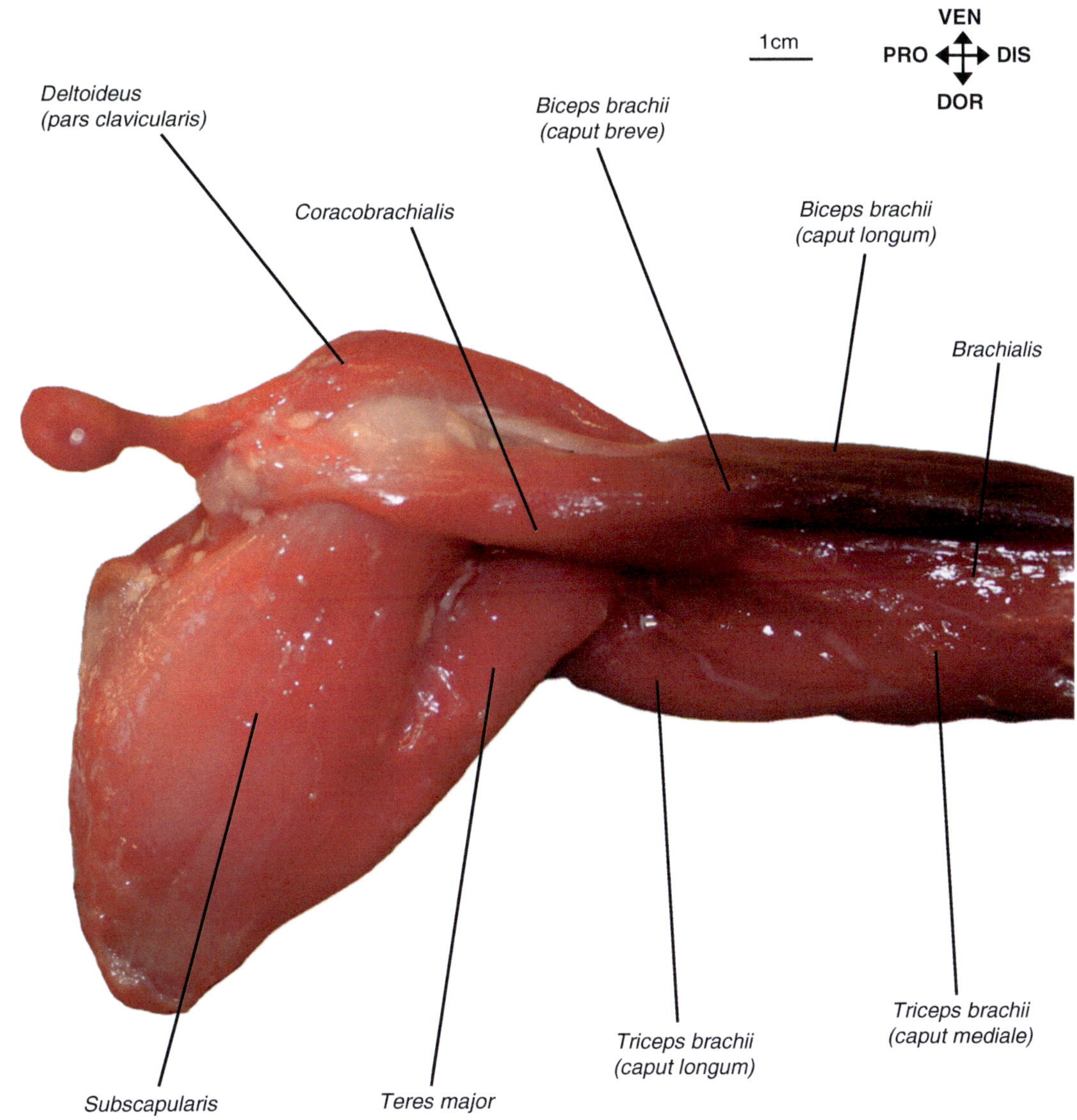


**Fig. 3.34** *Pan paniscus* (ZIMS 164052, Ano, fetus female): medial view of the shoulder and arm muscles after removal of latissimus dorsi and dorsoepitrochlearis

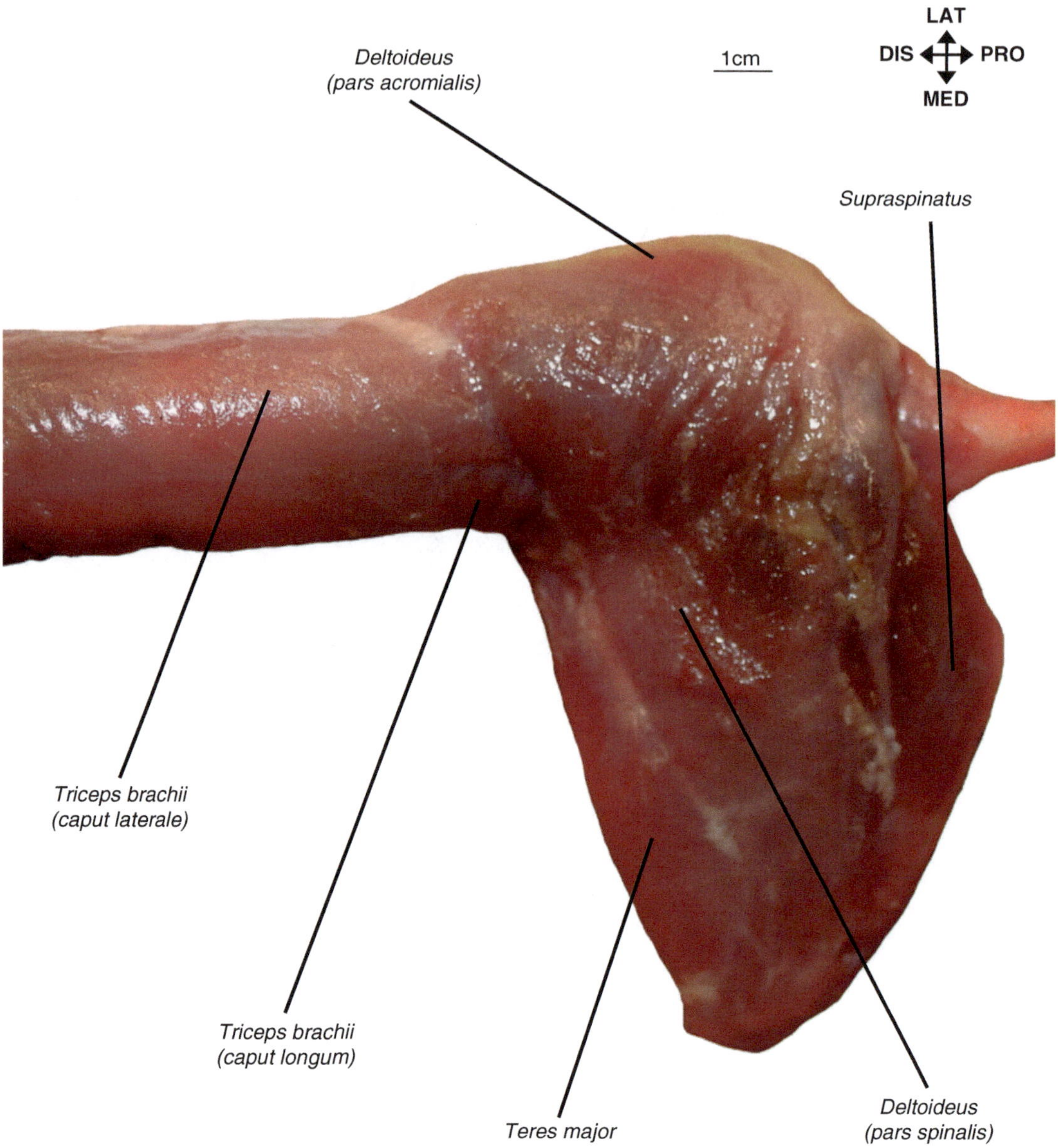


**Fig. 3.35** *Pan paniscus* (ZIMS 164052, Ano, fetus female): dorsal view of the shoulder and arm muscles

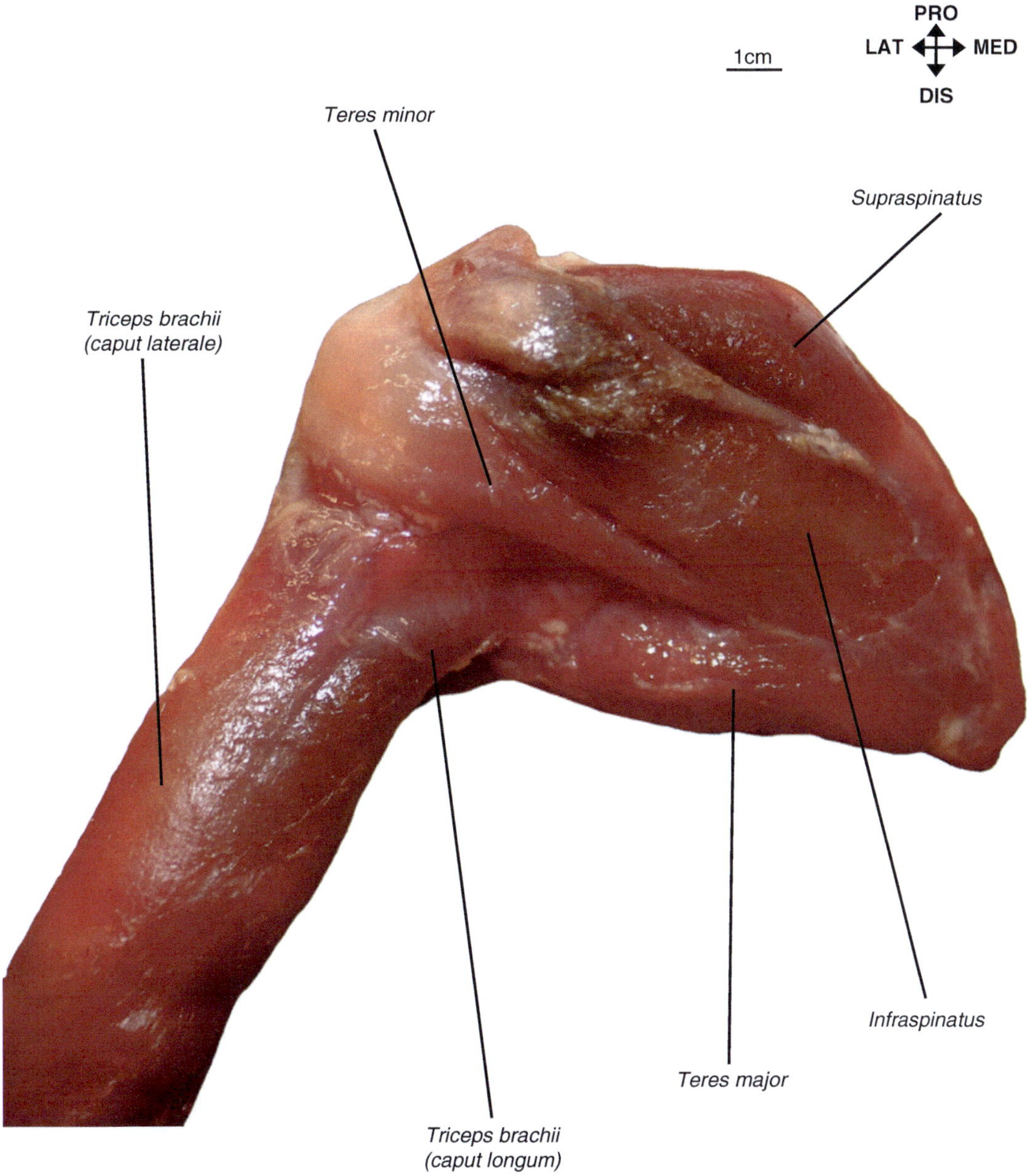


**Fig. 3.36** *Pan paniscus* (ZIMS 164052, Ano, fetus female): dorsal view of the shoulder and arm muscles after removal of deltoideus

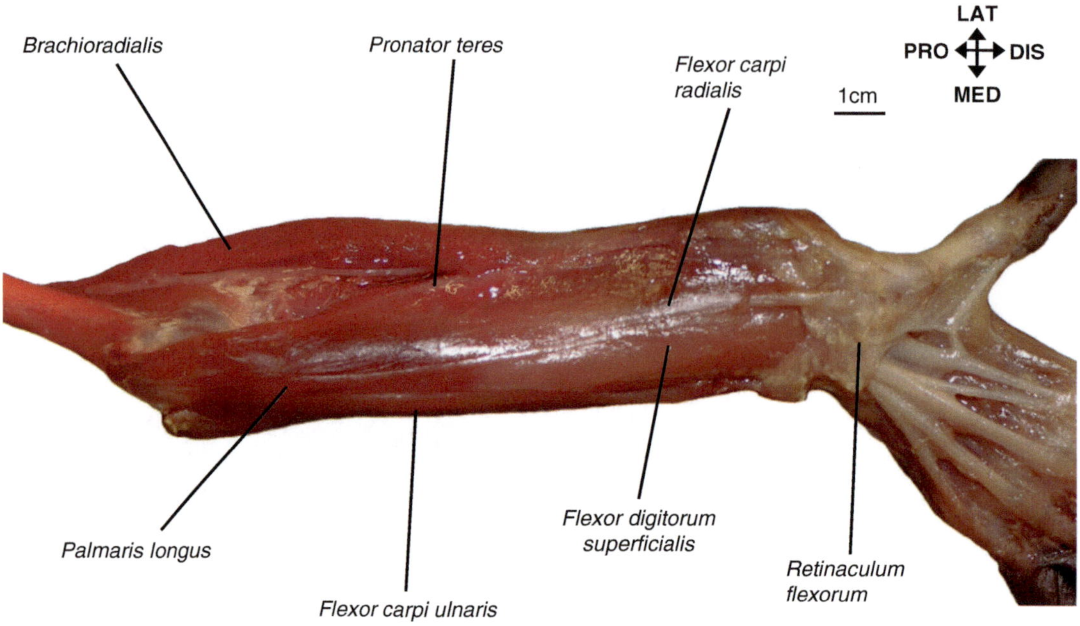


**Fig. 3.37** *Pan paniscus* (ZIMS 164052, Ano, fetus female): ventral view of the forearm muscles

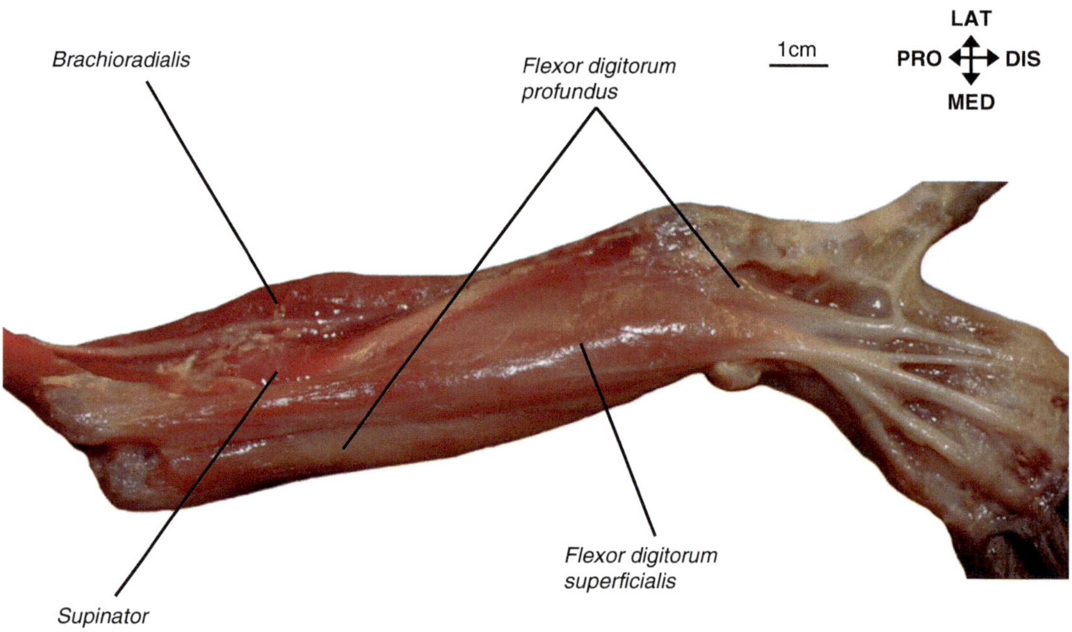


**Fig. 3.38** *Pan paniscus* (ZIMS 164052, Ano, fetus female): ventral view of the forearm muscles after removal of pronator teres, flexor carpi radialis, palmaris longus and flexor carpi ulnaris

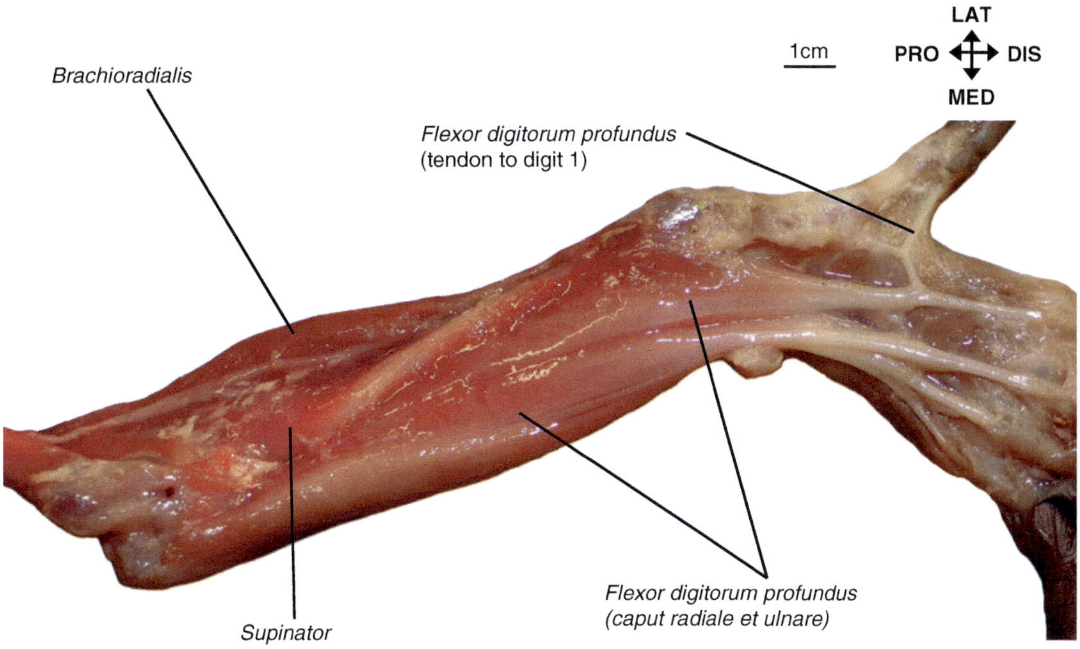


**Fig. 3.39** *Pan paniscus* (ZIMS 164052, Ano, fetus female): ventral view of the forearm muscles after removal of flexor digitorum superficialis

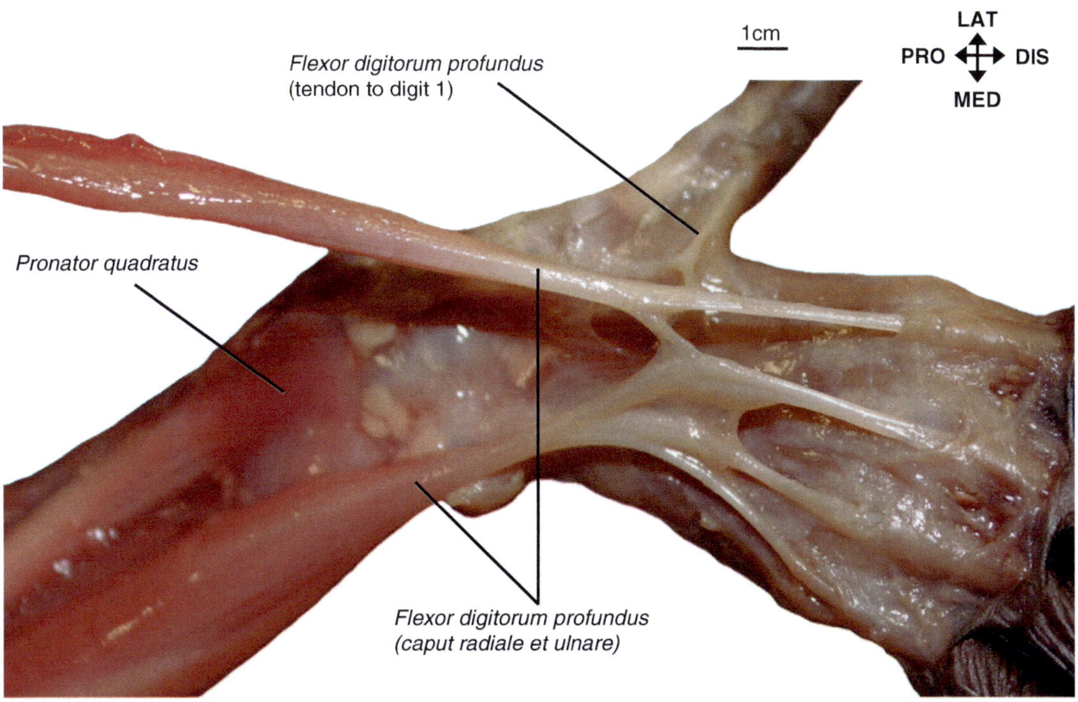


**Fig. 3.40** *Pan paniscus* (ZIMS 164052, Ano, fetus female): ventral view, detail of the insertion of flexor digitorum profundus showing the presence of a tendon to digit 1 and an anastomosis between the caput radiale and the caput ulnare

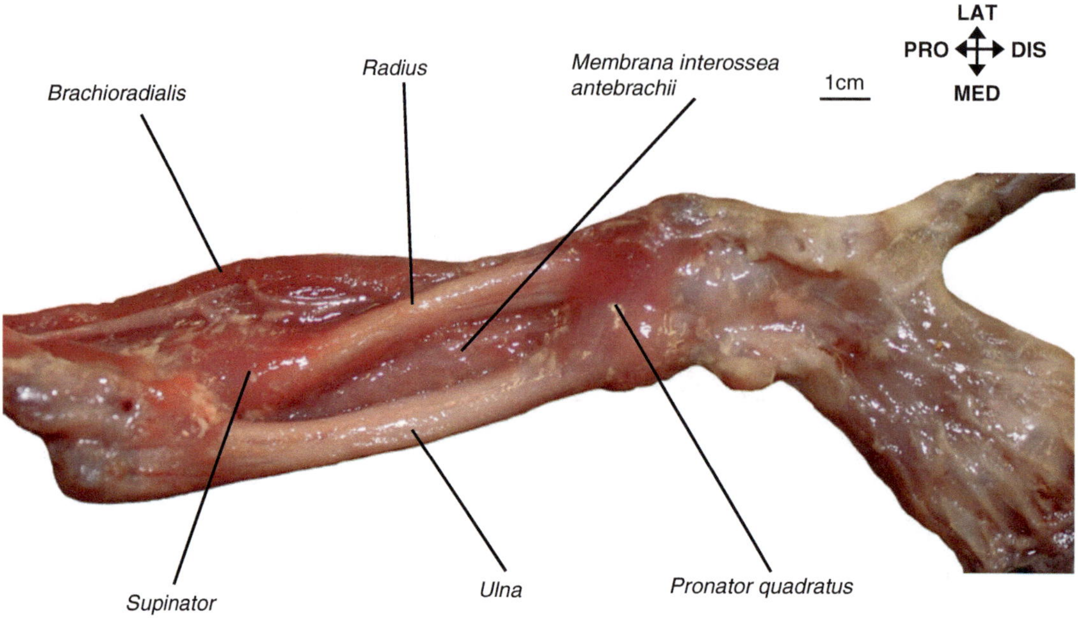


**Fig. 3.41** *Pan paniscus* (ZIMS 164052, Ano, fetus female): ventral view of the forearm muscles after removal of flexor digitorum profundus

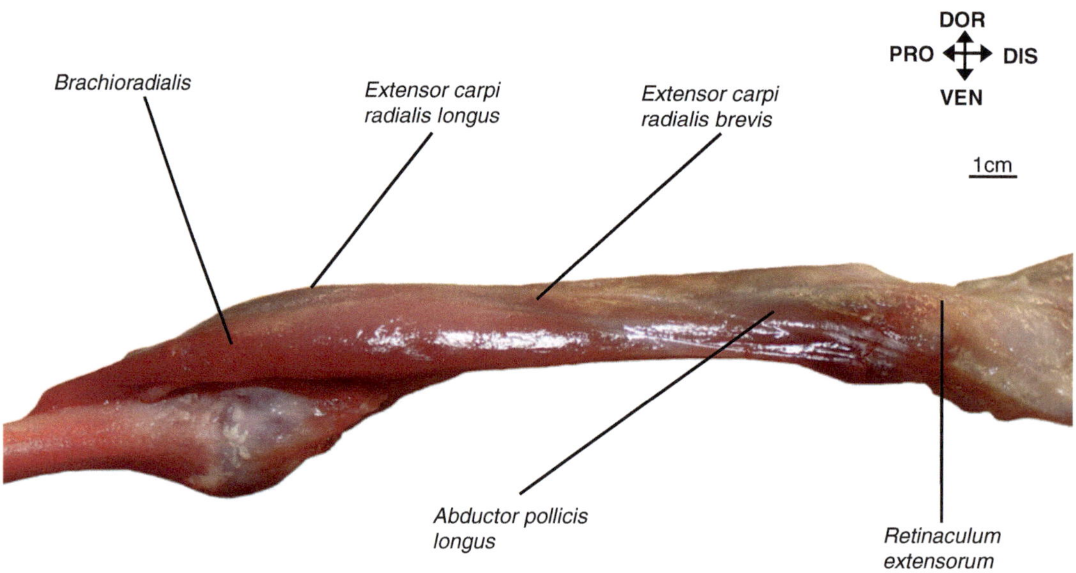


**Fig. 3.42** *Pan paniscus* (ZIMS 164052, Ano, fetus female): dorso-lateral view of the forearm muscles

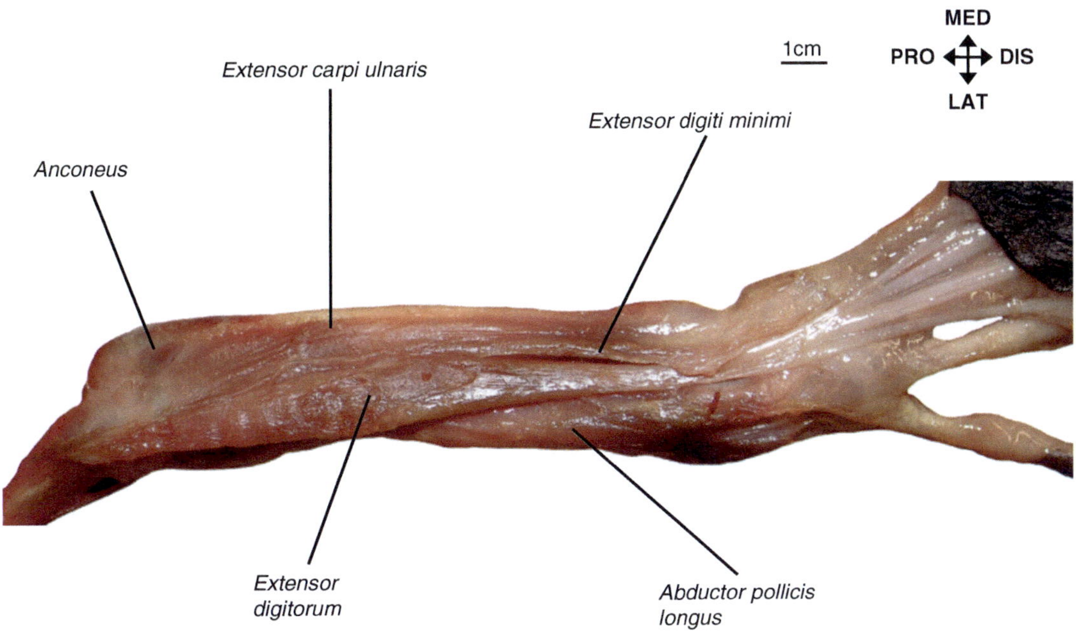


**Fig. 3.43** *Pan paniscus* (ZIMS 164052, Ano, fetus female): dorsal view of the forearm muscles

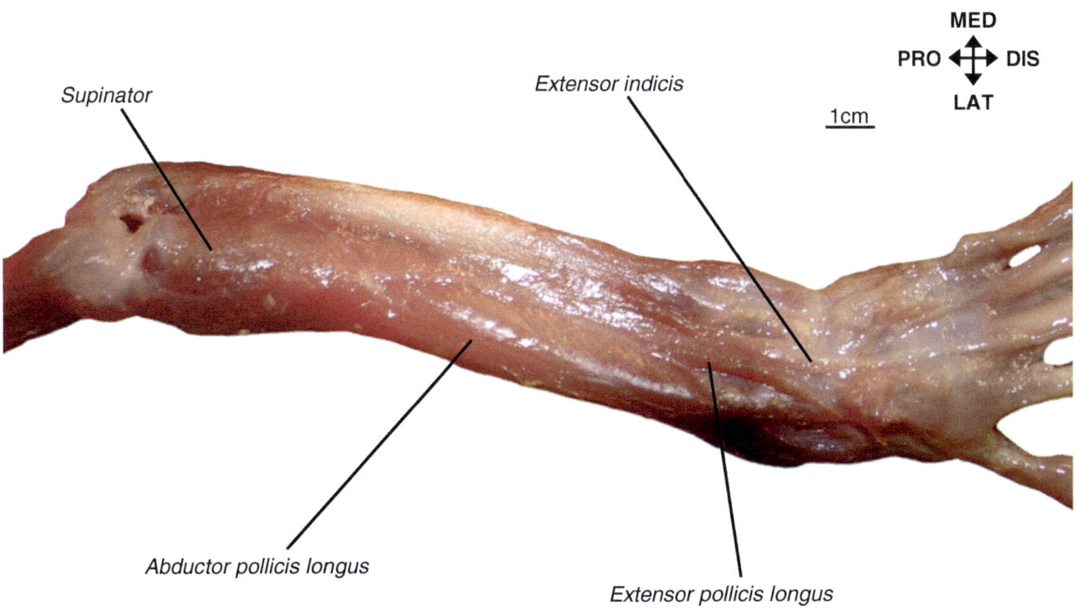


**Fig. 3.44** *Pan paniscus* (ZIMS 164052, Ano, fetus female): dorsal view of the forearm muscles after removal of brachioradialis, extensor carpi radialis longus et brevis, extensor digitorum, extensor digiti minimi and extensor carpi ulnaris

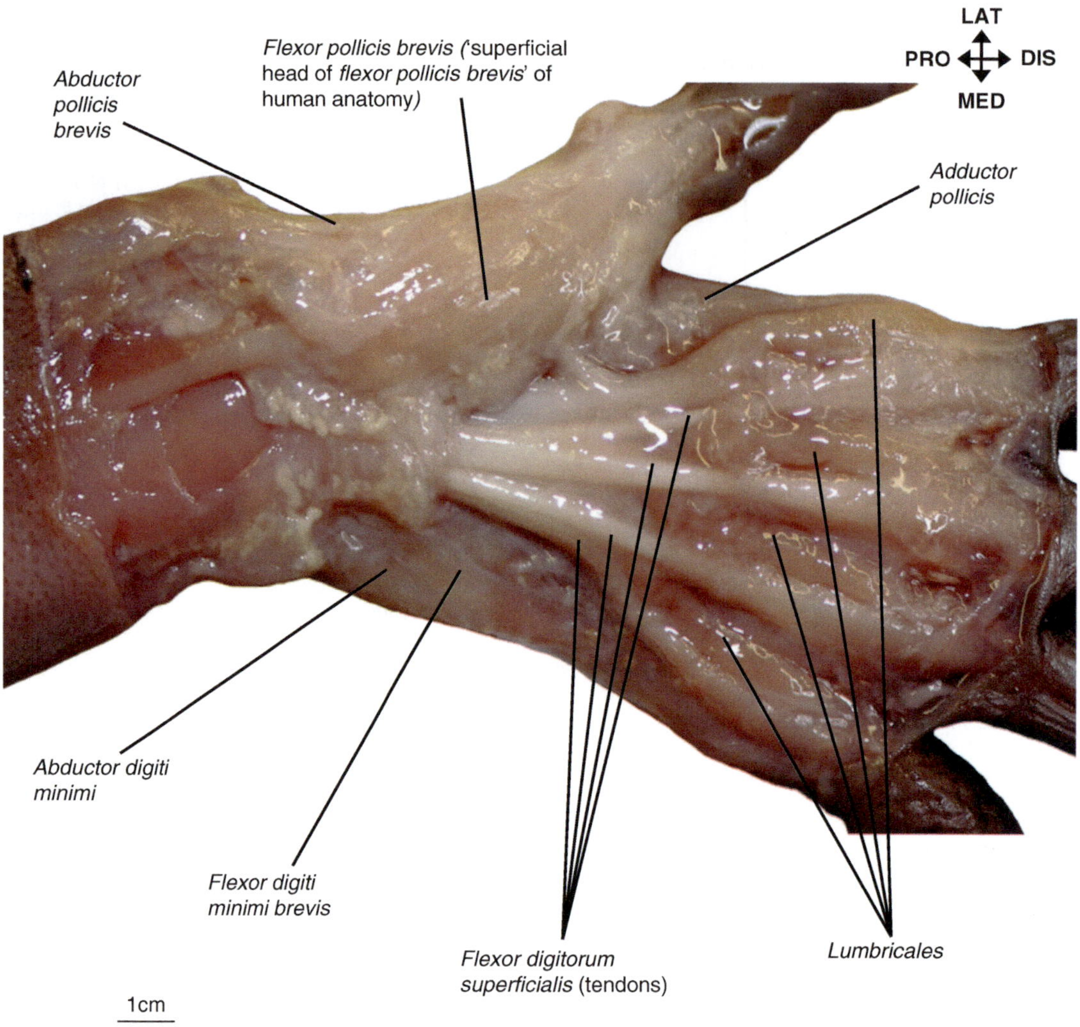


**Fig. 3.45** *Pan paniscus* (ZIMS 164052, Ano, fetus female): ventral view of the superficial muscles of the hand

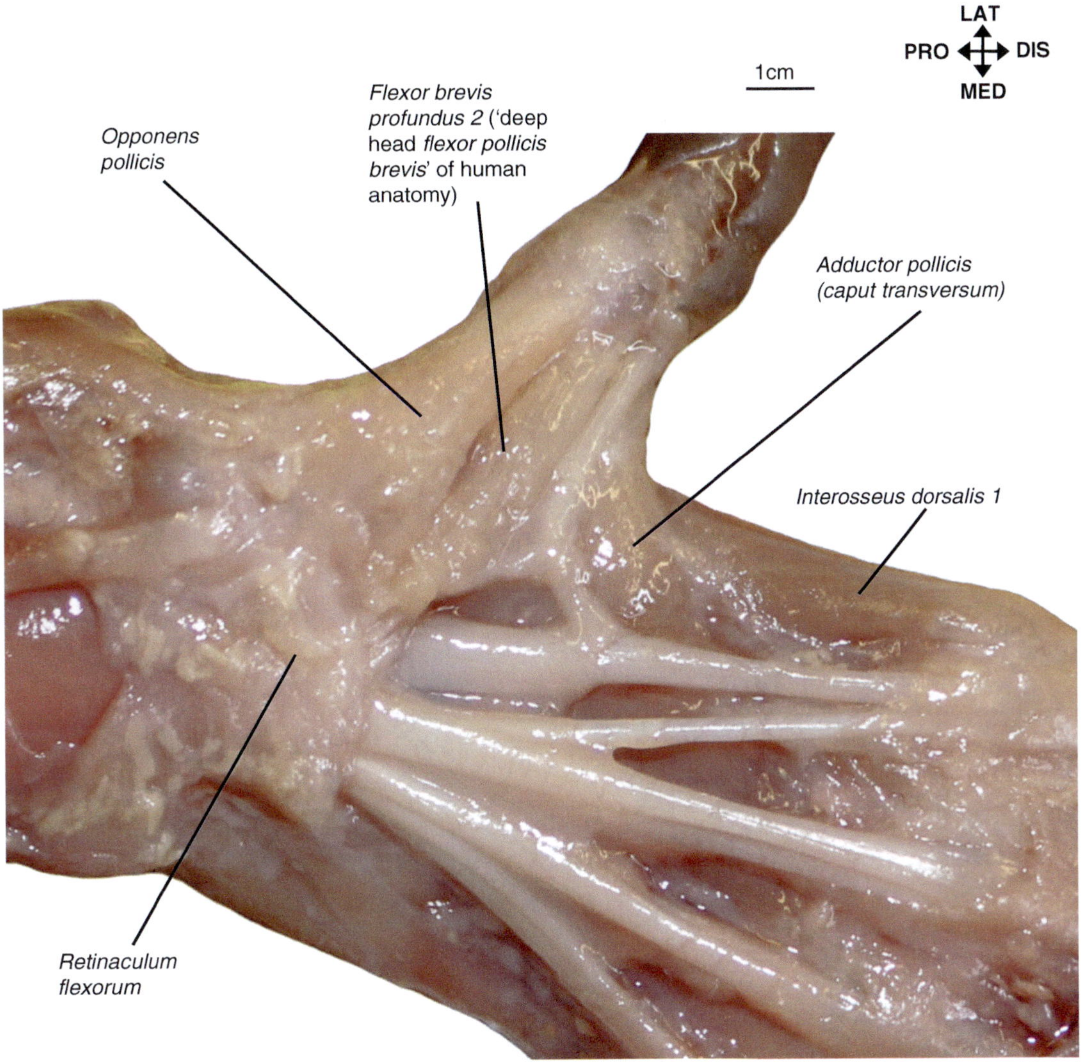


**Fig. 3.46** *Pan paniscus* (ZIMS 164052, Ano, fetus female): ventral view of the thenar muscles of the hand after removal of abductor pollicis brevis and flexor pollicis brevis

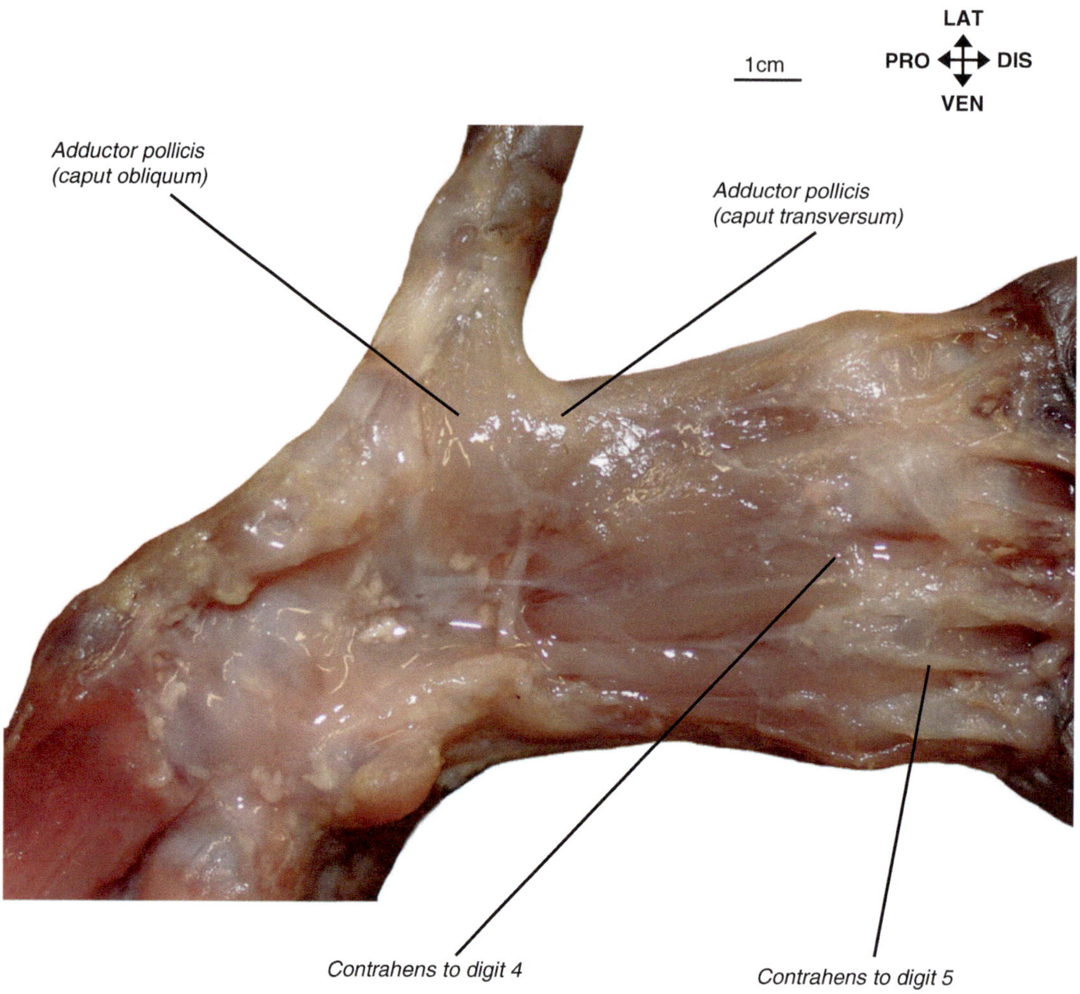


**Fig. 3.47** *Pan paniscus* (ZIMS 164052, Ano, fetus female): ventral view of the deep muscles of the hand after removal of thenar and hypothenar muscles, lumbricales and flexor digitorum superficialis and profundus

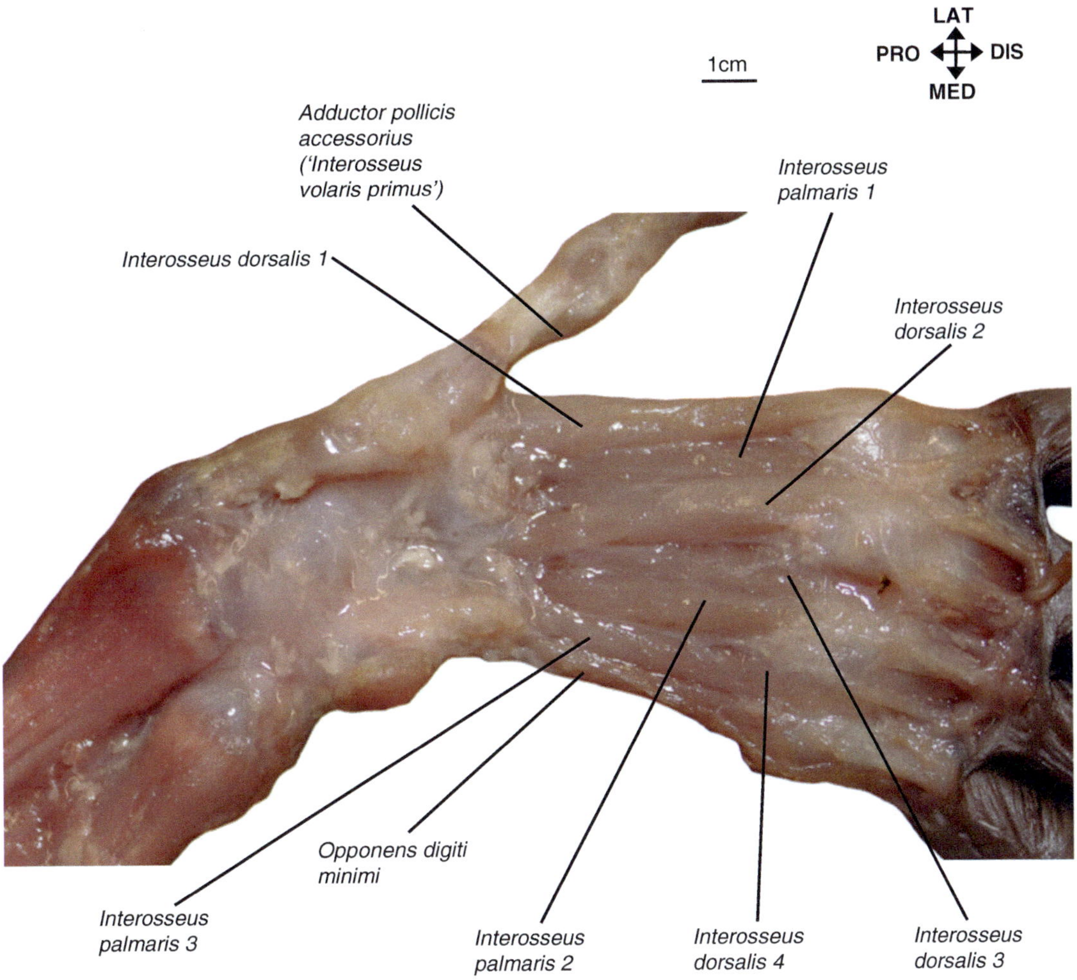


**Fig. 3.48** *Pan paniscus* (ZIMS 164052, Ano, fetus female): ventral view of the deep muscles of the hand after removal of contrahentes; note the presence of the adductor pollicis accessorius ('*interosseus volaris primus* or Henle's muscle')

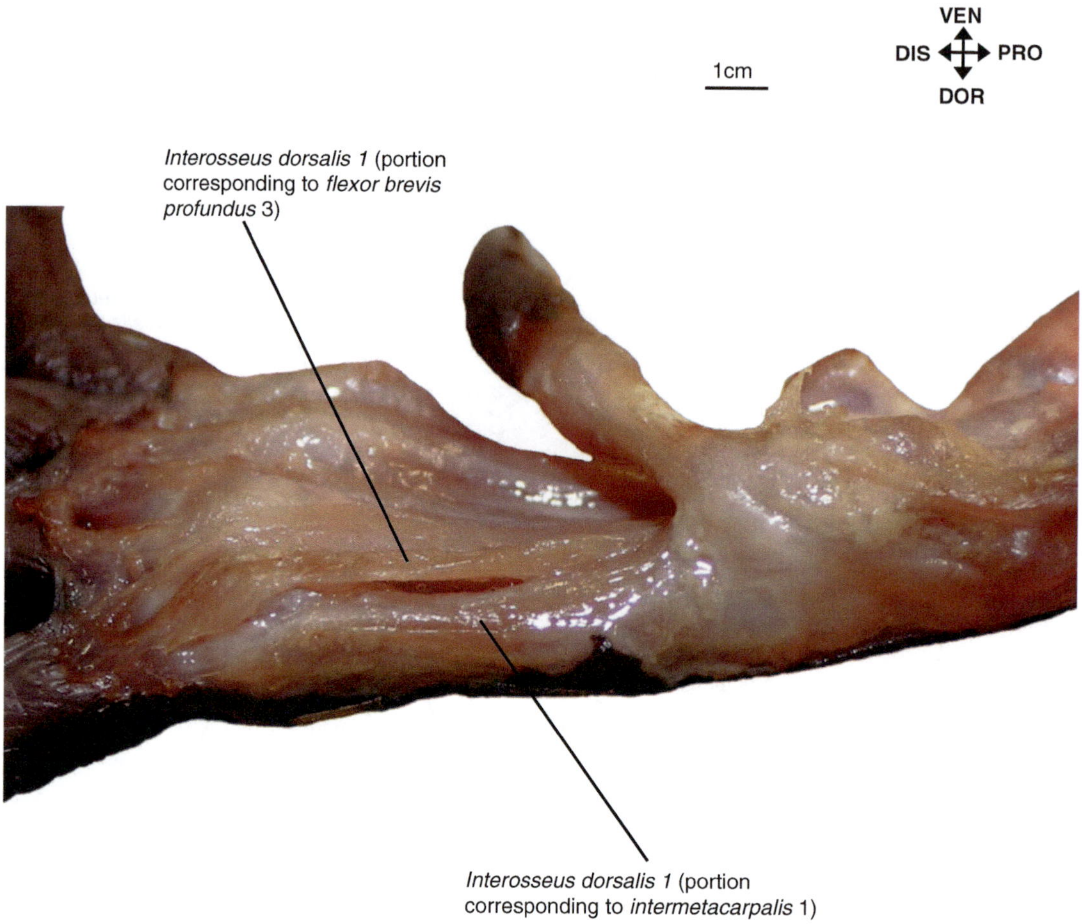


**Fig. 3.49** *Pan paniscus* (ZIMS 164052, Ano, fetus female): lateral view interosseus dorsalis 1, separated by dissection

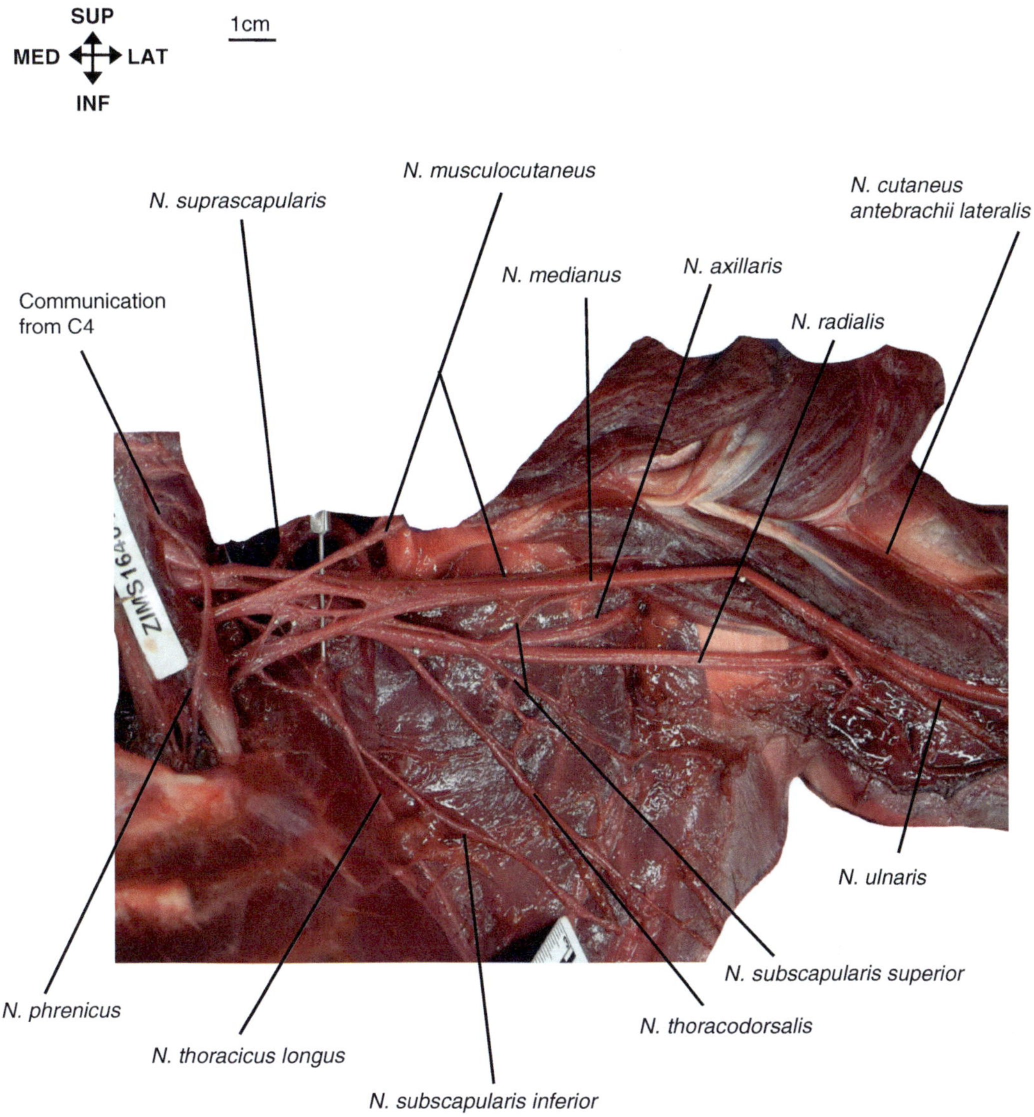


**Fig. 3.50** *Pan paniscus* (ZIMS 164047, Jasiri, adolescent female): ventral view of left *plexus brachialis*, pectoral musculature removed and clavicle retracted

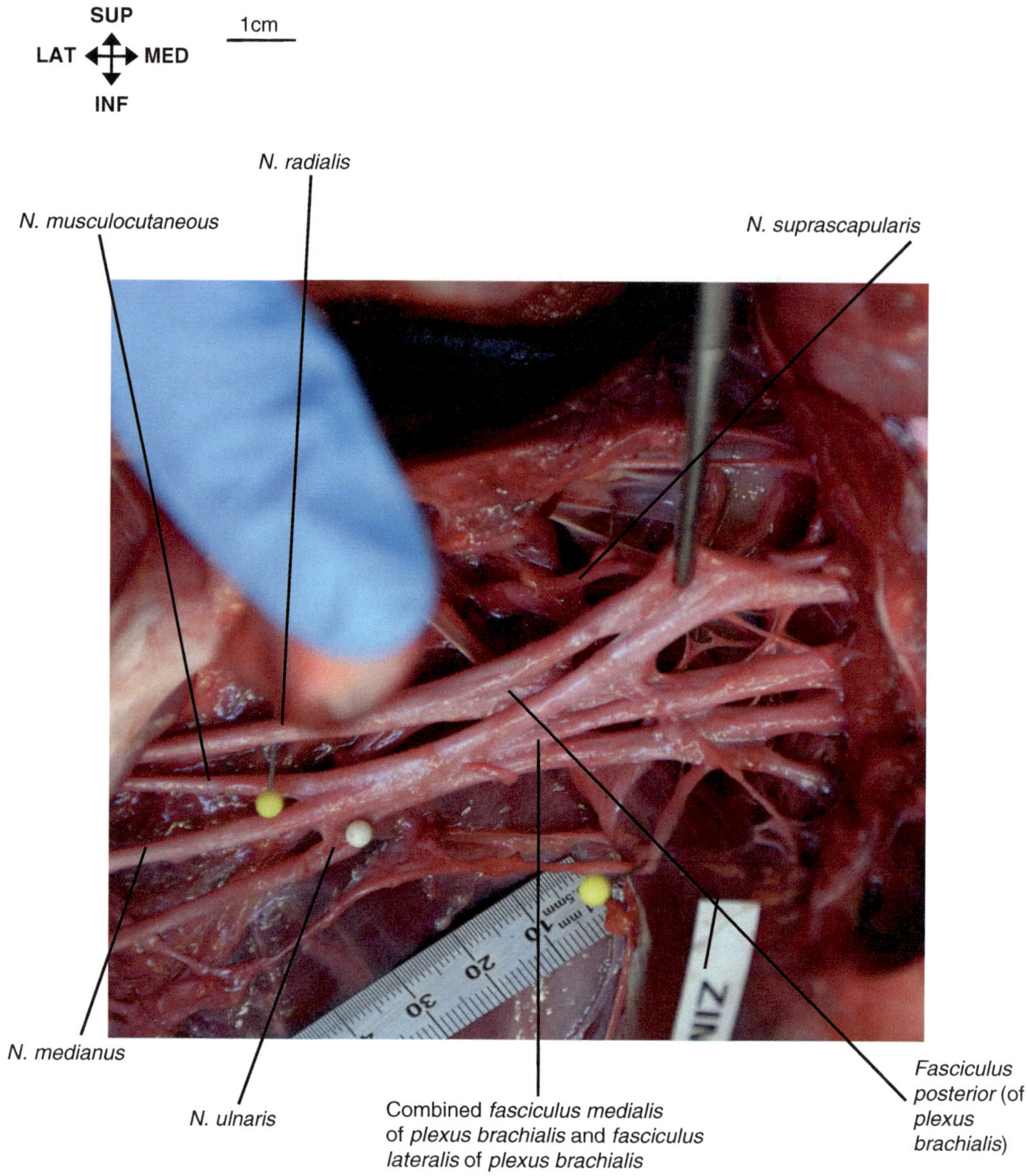


**Fig. 3.51** *Pan paniscus* (ZIMS 164047, Jasiri, adolescent female): ventral view of right *plexus brachialis*, clavicle retracted. Note the abnormal conjunction of the *fasciculus medialis plexus brachialis* and *fasciculus lateralis plexus brachialis*

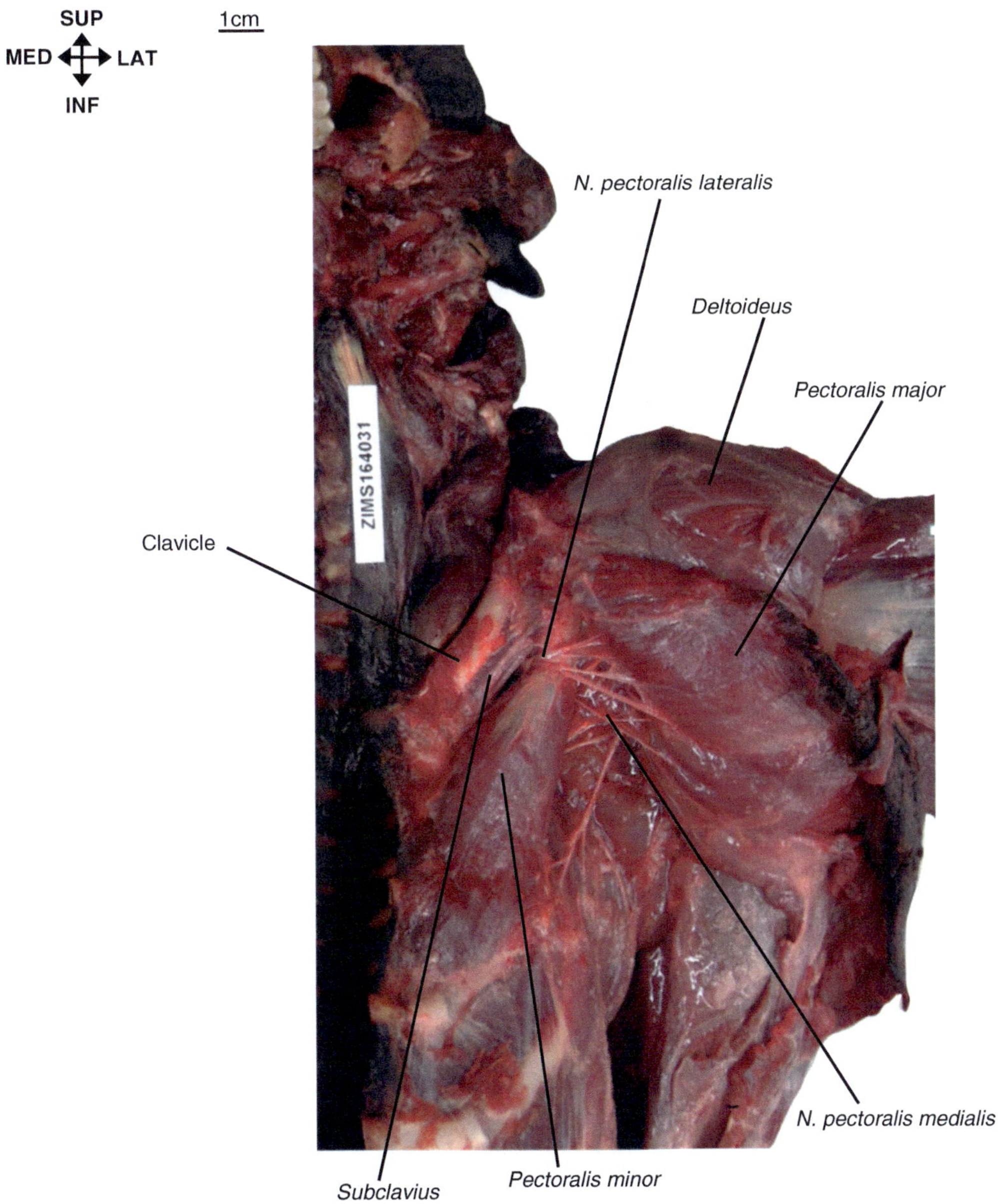


**Fig. 3.52** *Pan paniscus* (ZIMS 164031, Kidogo, adult male): ventral view of the left pectoral musculature and pectoral nerves. The pectoralis major was reflected laterally from its sternal and clavicular heads. Note the *n. pectoralis medialis* passing deep to the pectoralis minor, rather than piercing it, as is the usual condition in *Homo sapiens*

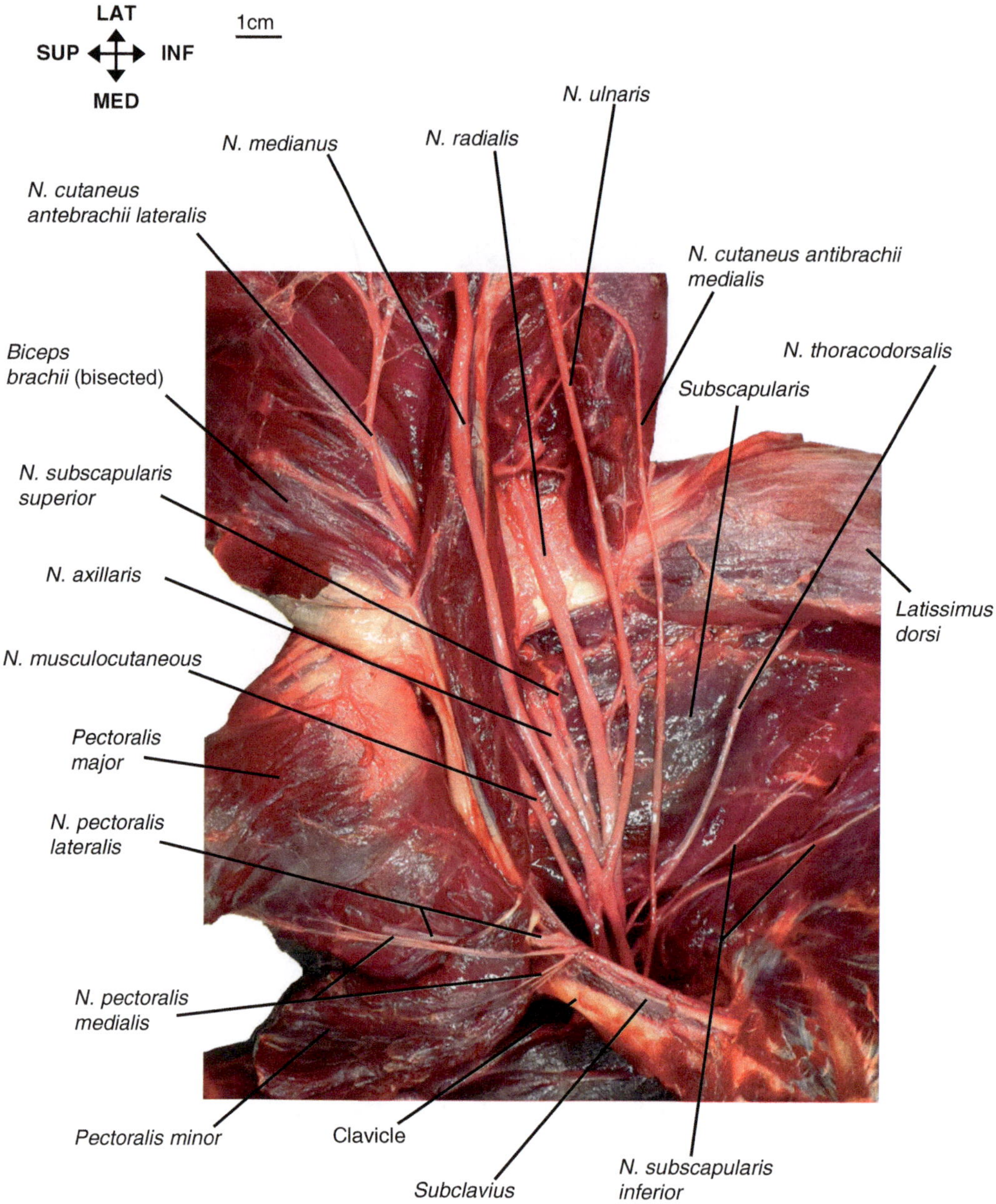


**Fig. 3.53** *Pan paniscus* (ZIMS 164031, Kidogo, adult male): ventral view of the left *plexus brachialis*; shoulder musculature removed from the axillary sheath. Pectoralis muscles reflected, biceps brachii bisected

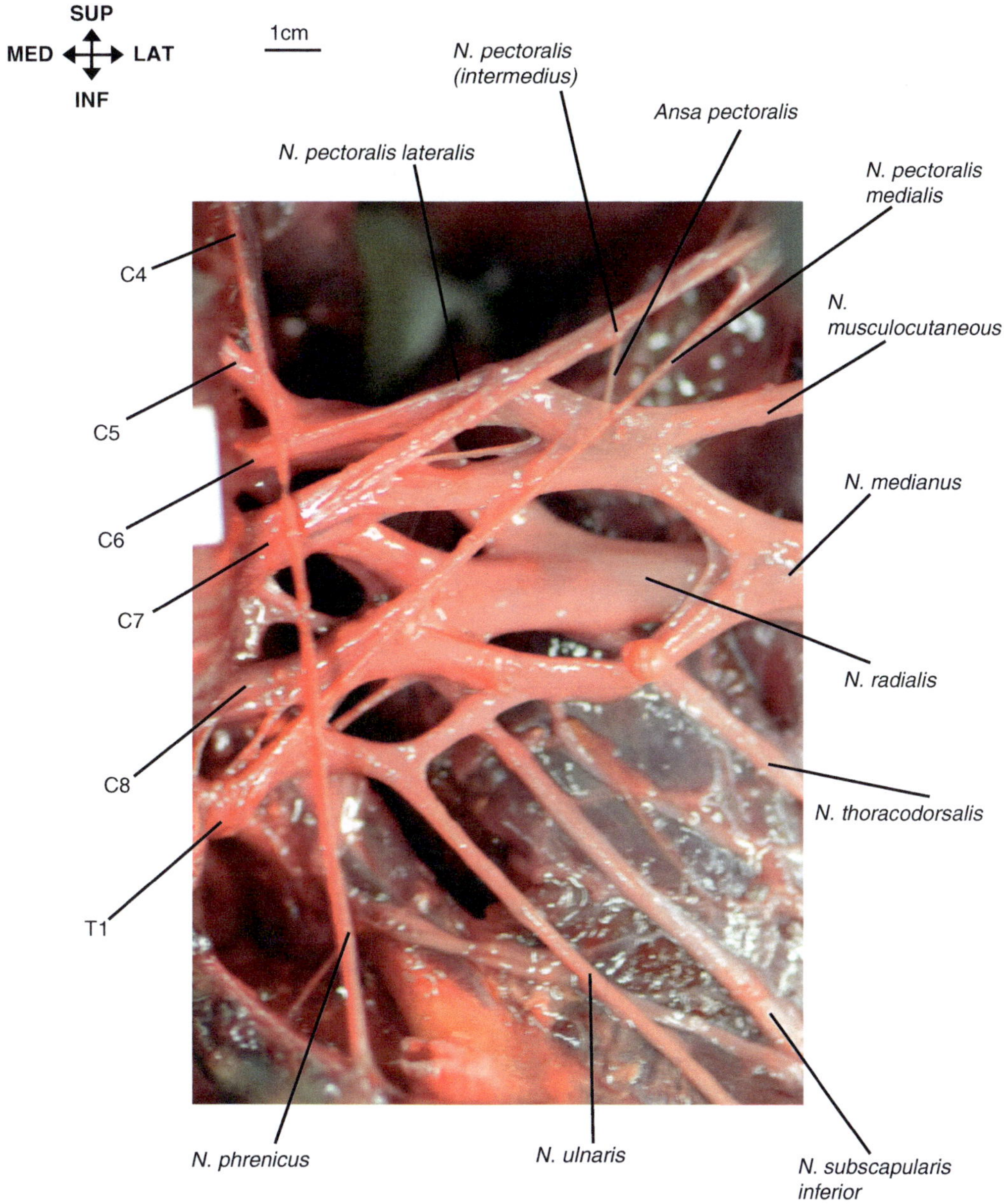


**Fig. 3.54** *Pan paniscus* (ZIMS 164031, Kidogo, adult male): ventral view of proximal portion of the left *plexus brachialis*, clavicle retracted

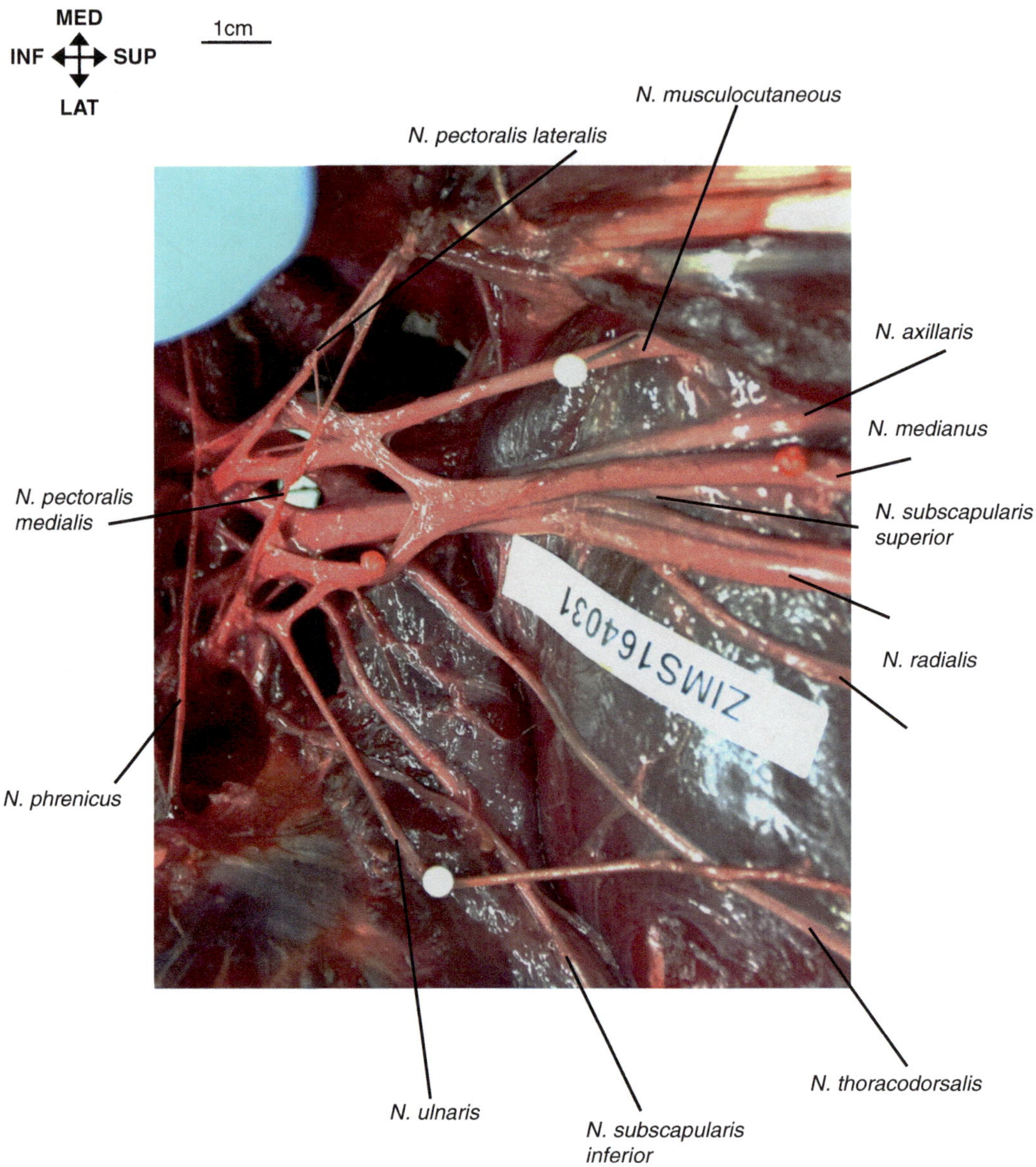


**Fig. 3.55** *Pan paniscus* (ZIMS 164031, Kidogo, adult male): ventral view of left *plexus brachialis*, clavicle retracted

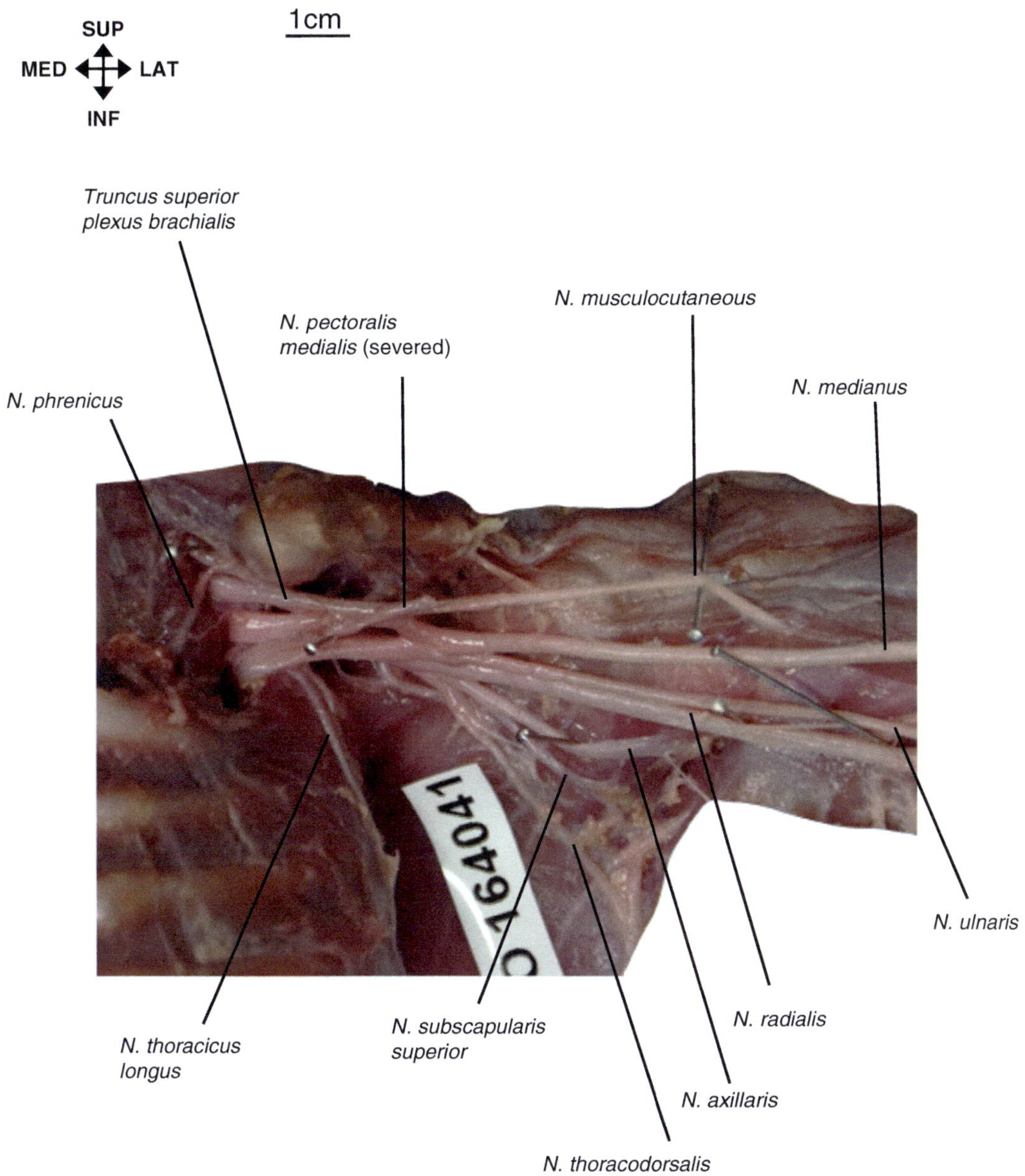


**Fig. 3.56** *Pan paniscus* (ZIMS 164041, Foyo, infant male): ventral view of left *plexus brachialis*, pectoral musculature removed and clavicle retracted

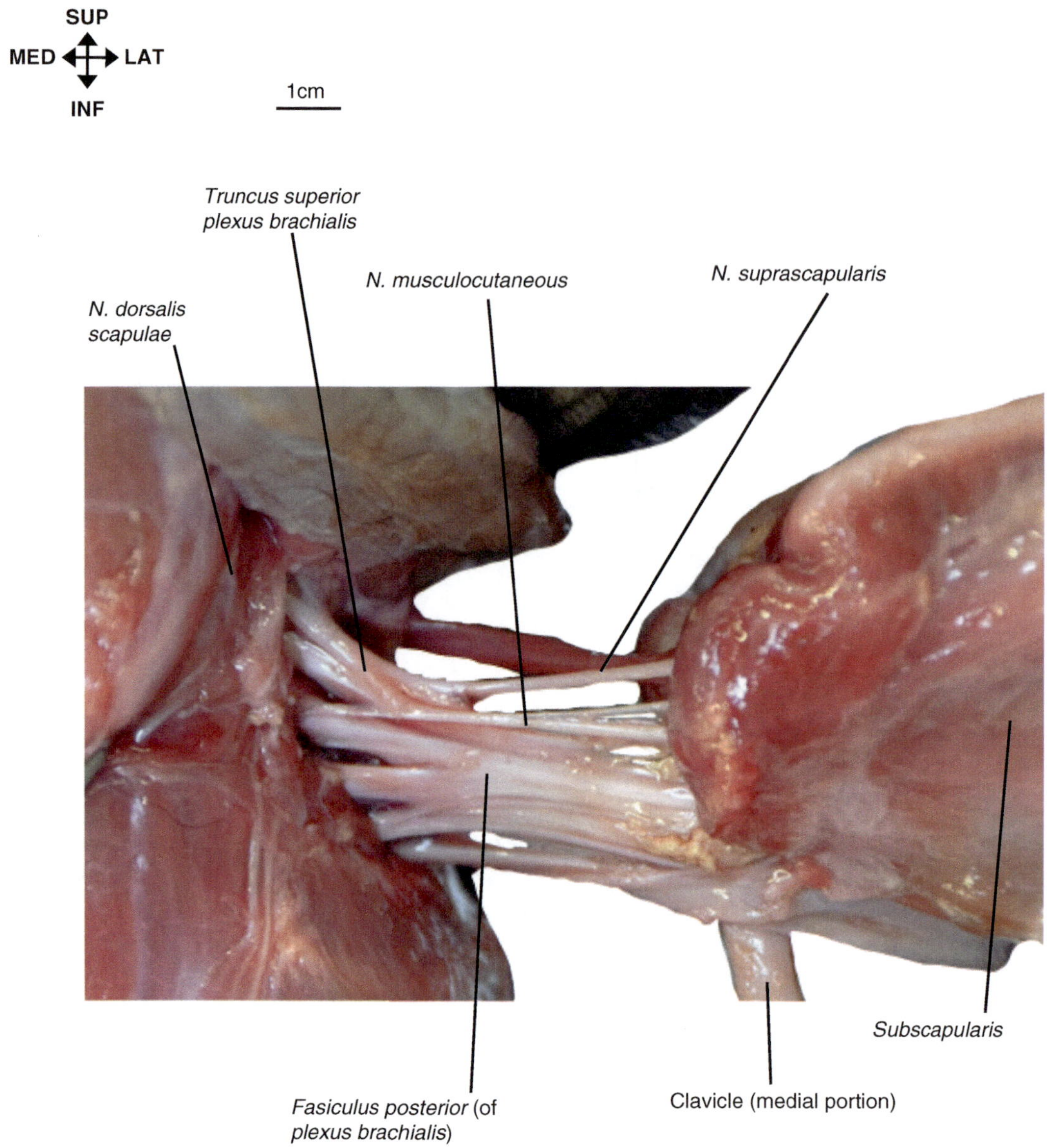


**Fig. 3.57** *Pan paniscus* (ZIMS 164041, Foyo, infant male): dorsal view of right *plexus brachialis*, clavicle separated and scapula rotated internally, about 180, and inferiorly, about 90°

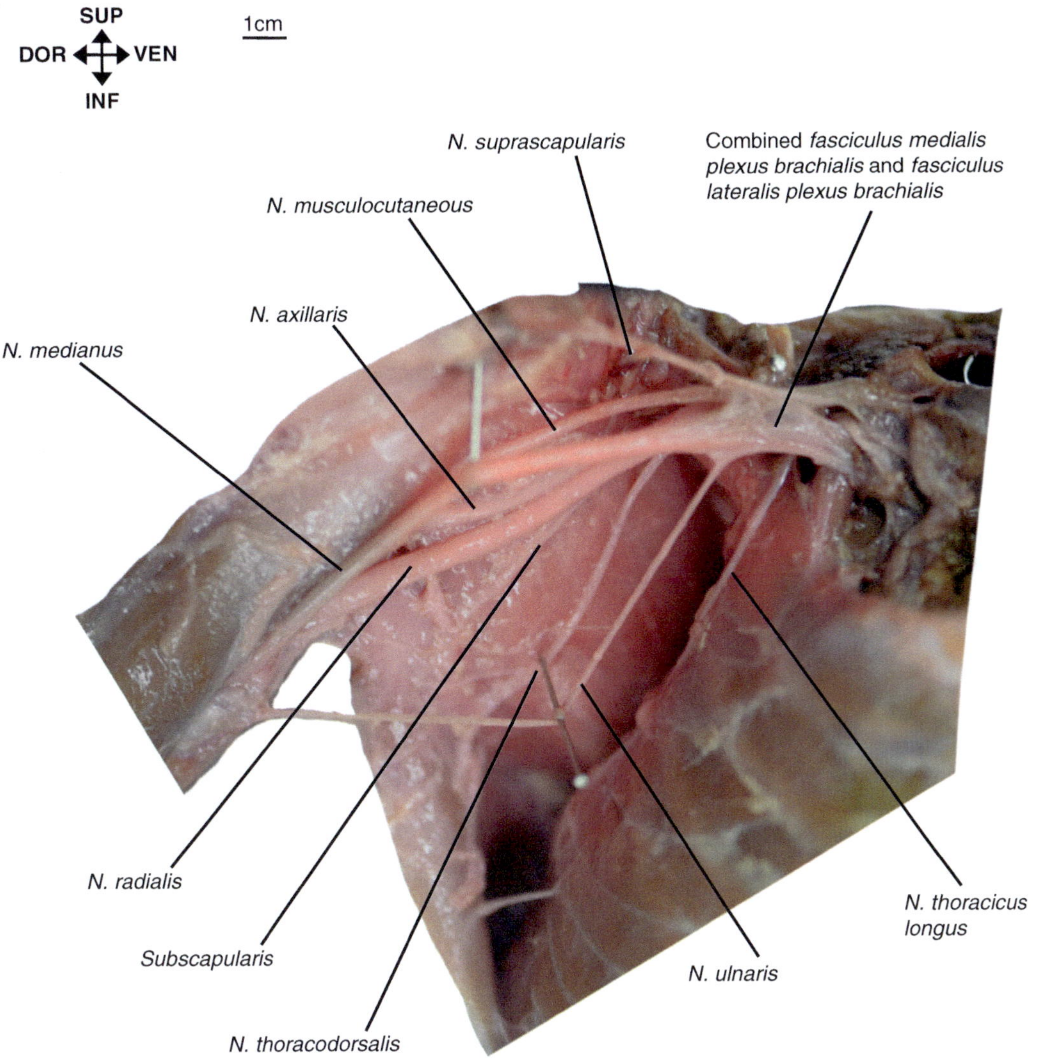


**Fig. 3.58** *Pan paniscus* (ZIMS 164040, Etje, infant male): ventral view of right *plexus brachialis*, pectoral musculature removed and scapula retracted

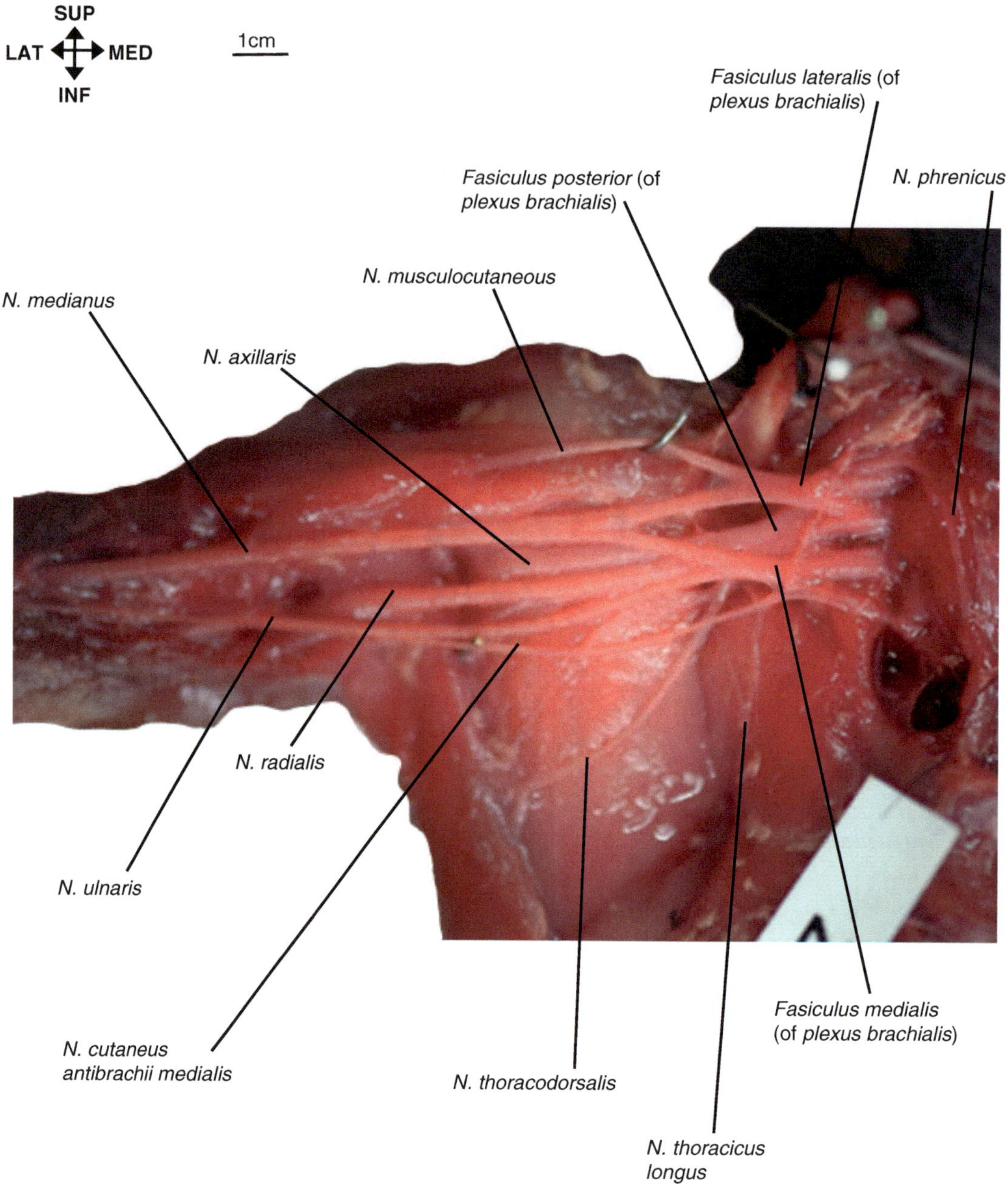


**Fig. 3.59** *Pan paniscus* (ZIMS 164052, Ano, female fetus): ventral view of right *plexus brachialis*, pectoral musculature removed and scapula retracted

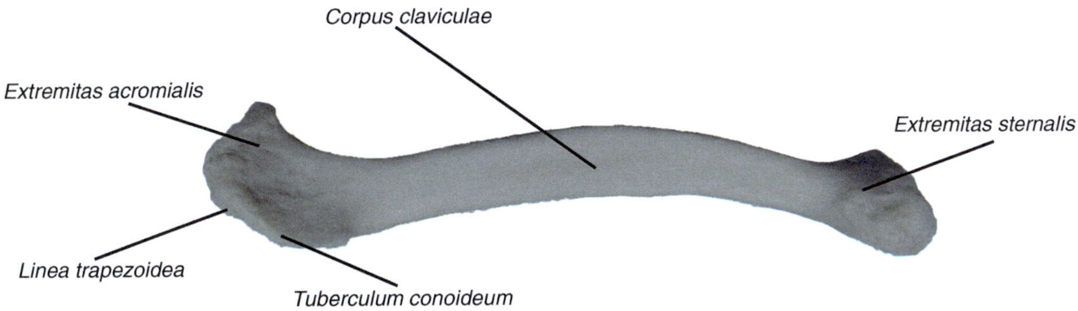


**Fig. 3.60** *Pan paniscus*: inferior view of clavicle

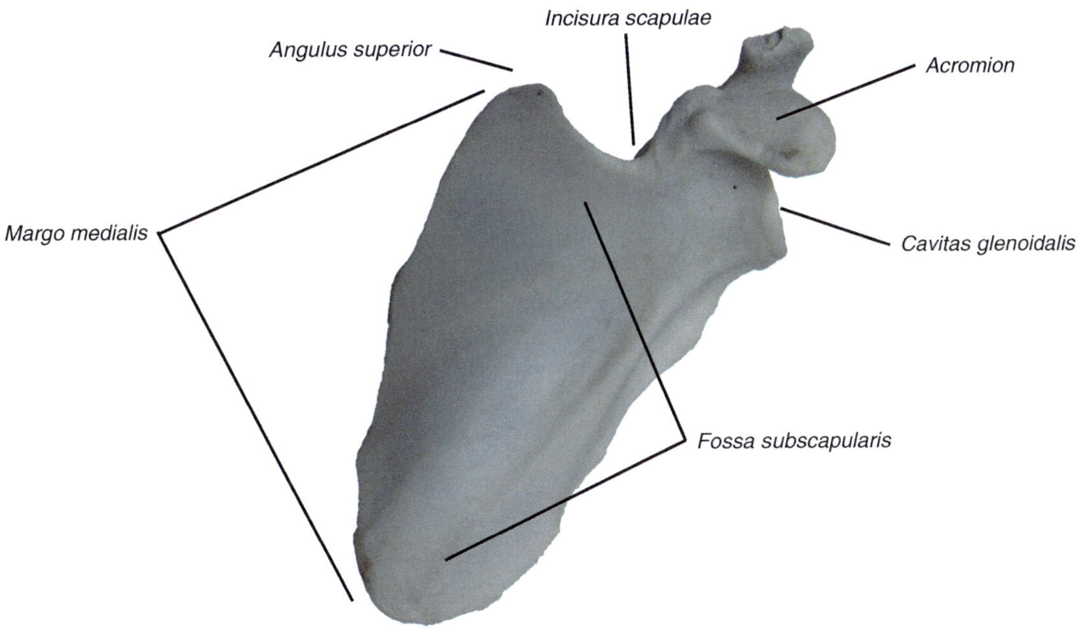


**Fig. 3.61** *Pan paniscus*: medical view of scapula

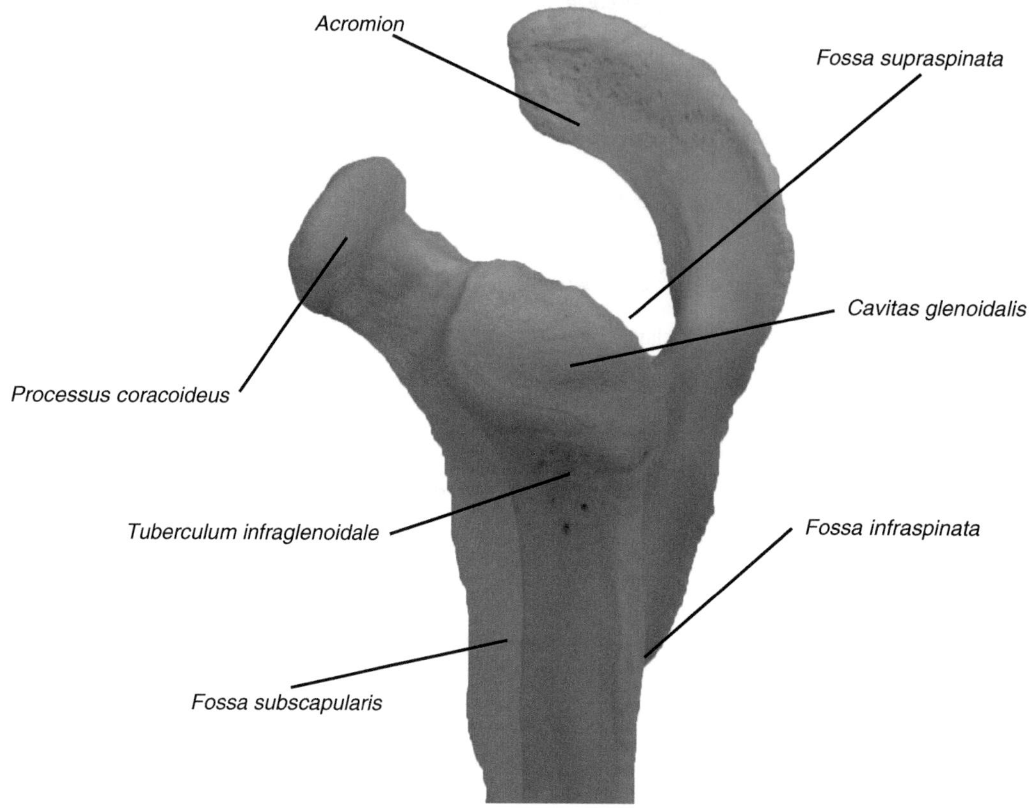


**Fig. 3.62** *Pan paniscus*: lateral view of the glenohumeral joint of scapula

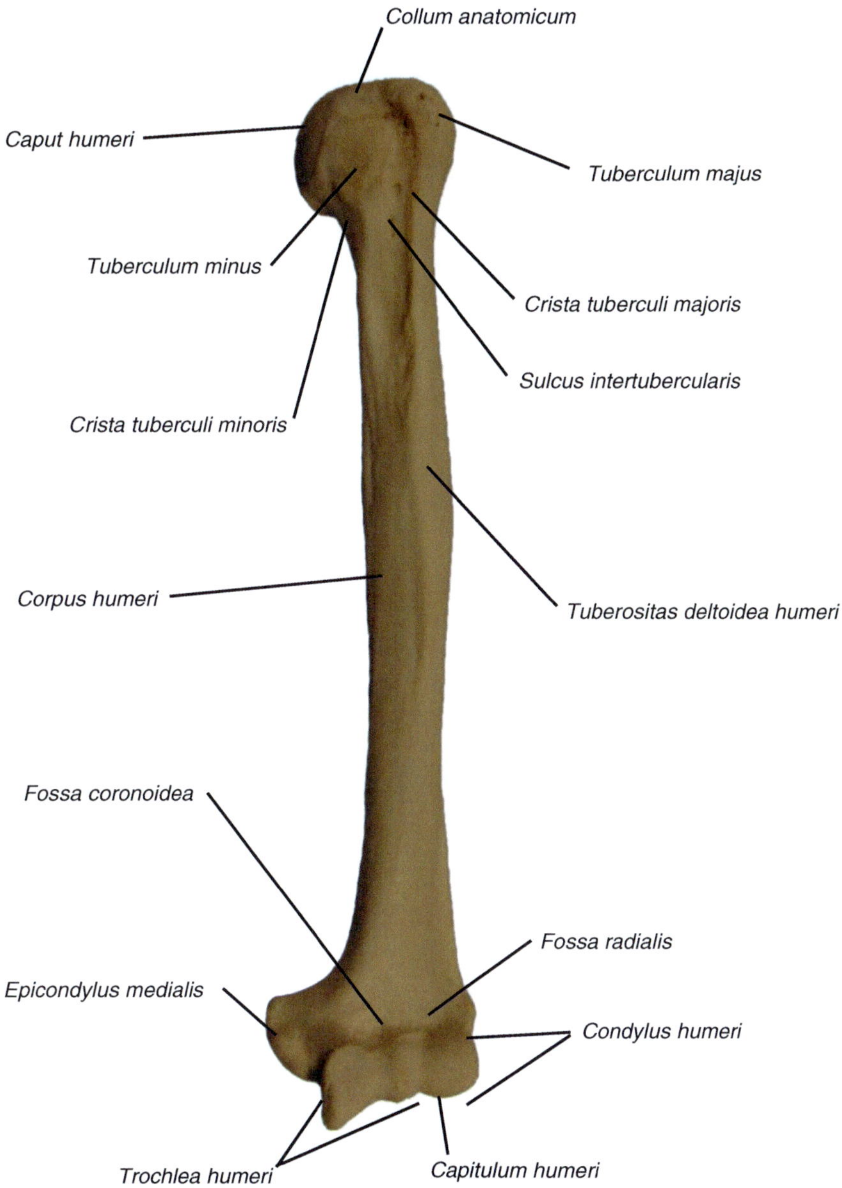


**Fig. 3.63** *Pan paniscus*: ventral (anterior in human anatomy) view of humerus

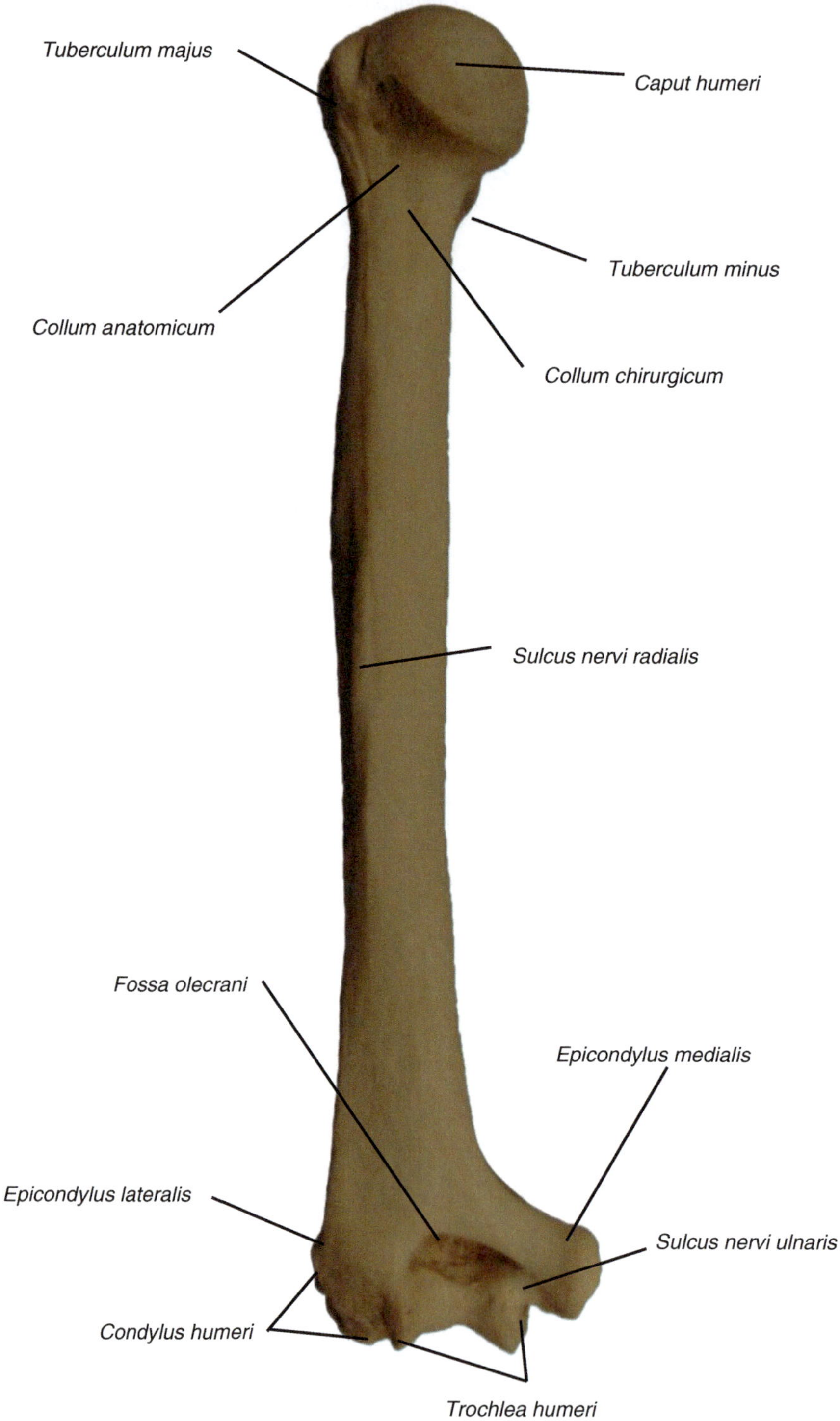


**Fig. 3.64** *Pan paniscus*: dorsal (posterior in human anatomy) view of humerus

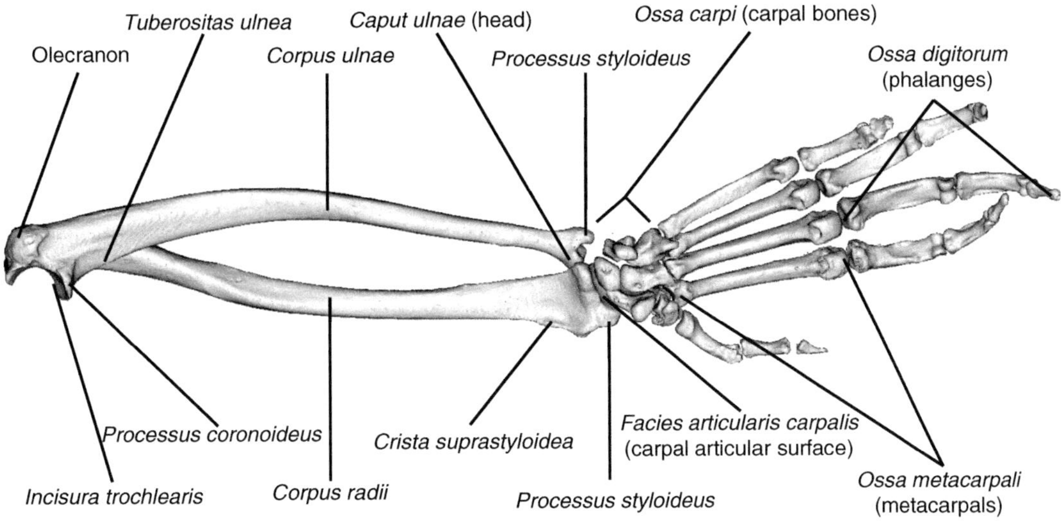


**Fig. 3.65** *Pan paniscus*: palmar view of forearm and hand

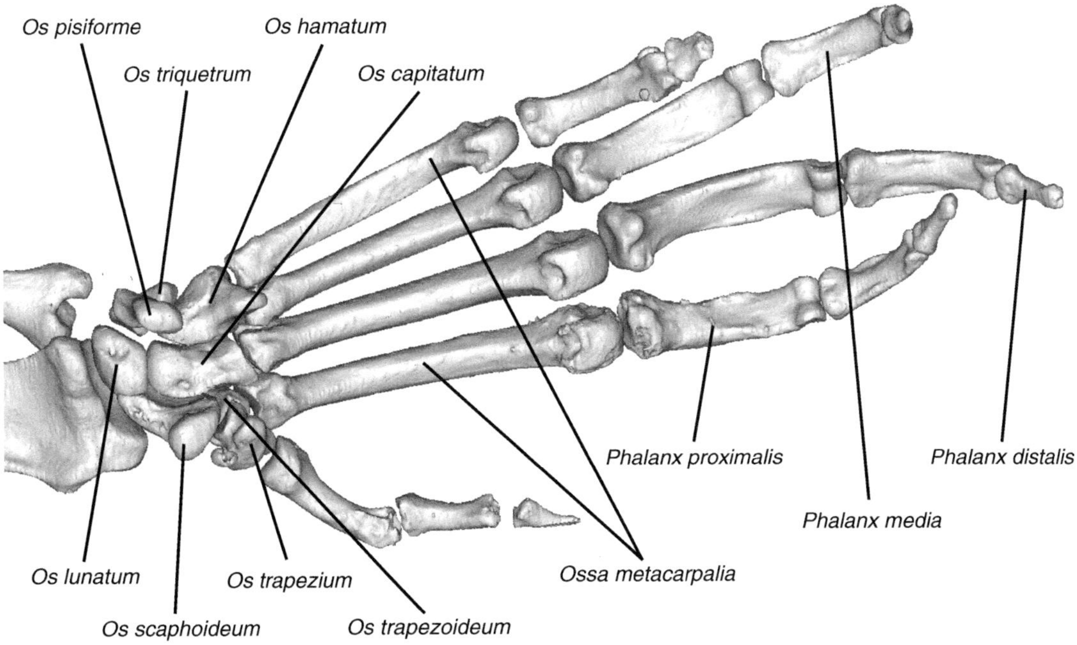


**Fig. 3.66** *Pan paniscus*: palmar view of hand

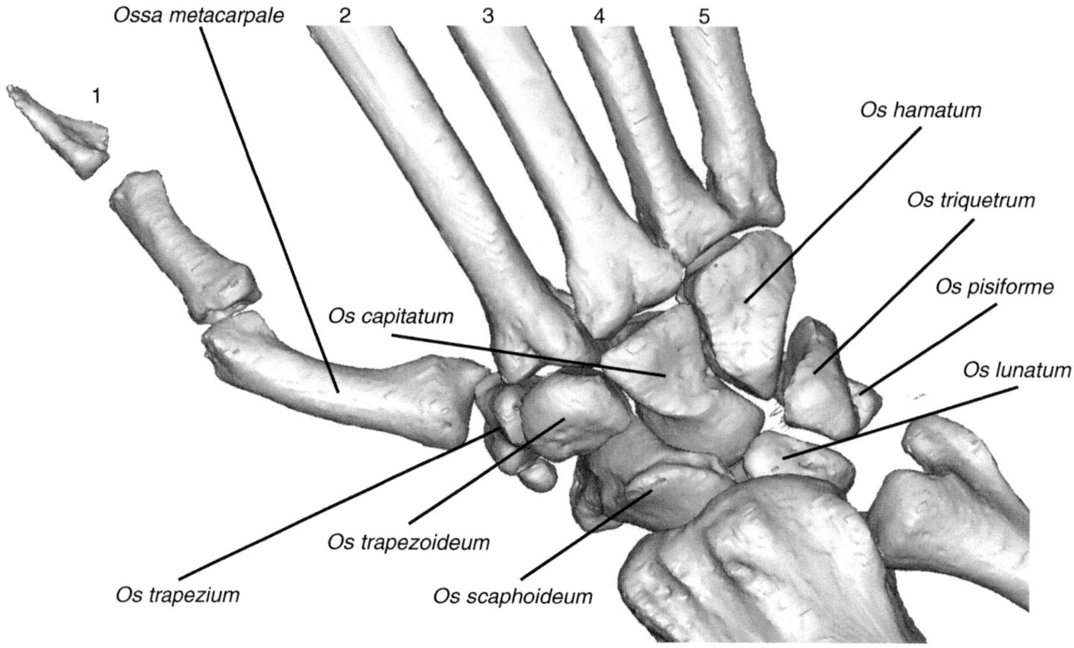


**Fig. 3.67** *Pan paniscus*: dorsal (posterior in human anatomy) view of hand

# Bibliography

## Appendix I: Literature Including Information about the Muscles of Bonobos and Common Chimpanzees


Abramowitz I (1955) On the existence of a palmar interosseous muscle in the thumb with particular reference to the Bantu-speaking Negro. South African J Sci 51:270–276

Aiello L, Dean C (1990) An introduction to human evolutionary anatomy. Academic Press, San Diego

Andrews P, Groves CP (1976) Gibbons and brachiation. In: Rumbaugh DM (ed) Gibbon and Siamang, vol 4. Karger, Basel, pp 167–218

Ashton EH, Oxnard CE (1963) The musculature of the primate shoulder. J Zool Soc Lond 29:553–650

Aziz MA, Dunlap SS (1986) The human extensor digitorum profundus muscle with comments on the evolution of the primate hand. Primates 27:293–319

Barnard WS (1875) Observations on the membral musculation of *Simia satyrus* (Orang) and the comparative myology of man and the apes. Proc Amer Assoc Adv Sci 24:112–144

Beddard FE (1893) Contributions to the anatomy of the anthropoid apes. Trans Zool Soc Lond 13:177–218

Bojsen-Møller F (1978) Extensor carpi radialis longus muscle and the evolution of the first intermetacarpal ligament. Am J Phys Anthropol 48:177–184

Bolk L (1902) Beiträge zur Affenanatomie, III, Der Plexus cervico-brachialis der Primaten. Petrus Campter 1:371–566

Broca P (1869) L'ordre des primates – parallele anatomique de l'homme et des singes. Bull Soc Anthropol Paris 4:228–401

Brooks HSJ (1886) On the morphology of the intrinsic muscles of the little finger, with some observations on the ulnar head of the short flexor of the thumb. J Anat Physiol 20:644–661

Brooks HSJ (1887) On the short muscles of the pollex and hallux of the anthropoid apes, with special reference to the opponens hallucis. J Anat Physiol 22:78–95

Carlson KJ (2006) Muscle architecture of the common chimpanzee (*Pan troglodytes*): perspectives for investigating chimpanzee behavior. Primates 47:218–229

Champneys F (1872) The muscles and nerves of a Chimpanzee (*Troglodytes Niger*) and a *Cynocephalus Anubis*. J Anat Physiol 6:176–211

Chapman HC (1879) On the structure of the chimpanzee. Proc Acad Nat Sci Phila 31:52–63

Day MH, Napier J (1963) The functional significance of the deep head of flexor pollicis brevis in primates. Folia Primatol 1:122–134

Deniker A, Boulart R (1885) Les sacs laryngiens, les excroissances adipeuses, les poumons, le cerveau, etc. des orang-utans. *Nouv Arch Mus Hist Nat, Ser. 3, 7,* 35–56.

Diogo R, Abdala V (2010) Muscles of vertebrates – comparative anatomy, evolution, homologies and development. Taylor & Francis, Oxford

Diogo R, Wood BA (2011) Soft-tissue anatomy of the primates: phylogenetic analyses based on the muscles of the head, neck, pectoral region and upper limb, with notes on the evolution of these muscles. J Anat 219:273–359

Diogo R, Wood BA (2012) Comparative anatomy and phylogeny of primate muscles and human evolution. Taylor & Francis, Oxford

Diogo R, Abdala V, Aziz MA, Lonergan N, Wood BA (2009) From fish to modern humans – comparative anatomy, homologies and evolution of the pectoral and forelimb musculature. J Anat 214:694–716

Duckworth WLH (1904) Studies from the anthropological laboratory, the Anatomy School, Cambridge. C. J. Clay & Sons, London

Dunlap SS, Thorington RW, Aziz MA (1985) Forelimb anatomy of New World monkeys: myology and the interpretation of primitive anthropoid models. Am J Phys Anthropol 68:499–517

Dylevsky I (1967) Contribution to the ontogenesis of the flexor digitorum superficialis and the flexor digitorum profundus in man. Folia Morphol (Praha) 15:330–335

Dwight T (1895) Notes on the dissection and brain of the chimpanzee 'Gumbo'. Mem Boston Soc Nat Hist 5:31–51

Fick R (1925) Beobachtungen an den muskeln einger schimpansen. Z Anat Entwicklungsgesch 76:117–141

Gibbs S (1999) Comparative soft tissue morphology of the extant Hominoidea, including Man. Unpublished PhD Thesis, The University of Liverpool, Liverpool

Glidden EM, De Garis CF (1936) Arteries of the chimpanzee (*Pan* spec. ?), I, the aortic arch, II, arteries of the upper extremity, III, the descending aorta, IV, arteries of the lower extremity. Am J Anat 58:501–527

Gratiolet LP, Alix PHE (1866) Recherches sur l'anatomie du *Troglodytes aubryi*. Nouv Arch Mus Hist Nat Paris 2:1–264

Grönroos H (1903) Die musculi biceps brachii und latissimocondyloideus bei der affengattung *Hylobates* im vergleich mit den ensprechenden gebilden der anthropoiden und des menschen. Abh Kön Preuss Akad Wiss Berlin 1903:1–102

Hartmann R (1886) Anthropoid apes. Keegan, London

Hepburn D (1892) The comparative anatomy of the muscles and nerves of the superior and inferior extremities of the anthropoid apes: I – myology of the superior extremity. J Anat Physiol 26:149–186

Howell AB, Straus WL (1932) The brachial flexor muscles in primates. Proc US Natl Mus 80:1–31

Humphry G (1867) On some points in the anatomy of the chimpanzee. J Anat Physiol 1:254–268

Huntington GS (1903) Present problems of myological research and the significance and classification of muscular variations. Am J Anat 2:157–175

Huxley TH (1864) The structure and classification of the Mammalia. Med Times Gazette 1864:398–468

Jouffroy FK (1971) Musculature des membres. In: Grassé PP (ed) Traité de Zoologie, XVI: 3 (Mammifères). Masson et Cie, Paris, pp 1–475

Jouffroy FK, Lessertisseur J (1959) Reflexions sur les muscles contracteurs des doigts et des orteils (contrahentes digitorum) chez les primates. Ann Sci Nat Zool, Ser 12(1):211–235

Jouffroy FK, Lessertisseur J (1960) Les spécialisations anatomiques de la main chez les singes à progression suspendue. Mammalia 24:93–151

Kaneff A (1979) Évolution morphologique des musculi extensores digitorum et abductor pollicis longus chez l'Homme. I. Introduction, méthodologie, M. extensor digitorum. Gegenbaurs Morphol Jahrb 125:818–873

Kaneff A (1980a) Évolution morphologique des musculi extensores digitorum et abductor pollicis longus chez l'Homme. II. Évolution morphologique des m. extensor digiti minimi, abductor pollicis longus, extensor pollicis brevis et extensor pollicis longus chez l'homme. Gegenbaurs Morphol Jahrb 126:594–630

Kaneff A (1980b) Évolution morphologique des musculi extensores digitorum et abductor pollicis longus chez l'Homme. III. Évolution morphologique du m. extensor indicis chez l'homme, conclusion générale sur l 'évolution morphologique des musculi extensores digitorum et abductor pollicis longus chez l'homme. Gegenbaurs Morphol Jahrb 126:774–815

Keith A (1894) Notes on a theory to account for the various arrangements of the flexor profundus digitorum in the hand and foot of primates. J Anat Physiol 28:335–339

Keith A (1899) On the chimpanzees and their relationship to the gorilla. Proc Zool Soc Lond 1899:296–312

Kikuchi Y (2010) Comparative analysis of muscle architecture in primate arm and forearm. Anat Histol Embryol 39:93–106

Koizumi M, Sakai T (1995) The nerve supply to coracobrachialis in apes. J Anat 186:395–403

Lander KF (1918) The pectoralis minor: a morphological study. J Anat 52:292–318

Landsmeer JM (1986) A comparison of fingers and hand in *Varanus*, opossum and primates. Acta Morphol Neerl Scand 24:193–221

Larson SG, Stern JT (1986) EMG of scapulohumeral muscles in the chimpanzee during reaching and 'arboreal' locomotion. Am J Anat 176:171–190

Larson SG, Stern JT, Jungers WL (1991) EMG of serratus anterior and trapezius in the chimpanzee: scapular rotators revisited. Am J Phys Anthropol 85:71–84

Lewis OJ (1989) Functional morphology of the evolving hand and foot. Clarendon Press, Oxford

Loth E (1931) Anthropologie des parties molles (muscles, intestins, vaisseaux, nerfs peripheriques). Mianowski-Masson et Cie, Paris

Macalister A (1871) On some points in the myology of the chimpanzee and others of the primates. Ann Mag Nat Hist 7:341–351

MacDowell EC (1910) Notes on the myology of *Anthropopithecus niger* and *Papio-thoth ibeanus*. Am J Anat 10:431–460

Miller RA (1932) Evolution of the pectoral girdle and forelimb in the primates. Amer J Phys Anthropol 17: 1–56

Miller RA (1952) The musculature of *Pan paniscus*. Am J Anat 91:182–232

Ogihara N, Kunai T, Nakatsukasa M (2005) Muscle dimensions in the chimpanzee hand. Primates 46:275–280

Oishi M, Ogihara N, Endo H, Ichihara N, Asari M (2009) Dimensions of forelimb muscles in orangutans and chimpanzees. J Anat 215:373–382

Owen R (1868) The anatomy of vertebrates, Mammals, vol 3. Longmans, Green & Co., London

Parsons FG (1898a) The muscles of mammals, with special relation to human myology, Lecture 1, The skin muscles and muscles of the head and neck. J Anat Physiol 32:428–450

Parsons FG (1898b) The muscles of mammals, with special relation to human myology: a course of lectures delivered at the Royal College of Surgeons of England – lecture II, the muscles of the shoulder and forelimb. J Anat Physiol 32:721–752

Payne RC (2001) Musculoskeletal adaptations for climbing in hominoids and their role as exaptations for the acquisition of bipedalism. Unpublished PhD thesis, The University of Liverpool, Liverpool

Pira A (1913) Beiträge Zür Anatomie des Gorilla, I, Das Extremitätenmuskelsystem. *Morphol Jahrb* 47;309–3.

Potau JM, Bardina X, Ciurana N, Camprubi D, Pastor JF, De Paz F, Barbosa M (2009) Quantitative analysis of the deltoid and rotator cuff muscles in humans and great Apes. Int J Primatol 30:697–708

Ribbing L, Hermansson K (1912) Kleinere muskelstudien, I11, die distale extremitatenmuskulatur eines schimpansen. Lunds Universitets Arsskrift NF Afd 2(8): 1–10

Sarmiento EE (1994) Terrestrial traits in the hands and feet of gorillas. *Am Mus Novit* 3091:1–56

Schück AC (1913a) Beiträge zur Myologie der Primaten, I – der m. lat. dorsi und der m. latissimo-tricipitalis. Morphol Jahrb 45:267–294

Schück AC (1913b) Beiträge zur Myologie der Primaten, II – 1 die gruppe sterno-cleido-mastoideus, trapezius, omo-cervicalis, 2 die gruppe levator scapulae, rhomboides, serratus anticus. Morphol Jahrb 46:355–418

Sonntag CF (1923) On the anatomy, physiology, and pathology of the chimpanzee. Proc Zool Soc Lond 23:323–429

Sonntag CF (1924) The morphology and evolution of the apes and man. John Bale Sons and Danielsson, Ltd., London

Stern JT, Larson SG (2001) Telemetered electromyography of the supinators and pronators of the forearm in gibbons and chimpanzees: implications for the fundamental positional adaptation of hominoids. Am J Phys Anthropol 115:253–268

Stern JT, Wells JP, Jungers WL, Vangor AK (1980) An electromyographic study of serratus anterior in atelines and *Alouatta*: implications for hominoid evolution. Am J Phys Anthropol 52:323–334

Stewart TD (1936) The musculature of the anthropoids, I, neck and trunk. Am J Phys Anthropol 21:141–204
Straus WL (1941a) The phylogeny of the human forearm extensors. Hum Biol 13:23–50
Straus WL (1941b) The phylogeny of the human forearm extensors (concluded). Hum Biol 13:203–238
Straus WL (1942a) The homologies of the forearm flexors: urodeles, lizards, mammals. Am J Anat 70:281–316
Straus WL (1942b) Rudimentary digits in primates. Q Rev Biol 17:228–243
Susman RL, Stern JT (1980) EMG of the interosseous and lumbrical muscles in the chimpanzee (*Pan troglodytes*) hand during locomotion. Am J Anat 157: 389–397
Susman RL, Nyati L, Jassal MS (1999) Observations on the pollical palmar interosseus muscle (of Henle). Anat Rec 254:159–165
Sutton JB (1883) On some points in the anatomy of the chimpanzee (*Anthropopithecus troglodytes*). J Anat Physiol 18:66–85
Swindler DR, Wood CD (1973) An atlas of primate gross anatomy: baboon, chimpanzee and men. University of Washington Press, Seattle
Testut L (1883) Le long fléchisseur propre du pouce chez l'homme et les singes. Bull Soc Zool Fr 8:164–185
Tocheri MW, Orr CM, Jacofsky MC, Marzke MW (2008) The evolutionary history of the hominin hand since the last common ancestor of *Pan* and *Homo*. J Anat 212:544–562
Tuttle RH (1969) Quantitative and functional studies on the hands of the Anthropoidea, I, the Hominoidea. J Morphol 128:309–363
Tuttle RH (1970) Postural, propulsive and prehensile capabilities in the cheiridia of chimpanzees and other great apes. In: Bourne GH (ed) The Chimpanzee, vol 2. Karger, Basel, pp 167–263
Tuttle RH, Basmajian JV (1976) Electromyography of pongid shoulder muscles and hominoid evolution I – retractors of the humerus and rotators of the scapula. Yearbook Phys Anthropol 20:491–497
Tuttle RH, Basmajian JV (1978a) Electromyography of pongid shoulder muscles II – deltoid, rhomboid and "rotator cuff". Am J Phys Anthropol 49:47–56
Tuttle RH, Basmajian JV (1978b) Electromyography of pongid shoulder muscles III – quadrupedal positional behavior. Am J Phys Anthropol 49:57–70
Tyson E (1699) Orang-Outang sive Homo sylvestris, or the anatomy of a pygmie compared to that of a monkey, an ape and a man. T. Bennett and D. Brown, London
Vrolik W (1841) Recherches d' anatornie comparé, sur le chimpanzé. Johannes Miller, Amsterdam
Wilder B (1862) Contributions to the comparative myology of the chimpanzee. Boston J Nat Hist 6:352–384
Wood Jones F (1920) The principles of anatomy as seen in the hand. J. & A Churchill, London
Wyman J (1855) An account of the dissection of a black chimpanzee. Proc Boston Soc Nat Hist 5:270–275
Ziegler AC (1964) Brachiating adaptations of chimpanzee upper limb musculature. Am J Phys Anthropol 22:15–32

## Appendix II: Literature Cited, Not Including Information About the Muscles of Bonobos and Common Chimpanzees

Terminologia Anatomica (1998) Federative committee on anatomical terminology. Georg Thieme, Stuttgart. New York: Columbia University Press

# 4 Trunk, Diaphragmatic, Perineal and Coccygeal Musculature

## 4.1 Trunk and Back Musculature

***Obliquus capitis inferior*** (adult Kidogo 2.61 g; Fig. 4.5)

- Attachments in bonobos: From spinous processes of C2 to transverse process of C1 (Miller's 1952 adult bonobo and our adult Kidogo).
- Usual innervation: Branch of the dorsal ramus of the first cervical nerve (Miller 1952: *P. paniscus*).
- Comparison with common chimpanzees: Similar to bonobos.

***Obliquus capitis superior*** (adult Kidogo 1.18 g; Fig. 4.5)

- Attachments in bonobos: From transverse process of C1 to occipital bone (Miller's 1952 adult bonobo and our adult Kidogo).
- Usual innervation: Branch of the dorsal ramus of the first cervical nerve (Miller 1952: *P. paniscus*).
- Comparison with common chimpanzees: Similar to bonobos.

***Rectus capitis anterior*** (adult Kidogo 0.42 g)

- Attachments in bonobos: Mainly from C1 to basiocciput (our adult Kidogo).
- Usual innervation: Data are not available.
- Comparison with common chimpanzees: In *P. troglodytes* rectus capitis anterior major runs from C3 to C6 to the basiocciput, receiving a slip from the scalenus anterior, while the rectus capitis anterior minor runs from the anteroventral surface of the lateral mass of C1 to the basiocciput, ventral to the foramen magnum and occipital condyle, and dorsolateral to the insertion of the longus capitis (Sonntag 1923, 1924; Dean 1985).

***Rectus capitis lateralis*** (adult Kidogo 0.75 g)

- Attachments in bonobos: From transverse process of C1 to jugular process of occipital bone (our adult Kidogo).
- Usual innervation: Data are not available.
- Comparison with common chimpanzees: Similar to bonobos.

***Rectus capitis posterior major*** (adult Kidogo 3.27 g; Fig. 4.5)

- Attachments in bonobos: From spinous process of C2 to occipital bone, between inferior nuchal line and foramen magnum, lateral to insertion of rectus capitis posterior minor (our adult Kidogo and Miller's 1952 adult bonobo).
- Usual innervation: Branches of the dorsal ramus of the first cervical nerve (Miller 1952: *P. paniscus*).



R. Diogo et al., *Photographic and Descriptive Musculoskeletal Atlas of Bonobos*,
DOI 10.1007/978-3-319-54106-8_4

- Comparison with common chimpanzees: Similar to bonobos.

***Rectus capitis posterior minor*** (adult Kidogo 3.00 g; Fig. 4.5)

- Attachments in bonobos: From dorsal tubercle process of C1 to medial region of inferior nuchal line (our adult Kidogo and Miller's 1952 adult bonobo).
- Usual innervation: Branch of the dorsal ramus of the first cervical nerve (Miller 1952: *P. paniscus*).
- Comparison with common chimpanzees: Similar to bonobos.

***Longus capitis*** (adult Kidogo 7.11 g; Fig. 4.1)

- Attachments in bonobos: From transverse processes of C4 to C7 (Miller's 1952 adult bonobo) or C2-C5 (our adult Kidogo), to basiocciput.
- Usual innervation: Branches of the ventral rami of the upper cervical nerves (Miller 1952: *P. paniscus*).
- Comparison with common chimpanzees: In *P. troglodytes* it runs from transverse processes of C4 to C7, to basiocciput (Dean 1985).

***Longus colli*** (adult Kidogo 9.59 g)

- Attachments in bonobos: Mainly from body of C1-T4 (medially) to transverse processes of C2-C7 (laterally) in our adult Kidogo. In Miller's 1952 adult bonobo it has three components: (1) the portion arising from the transverse processes of the 3rd to the 6th cervical vertebrae is directed obliquely medially to be inserted upon the anterior arch of the atlas; (2) another component arises from the bodies of the 5th cervical to the 3rd thoracic vertebrae, passes vertically upward, and is inserted upon the bodies of the 2nd to the 4th cervical vertebrae; (3) a third component arises from the bodies of the first three thoracic vertebrae, passing obliquely laterally to insert upon the transverse processes of the 5th and 6th cervical vertebrae.
- Usual innervation: Branches of the ventral rami of the cervical nerves (Miller 1952: *P. paniscus*).
- Comparison with common chimpanzees: In *P. troglodytes* the superior oblique division originates from the transverse processes of C3-C6, while the inferior oblique division and the vertical division originate from the body of C6-T3 or C6-T4; the muscle inserts onto C1, C5 and C6 (Sonntag 1923).

***Scalenus anterior*** (adult Kidogo 4.81 g; Fig. 4.1)

- Attachments in bonobos: From transverse processes of C4, C5 and C6, to first rib (our adult Kidogo and Miller's 1952 adult bonobo).
- Usual innervation: Lower cervical nerves (Miller 1952: *P. paniscus*) [*P. troglodytes*: C8 (Champneys 1872)].
- Comparison with common chimpanzees: In *P. troglodytes* it runs from transverse processes of C4, C5 and C6, and occasionally of C7 and/or C3, to the anterior scalene tubercle of the first rib (Champneys 1872; Sonntag 1923). Stewart (1936) stated that common chimpanzees and other apes usually have a scalenius anterior and a 'scalenus medius' (which corresponds to the scalenus medius + scalenus posterior sensu the present study). According to him, the scalenius anterior usually inserts onto the first rib in primates, its origin being limited to C5 and C6 in 'lower primates', but in hominoids the insertion included some of the upper cervical vertebrae. Also according to him, within primates the 'scalenus medius' (i.e., medius + posterior sensu the present study) shows a progressive decrease in the number of ribs supplying attachment, usually inserting in hominoids upon the first rib, but occasionally also onto the second rib.

***Scalenus medius*** (adult Kidogo 3.44 g; Fig. 4.1)

- Attachments in bonobos: The combined scalenus medius-scalenus posterior runs from C2-C7 to first rib with a tendinous expansion to second rib (Miller's 1952 adult bonobo), or has a unique insertion onto second rib (our adult Kidogo). That is, the *scalenus posterior* is not

present as a distinct muscle but is probably not completely missing, being instead fused with the scalenus medius. The *scalenus minimus* is usually not present as a distinct muscle in common chimpanzees, and was also not described in Miller's 1952 adult bonobo, but it was seemingly present in our adult Kidogo (having 0.87 g; Fig. 4.1), extending from the transverse process of C7 to the first rib.
- Usual innervation: Lower cervical nerves (Miller 1952: *P. paniscus*).
- Comparison with common chimpanzees: In *P. troglodytes* the combined scalenus medius-scalenus posterior runs from C2-C7 to the first and second ribs to first (Champneys 1872) or second (Macalister 1871) ribs, and its insertion may extend as an aponeurotic sheet to the fifth rib (Sonntag 1923). See notes about scalenus anterior above.

***Levatores costarum***

- Attachments in bonobos: From transverse processes of C7-T12 to costal angle of the next ribs (Miller's 1952 adult bonobo).
- Usual innervation: Branches of the intercostal nerves (Miller 1952: *P. paniscus*).
- Comparison with common chimpanzees: Similar to bonobos.

***Intercostales externi*** (Figs. 4.2, 4.3, 4.4 and 4.6)

- Attachments in bonobos: Attached to the adjacent margins of each pair of ribs, being membranous between the coastal cartilages and the sternum in Miller's 1952 adult bonobo, while in *P. troglodytes* they are replaced by membranous fibrous tissue in all but the first three and last two intercostal spaces (Sonntag 1923).
- Usual innervation: Intercostal nerves (Miller 1952: *P. paniscus*).
- Comparison with common chimpanzees: Similar to bonobos.

***Intercostales interni***

- Attachments in bonobos: Attached to the adjacent margins of each pair of ribs, being membranous between the angles of the ribs and the vertebral column.
- Usual innervation: Intercostal nerves (Miller 1952: *P. paniscus*).
- Comparison with common chimpanzees: Similar to bonobos.

***Transversus thoracis***

- Attachments in bonobos: From internal surface of xiphoid process and body of sternum to costal cartilages of second to seventh ribs (Miller's 1952 adult bonobo).
- Usual innervation: Intercostal nerves (Miller 1952: *P. paniscus*).
- Comparison with common chimpanzees: In *P. troglodytes* it runs from the internal surface of the xiphoid process and the body of the sternum (Sonntag 1923; Swindler and Wood 1973), the sternal origin reaching as far anteriorly as the first or second costal spaces (Swindler and Wood 1973), to the inferior border of the costal cartilages of the second to sixth ribs (Sonntag 1923; Swindler and Wood 1973).

***Splenius capitis*** (adult Kidogo 33.90 g; Figs. 4.1 and 4.2)

- Attachments in bonobos: From the spinous processes of the lower cervical and the first four thoracic vertebrae (Miller's 1952 adult bonobo), or of C2-T4 (our adult Kidogo), to mastoid process and occipital bone beneath the superior occipital line.
- Usual innervation: Dorsal rami of the cervical nerves (Miller 1952: *P. paniscus*).
- Comparison with common chimpanzees: In *P. troglodytes* it runs from the spinous processes of C5-T4 (e.g., Sonntag 1923) and sometimes of T5-T7 (e.g., Sonntag 1923), to mastoid process and the occipital bone beneath the superior occipital line (Sonntag 1923); occasionally there is a slip to the levator scapulae (Sonntag 1923).

***Splenius cervicis*** (Figs. 4.1 and 4.2)

- Attachments in bonobos: From T5, receiving a slip from the longissimus cervicis at the level

of C2, and inserting onto the transverse process of C1, in Miller's 1952 adult bonobo. From spinous processes of T5-T7 to transverse processes of C1 and C2 in our adult Kidogo.
- Usual innervation: Dorsal rami of the lower cervical nerves (Miller 1952: *P. paniscus*).
- Comparison with common chimpanzees: Similar to bonobos.

***Serratus posterior superior*** (Infant Foyo 0.33 g; adult Kidogo 4.24 g)

- Attachments in bonobos: From spinous processes of C6-T2 (our adolescent Jasiri) or C3-T2 (Miller's 1952 adult bonobo) or C4-T1 (our infant Foyo) or C5-T1 (our adult Kidogo) to ribs 2–5 (Miller's 1952 adult bonobo, our adolescent Jasiri) or 2–4 (our infant Foyo and our adult Kidogo).
- Usual innervation: Branches of the first four intercostal nerves (Miller 1952: *P. paniscus*).
- Comparison with common chimpanzees: In *P. troglodytes* it runs from the spinous processes of C7-T1 (Sonntag 1923) to ribs 2–5 (Macalister 1871; Sonntag 1923) and occasionally to rib 1 (Sonntag 1923). The iliocostalis, longissimus and spinalis form the *erector spinae*.

***Serratus posterior inferior*** (infant Foyo 0.83 g; adult Kidogo 17.65 g)

- Attachments in bonobos: From thoracolumbar fascia (Miller's 1952 adult bonobo, our infant Foyo and adult Kidogo) and vertebrae T8-T10 (our infant Foyo) or T9-L3 (our adult Kidogo), to the last four ribs (Miller's 1952 adult bonobo) or ribs 9–12 (our infant Foyo) or ribs 9–13 (our adult Kidogo).
- Usual innervation: Ventral rami of the ninth to twelfth thoracic spinal nerves (Miller 1952: *P. paniscus*).
- Comparison with common chimpanzees: In *P. troglodytes* it runs from the thoracolumbar fascia (Sonntag 1923) to the inferior four or five ribs, lateral to their angles (Sutton 1883; Sonntag 1923).

***Iliocostalis*** (adult Kidogo 65.11 g; Figs. 4.2 and 4.3)

- Attachments in bonobos: Mainly extending from medial 1/3 of iliac crest to all ribs and transverse processes of vertebrae in our adult Kidogo. In Miller's 1952 adult bonobo the iliocostalis lumborum runs from sacral crest, spinous processes of lumbar vertebrae, and medial iliac crest to angles of ribs 7–14; iliocostalis thoracis (dorsi) from angles of 6th to 14th ribs to transverse processes of C7-T1 vertebrae; iliocostalis cervicis can scarcely be distinguished as a distinct muscle, running from ribs 5–6 to merge with the longissimus cervicis.
- Usual innervation: Dorsal rami of the spinal nerves (Miller 1952: *P. paniscus*).
- Comparison with common chimpanzees: In *P. troglodytes* the costal attachments are onto all the ribs (Sonntag 1923; our specimen VU PT2), the muscle being fused with the longissimus (Sonntag 1923); in our *P. troglodytes* specimen VU PT2, the muscle attaches to the iliac crest and some cervical vertebrae, being differentiated into a iliocostalis lomborum, iliocostalis thoracis and iliocostalis cervicis. The iliocostalis, longissimus and spinalis form the *erector spinae*.

***Longissimus*** (adult Kidogo 149.00 g; Figs. 4.2, 4.3, 4.4 and 4.6)

- Attachments in bonobos: Extending from medial 1/3 of iliac crest and sacrum to angle of all ribs, the transverse processes of C2-C5 and the mastoid process, in our adult Kidogo. In Miller's 1952 adult bonobo, the longissimus capitis runs from the transverse processes of C3-T1 to the occiput and mastoid process, being blended with the longissimus cervicis, which runs from the transverse processes of T1-T6, receiving a contribution from the iliocostalis cervicis at the level of T5, to the transverse processes of C2-C6 and blended with the splenius cervicis and longissimus capitis; the longissimus thoracis (dorsi) runs from the transverse processes of the lumbar vertebrae in common with the origin of the iliocostalis, to the angles of ribs 4–14.

- Usual innervation: Branches of the dorsal rami of the spinal nerves (Miller 1952: *P. paniscus*).
- Comparison with common chimpanzees: In our *P. troglodytes* specimen VU PT2 the muscle mainly runs from iliac crest, sacrum and adjacent structures, to the ribs (longissimus thoracis), the cervical vertebrae (longissimus cervicis) and the skull (longissimus capitis). The cranial insertion of the longissimus in *P. troglodytes* is to the occiput according to Sonntag (1923); the cervical insertion is onto the transverse processes of C3 to C5 (Sonntag 1923); in the thoracic region, longissimus inserts onto the transverse processes of the thoracic vertebrae according to Sonntag (1923) and inserts onto the ribs between the costal angle and the transverse processes of the thoracic vertebrae, the extent of the costal insertion being from the fourth rib to the last rib according to this author. The iliocostalis, longissimus and spinalis form the *erector spinae* (see Fig. 23). In common chimpanzees (e.g., Gratiolet and Alix 1866), modern humans, gorillas and a few other primates the *atlantomastoideus* might be occasionally present as a distinct muscle, running from the atlas to the mastoid process (Diogo and Wood 2012).

***Spinalis*** (adult Kidogo spinalis thoracis 125.22 g, spinalis cervicis 26.67 g; Figs. 4.2 and 4.3)

- Attachments in bonobos: In our adult Kidogo the spinalis thoracis seemingly connected the spinous processes from T2 to the last lumbar vertebra, while the spinalis cervicis connects the spinous processes from C2 to T4. In Miller's 1952 adult bonobo the spinalis thoracis (dorsi) runs from the spinous processes of L1-L3 and the last three thoracic vertebrae, to the spinous processes of C7-T7.
- Usual innervation: Branches from the dorsal rami of the spinal nerves (Miller 1952: *P. paniscus*).
- Comparison with common chimpanzees: In our *P. troglodytes* specimen VU PT2 the muscle mainly connects the spinous processes of different vertebrae; in specimens dissected by other authors the muscle originates from the spinous processes of T11 and T12, the insertion being to the spinous processes of T1 to T8 (Sonntag 1923). The iliocostalis, longissimus and spinalis form the *erector spinae* (see Fig. 4.3).

***Semispinalis thoracis***

- Attachments in bonobos: The origin of the muscle is not separable from the origin of semispinalis cervicis in common chimpanzees (Sonntag 1923; Winckler 1947; see semispinalis cervicis below), and was not described in Miller's 1952 adult bonobo; its insertion in *P. troglodytes* may reach as far posteriorly as T8 (Winckler 1947).
- Usual innervation: Dorsal rami of the cervical nerves (Miller 1952: *P. paniscus*).
- Comparison with common chimpanzees: See above, and semispinalis cervicis below.

***Semispinalis cervicis*** (Figs. 4.3, 4.4 and 4.6)

- Attachments in bonobos: The combined semispinalis thoracis and cervicis runs from transverse processes of T1-T4 to spinous processes or C1-C6.
- Usual innervation: Dorsal rami of the cervical nerves (Miller 1952: *P. paniscus*).
- Comparison with common chimpanzees: In *P. troglodytes* it runs from the transverse processes of T1 to T4 (Sonntag 1923), sometimes extending to T10 and also taking origin from the spinalis muscles (Winckler 1947). The combined origin of semispinalis thoracis and cervicis in common chimpanzees involves the transverse processes of all the thoracic vertebrae and the articular processes of C2 to C7 according to Sonntag (1923); Winckler (1947) puts the division between thoracis and cervicis at the level of its seventh fascicle, which arises from T7 to T10 and inserts onto T1; the insertion of the semispinalis cervicis is onto the

spinous processes of C1 to C6 (Sonntag 1923), reaching posteriorly as far as T1 according to Winckler (1947).

***Semispinalis capitis*** (adult Kidogo 25.05 g; Figs. 4.3, 4.4, 4.5 and 4.6)

- Attachments in bonobos: From transverse processes of C5-T4 (Miller's 1952 adult bonobo) or of C3-T5 (our adult Kidogo) to occipital bone between the superior and inferior nuchal lines.
- Usual innervation: Dorsal rami of the cervical nerves (Miller 1952: *P. paniscus*).
- Comparison with common chimpanzees: In *P. troglodytes* origin often extends cranially to C3 (Sonntag 1923, Winckler 1947) and posteriorly to T6 (Sonntag 1923) or T7 (Winckler 1947), and the muscle often fuses with the longissimus thoracis and the spinalis thoracis at its origin (Sonntag 1923), its insertion being onto the occipital bone between the superior and inferior nuchal lines.

***Multifidus*** (adult Kidogo 114.06 g)

- Attachments in bonobos: Mainly from sacrum to C2 in our adult Kidogo. In Miller's 1952 adult bonobo it runs from sacrum, posterior superior iliac spine, transverse processes of lumbar, thoracic and caudal five cervical vertebrae, to spinous processes of the lumbar and thoracic vertebrae of the cervical vertebrae as far as, and including, C1.
- Usual innervation: Dorsal rami of the spinal nerves (Miller 1952: *P. paniscus*).
- Comparison with common chimpanzees: In *P. troglodytes* origin is from the sacrum and sacroiliac ligaments, the mammillary processes of the lumbar vertebrae, the transverse processes of the thoracic vertebrae and the articular processes of C7 to C4 (Sonntag 1923, Winckler 1947), sometimes extending anteriorly to C2 (Winckler 1947). Insertion is onto the lamina and the spinous processes of the lumbar and thoracic vertebrae and to the cervical vertebrae as far as, and including, C2 (Sonntag 1923; Winckler 1947).

***Rotatores***

- Attachments in bonobos: From transverse processes of thoracic vertebrae to lamina and base of spinous process of vertebra above.
- Usual innervation: Data are not available.
- Comparison with common chimpanzees: Similar to bonobos (the 'rotatores breves' and 'rotatores longi' being seemingly absent in common chimpanzees; see Sonntag 1923).
- Synonymy: Rotatores breves and longi (Gibbs 1999).

***Interspinales***

- Attachments in bonobos: To our knowledge, there are no detailed descriptions of these muscles in *Pan*, the only mention to them being Sonntag (1923), who stated that in common chimpanzees they are similar to those of modern humans.
- Comparison with common chimpanzees: Similar to bonobos.
- Usual innervation: Data are not available.

***Intertransversarii***

- Attachments in bonobos: To our knowledge, there are no detailed descriptions of these muscles in *Pan*, the only mention to them being Sonntag (1923), who stated that in common chimpanzees they are similar to those of modern humans.
- Comparison with common chimpanzees: Similar to bonobos.
- Usual innervation: Data are not available.

## 4.2 Diaphragmatic and Abdominal Musculature

***Diaphragma***

- Attachments in bonobos: See notes below.
- Usual innervation: Phrenic nerve (Miller 1952: *P. paniscus*).
- Comparison with common chimpanzees: The position of the aortic aperture is higher

than in lower catarrhines and is shorter by two-thirds of a vertebra in modern humans than in *Pan* (Juraniec 1972). The hiatus extends posteriorly on average to the level of the upper part of L3, and anteriorly to T11 or T12 (Juraniec 1972). The division of the central tendon into folia is as distinct in *Pan* as it is in modern humans. In 1/3 *Pan* the central tendon is not divided into leaflets, being kidney-shaped (Juraniec and Szostakiewicz-Sawicka 1968). In 2/3 *Pan* the overall shape of the central tendon is similar to that of modern humans, with lateral and anterior leaflets of approximately the same size (Juraniec & Szostakiewicz-Sawicka 1968). The central point of decussation is present (Blair 1923), and muscle fibres are found in the anterior leaflet in 1/3 *Pan* (Juraniec and Szostakiewicz-Sawicka 1968). The oesophageal hiatus is situated in the muscular part of the diaphragm at the level of T11, and the middle of T10 (Juraniec 1972). It is elliptical in shape (Juraniec 1972), and is formed by the splitting of the medial fibres of the right crus in common chimpanzees (Juraniec 1972; Körner 1929; Low 1907) and in bonobos (Miller 1952). The oesophageal and aortic hiatuses are separated by half a vertebra (Juraniec 1972). Juraniec (1972) distinguished four types of fibres crossing at the oesophageal hiatus in primates (in type I, only muscular crossing is present, with 6 subtypes distinguished). Type I4, I5 and I6 represent various types of crossing of the fibres of the crura to the opposite side in bundles. These account for 1/4 *Pan*. In type III, not found in modern humans, a tendinous strand at the site of the crossing of the fibres is present on the right side of a *P. troglodytes* specimen dissected by Juraniec (1972) (subtype III2). The pars costalis of the diaphragm originates from the lower seven ribs and interdigitates with the transverses abdominis in common chimpanzees (Sonntag 1923) and bonobos (Miller 1952). The pars lumbaris originates from the lumbar vertebrae by two crura in common chimpanzees (Sonntag 1923) and bonobos (Miller 1952); additional slips arise from the transverse process of L2 and from the side of the body or L1 in common chimpanzees (Sonntag 1923). This pars lumbaris also originates from two medial and lateral arcuate ligaments and is divided into a right crus arising from the bodies and intervertebral discs as far posteriorly as L2 in common chimpanzees (Sonntag 1923) and bonobos (Miller 1952), and a left crus arising from L1 and L2 in *P. paniscus* (Miller 1952) and from L1 in *P. troglodytes* (Sonntag 1923). The lateral arcuate ligament is a thickened band in the quadrates lumborum fascia that attaches to the transverse process of L1 in bonobos (Miller 1952). It also attaches to the inferior margin to the last two ribs (ribs thirteen and fourteen) in bonobos (Miller 1952). The medial arcuate ligament is a tendinous arch in the psoas major fascia and attaches to the side of the body of L1, being laterally fixed to the front of the transverse process of L1 in bonobos (Miller 1952). The pars sternalis originates by two slips from the inner surface of the xiphoid process in *P. paniscus* (Miller 1952), while in *P. troglodytes* it takes origin by two slips from the inner surface of the sternum (Sonntag 1923).

***Rectus abdominis*** (Fig. 4.7)

- Attachments in bonobos: From outer surface of costal cartilages of ribs 5–7 to pubic crest and symphysis; there are four tendinous intersections, of which two are caudal to the umbilicus (Miller's 1952 adult bonobo).
- Usual innervation: Lower intercostals nerves in *Pan paniscus* Miller 1952.
- Comparison with common chimpanzees: In common chimpanzees there are four tendinous intersections (Champneys 1872; Sonntag 1923); the superficial rectus sheath is formed by the external and internal oblique, and the deep sheath by the transversus abdominis alone (Champneys 1872; Lunn 1949). To our knowledge, the presence of a *supracostalis* and *tensor linea semilunaris* has not been reported in common chimpanzees.

***Pyramidalis***

- Attachments in bonobos: It was present on the left side of the adult bonobo reported by Miller (1952) – it ran from the pubic crest between the rectus abdominis and the rectus sheath, radiating into the linea alba – but it was absent on the right side.
- Usual innervation: Twelfth thoracic nerve (Miller 1952: *P. paniscus*).
- Comparison with common chimpanzees: In her review of the literature, Gibbs (1999) suggested that the pyramidalis is probably usually missing in common chimpanzees, being absent in the *P. troglodytes* specimens reported by Champneys (1872), Sonntag (1923), Vrolik (1841). In the single *P. troglodytes* in which we analysed this feature in detail (VU PT1), the muscle was present as a small and triangular structure that lies superficial (ventral) to the rectus abdominis running from the pelvic region to the region of the linea alba. This muscle is present in about 20% of modern humans, which in general display a configuration similar to that found in our *P. troglodytes* specimen.

***Cremaster***

- Attachments in bonobos: A thin layer of cremaster fibres from the obliquas internus abdominis enclose the round ligament as it emerges from the subcutaneous ring (Miller's 1952 adult bonobo).
- Usual innervation: Data are not available.
- Comparison with common chimpanzees: In common chimpanzees it runs from the internal abdominal oblique to the 'inguinal ligament', also receiving a contribution from the transversus abdominis (Mijsberg 1923; Miller 1947, 1952; Lunn 1949). According to the literature review of Gibbs (1999), there is no true inguinal ligament in any ape, only a series of tendinous arches over the sartorius and the femoral vessels and nerves, merging with the fascia lata (Glidden and De Garis 1936; Miller 1947, 1952; Lunn 1948). Schultz (1936), Miller (1947) and Lunn (1948) suggest that a true inguinal ligament is only found in modern humans, but ligament-like connective tissue in a plane ventral to the pelvic girdle has been identified in all of the mammals dissected by Lunn (1948).

***Obliquus externus abdominis*** (Fig. 4.7)

- Attachments in bonobos: Mainly from last seven ribs to iliac crest and to linea alba and crest of pubis (Miller's 1952 adult bonobo).
- Usual innervation: Branches of the lower intercostal nerves and also by the iliohypogastric and ilioinguinal nerves (Miller 1952: *P. paniscus*).
- Comparison with common chimpanzees: In *P. troglodytes* it originates by fleshy slips from the external surface of the sixth to eleventh ribs (Champneys 1872; Sonntag 1923; Winckler 1950). The anterior extent of the origin is also variable in *P. troglodytes* (Champneys 1872; Sonntag 1923; Winckler 1950); the anterior part of the muscle interdigitates with slips of the serratus anterior and the posterior part with slips of the latissimus dorsi (Champneys 1872; Sonntag 1923), while its medial part interdigitates with the pectoralis major (Sonntag 1923). Insertion of the fibres in *P. troglodytes* from the last ribs is onto the iliac crest and its lateral lip, as far as the anterior superior iliac spine (Champneys 1872; Kohlbrügge 1897; Sonntag 1923; Waterman 1929); from the anterior superior iliac spine to the pubic tubercle, the muscle has a free posterior aponeurotic border (Kohlbrügge 1897; Sonntag 1923; Lunn 1949).

***Obliquus internus abdominis***

- Attachments in bonobos: From iliac crest and deep layer of thoracolumbar fascia and aponeurosis of the external oblique, or 'inguinal ligament'; the cranial part of the muscle inserts onto the inferior margins of the cartilage of the last four ribs, while the middle fibres become aponeurotic medially, forming the anterior wall of the rectus sheath (Miller's 1952 adult bonobo).
- Usual innervation: Branches of the lower intercostal nerves and the first lumbar nerve (Miller 1952: *P. paniscus*).

- Comparison with common chimpanzees: In *P. troglodytes* its origin includes the aponeurosis of the external oblique, or the 'inguinal ligament' (Sonntag 1923; Miller 1947; Lunn 1949). The cranial part of the muscle inserts onto the inferior margins of the cartilage of the last four ribs (Sonntag 1923); the middle fibres become aponeurotic medially, forming the anterior wall of the rectus sheath (Champneys 1872; Lunn 1949).

***Transversus abdominis***

- Attachments in bonobos: From last six ribs with the diaphragm, and from lumbodorsal fascia, iliac crest and obliquus externus, merging with other abdominal muscles and inserting also upon the crest of the pubis (Miller's 1952 adult bonobo).
- Usual innervation: Lower intercostal nerves and first lumbar nerve (Miller 1952: *P. paniscus*).
- Comparison with common chimpanzees: In *P. troglodytes* this muscle also forms the posterior layer of the rectus sheath (Champneys 1872; Lunn 1949); its fibres decussate in the linea alba, and the aponeurotic part forms, along with the internal oblique, the conjoined tendon, which inserts into the superior pubic surface in the region of the pubic crest (Sonntag 1923; Miller 1947; Lunn 1949).

## 4.3 Perineal, Coccygeal and Anal Musculature

***Coccygeus***

- Attachments in bonobos: From sacrosciatic ligament to coccyx, being blended with the levator ani (Miller's 1952 adult bonobo).
- Usual innervation: Third sacral nerve (Champneys 1872: *P. troglodytes*); fourth and fifth sacral nerves (Miller 1952: *P. paniscus*).
- Comparison with common chimpanzees: In *P. troglodytes* the muscle is mostly tendinous with few muscle fibres inserting onto the side of the coccyx, its posterior border being continuous with the pubococcygeus (Elftman 1932), or its insertion with the coccygeal part of the origin of the gluteus maximus (Champneys 1872). According to Elftman (1932) in common chimpanzees there is no trace of the *flexor caudae* and according to Elftman (1932) and Sonntag (1923) the *iliococcygeus* is also usually absent.

***Levator ani***

- Attachments in bonobos: From obturator fascia on the inner wall of the lesser pelvis that extends from the pubic symphysis to the ischial ramus, the posteriorly directed fibres converging in the midline; the more anterior fibres encircle the rectum and the more posterior ones insert onto the anococcygeal raphe behind the anus as well as onto the tip of the coccyx (Miller's 1952 adult bonobo).
- Usual innervation: Branches of the rectal nerve from the fourth and fifth sacral nerves (Miller 1952: *P. paniscus*).
- Comparison with common chimpanzees: In *P. troglodytes* the muscle is formed by bilateral 'plates' of muscle and fibrous tissue that have a linear attachment to the obturator fascia on the inner wall of the lesser pelvis that extends from the pubic symphysis to the ischial ramus (Thompson 1901; Sonntag 1923); the posteriorly directed fibres converge in the midline; the more anterior fibres encircle the rectum and the more posterior ones insert onto the anococcygeal raphe behind the anus as well as onto the tip of the coccyx (Sonntag 1923). To our knowledge, the *pubovesicalis* ('ligamentum puboprostaticum' or 'puboampullaris': Gibbs 1999) has never been reported in common chimpanzees.

***Pubococcygeus***

- Attachments in bonobos: Data are not available.
- Usual innervation: Data are not available.
- Comparison with common chimpanzees: In *P. troglodytes* origin is from the posterior surface of the pubic body (Thompson 1901; Elftman 1932), extending to the ischial spine (Smith 1923; Elftman 1932) and also originating from

the obturator fascia (Thompson 1901; Van den Broek 1908); insertion is onto the rectal wall (Thompson 1901; variably in *Pan* according to Elftman 1932) as well as onto the tip of the coccyx (Smith 1923; Elftman 1932), the anococcygeal raphe (Elftman 1932) and/or the sacrum (Smith 1923); the pubococcygeus muscles of the two sides unite at the midline (Thompson 1901; Smith 1923; Elftman 1932). According to Elftman (1932) and Thompson (1901) in great apes the *puborectalis* is homologous with the inferior fibres of the pubococcygeus.

***Sphincter ani externus***

- Attachments in bonobos: The sphincter encircles the anus, extending from the anococcygeal raphe to the perineal body, its superficial fibres attaching to the skin around the anus ((Miller's 1952 adult bonobo).
- Usual innervation: Branches of the fourth and fifth sacral nerves (Miller 1952: *P. paniscus*).
- Comparison with common chimpanzees: In *P. troglodytes* the sphincter encircles the anus, extending from the anococcygeal raphe (Sonntag 1923) to the perineal body (Elftman 1932) and forming a muscular basin (less marked degree in *Pan* than in *Gorilla* and Pongo) for the support of both alimentary and urogenital viscera according to Elftman 1932; some deeper fibres blend with the inferior margin of the levator ani (Sonntag 1923; Elftman 1932).

***Bulbospongiosus***

- Attachments in bonobos: Data are not available.
- Usual innervation: Data are not available.
- Comparison with common chimpanzees: In *P. troglodytes* some fibres of the bulbospongiosus that lie in contact with the pelvic diaphragm and insert into the ischium in chimpanzees may be homologous with the *transversus perinei profundus* according to Elftman (1932). According to Sonntag (1923) the *transversus perinei superficialis* is absent in *Pan*.
- Synonymy: Bulbocavernosus (Gibbs 1999).

***Ischiocavernosus***

- Attachments in bonobos: Data are not available.
- Usual innervation: Data are not available.
- Comparison with common chimpanzees: In *P. troglodytes* its origin is from the ascending pubic ramus (Sonntag 1923); no information about insertion.

***Sphincter urethrae***

- Attachments in bonobos: Data are not available.
- Usual innervation: Data are not available.
- Comparison with common chimpanzees: In female *P. troglodytes*, its fibres interlace in the vaginal wall (Roberts and Seibold 1971).

***Rectococcygeus***

- Attachments in bonobos: To our knowledge there are no detailed descriptions of this muscle in chimpanzees.
- Usual innervation: Data are not available.
- Synonymy: Caudoanalis (Gibbs 1999).

***Rectourethralis***

- Attachments in bonobos: To our knowledge there are no detailed descriptions of this muscle in chimpanzees.
- Usual innervation: Data are not available.
- Comparison with common chimpanzees: To our knowledge the muscles *rectovesicalis*, *regionis analis*, *regionis urogenitalis*, *compressor urethrae*, *sphincter urethrovaginalis*, *sphincter pyloricus*, *suspensori duodeni*, *sphincter ani internus*, *sphincter ductus choledochi*, *sphincter ampullae*, *detrusor vesicae*, *trigoni vesicae*, *vesicoprostaticus*, *vesicovaginalis*, *puboprostaticus* and *rectouterinus* have not been described in detail in chimpanzees. We were unable to examine these muscles in detail in the chimpanzees dissected by us.

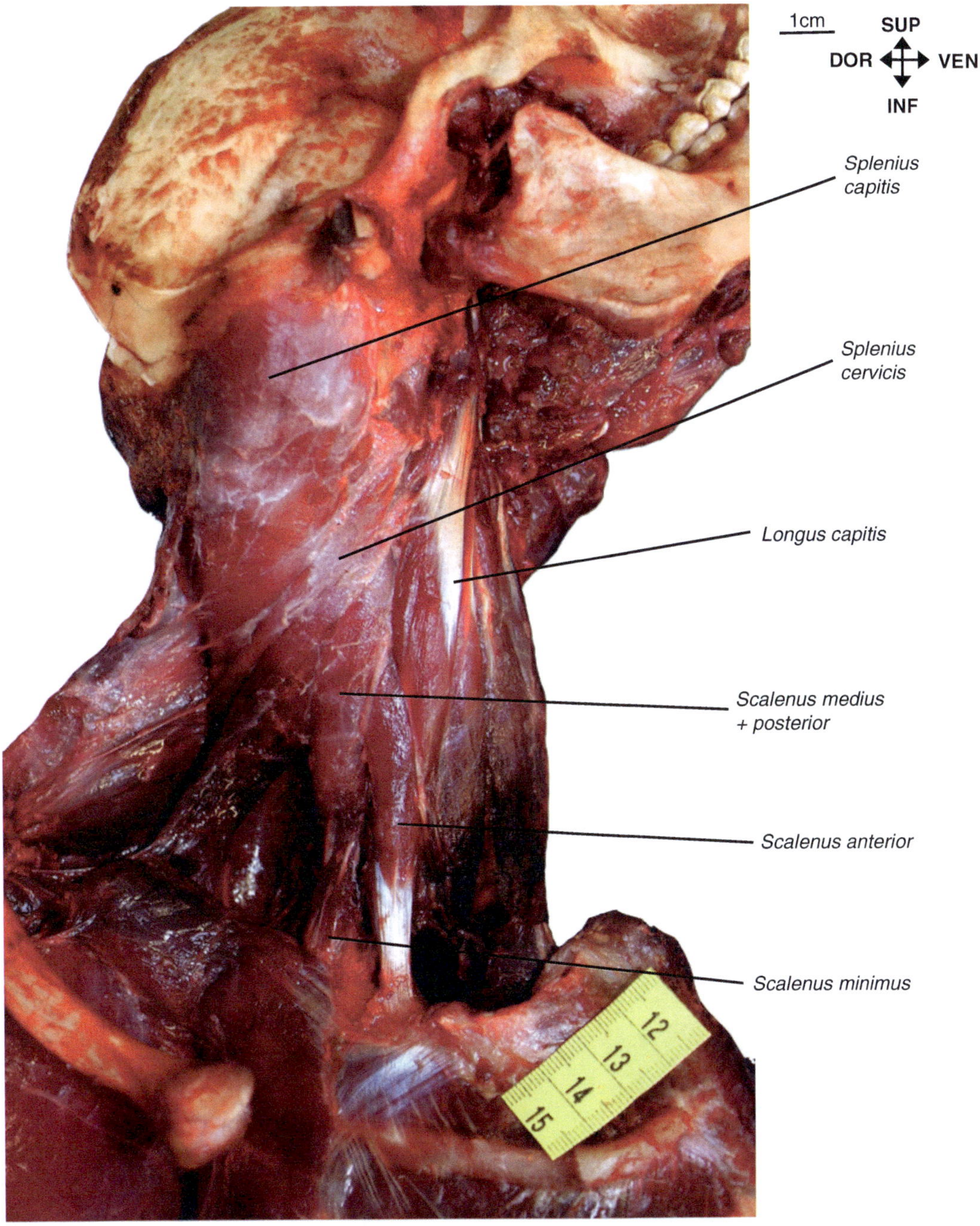


**Fig. 4.1** *Pan paniscus* (ZIMS 164031, Kidogo, adult male): ventrolateral view of the deep muscles of the neck, showing the *scalenus minimus* and the absence of a distinct, separate *scalenus posterior* muscle

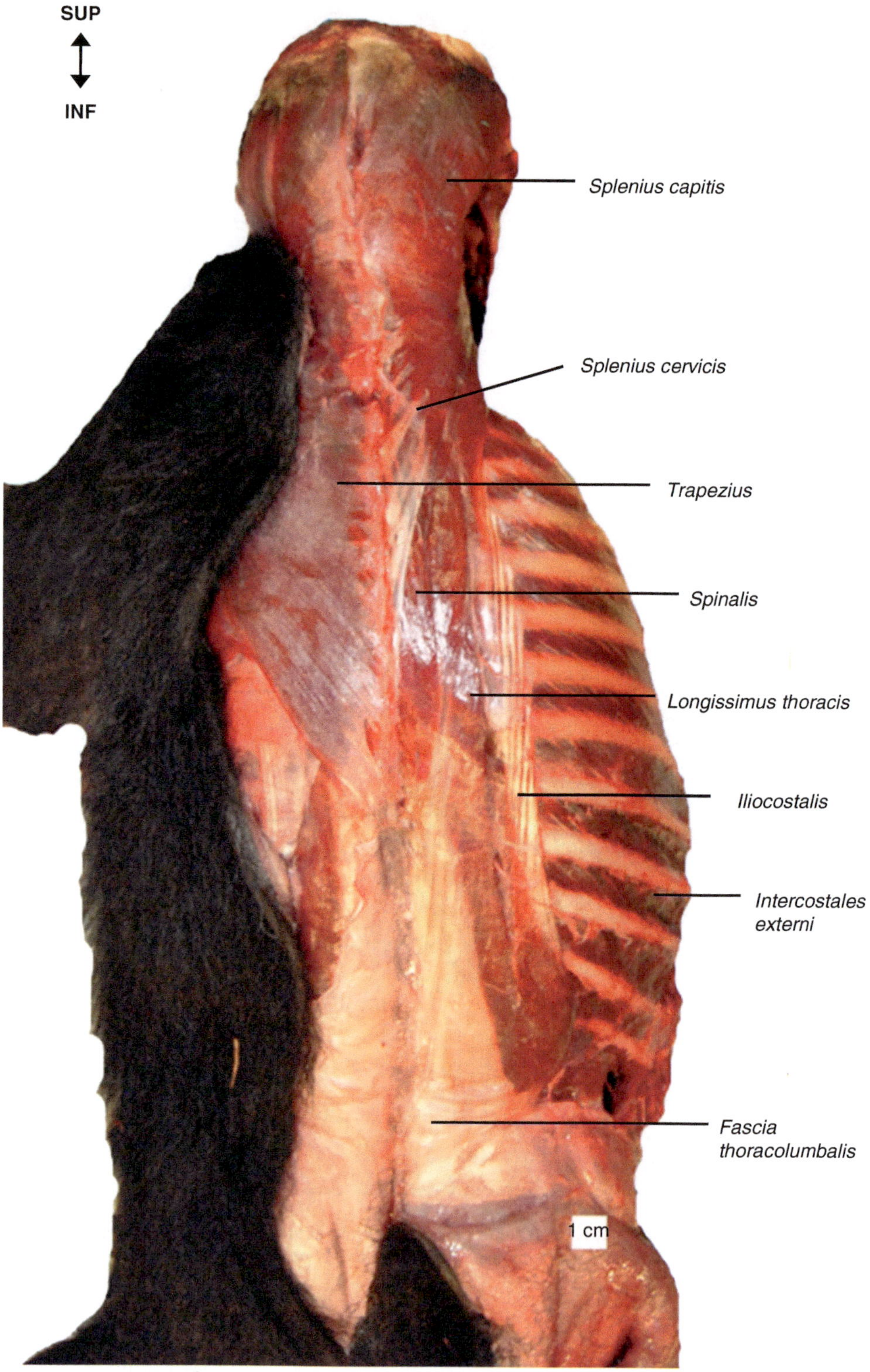


**Fig. 4.2** *Pan paniscus* (ZIMS 164031, Kidogo, adult male): dorsal view of the back muscles, after removal of *trapezius*, latissimus dorsi and rhomboideus on the right side

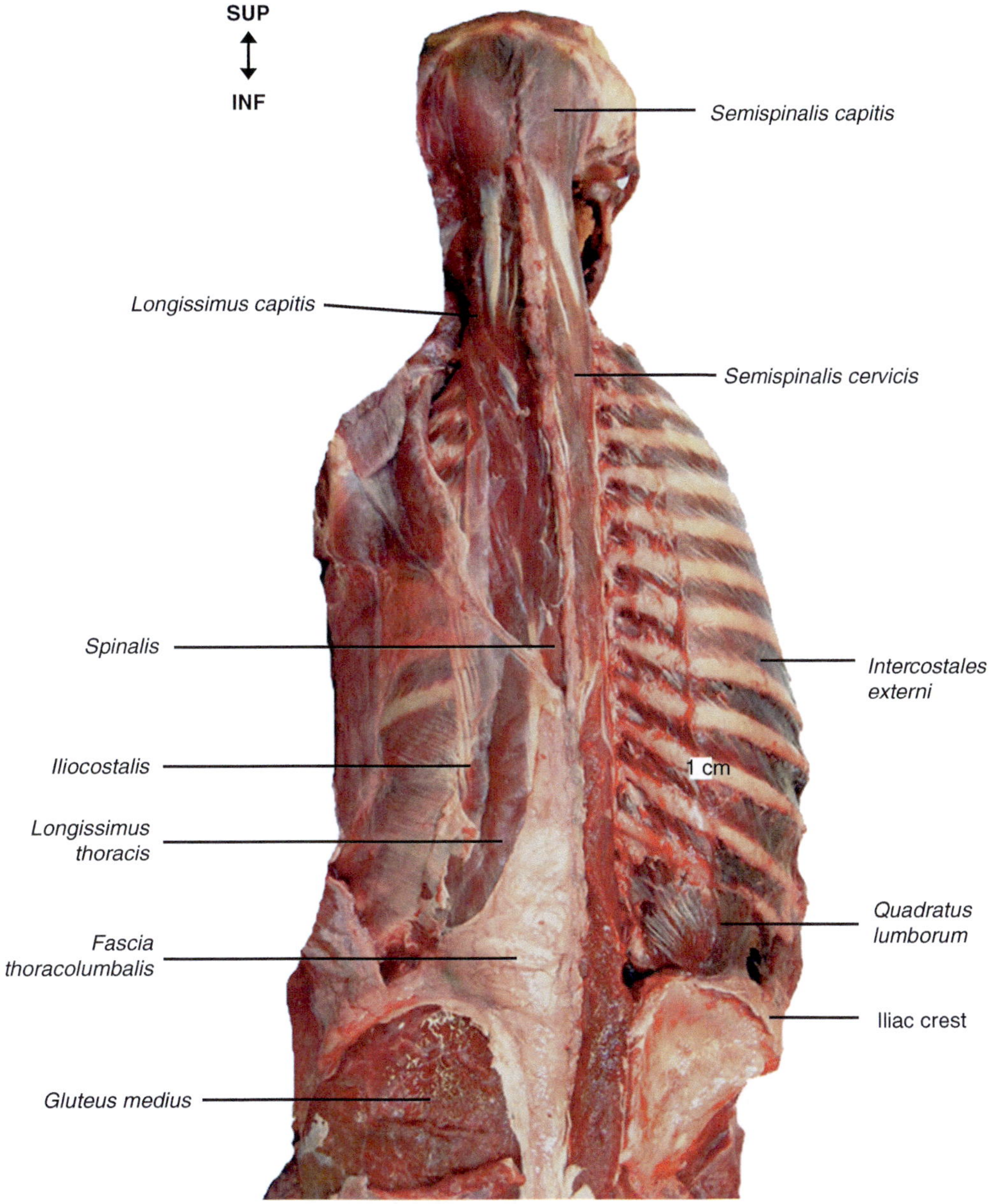


**Fig. 4.3** *Pan paniscus* (ZIMS 164031, Kidogo, adult male): dorsal view of the deeper back muscles

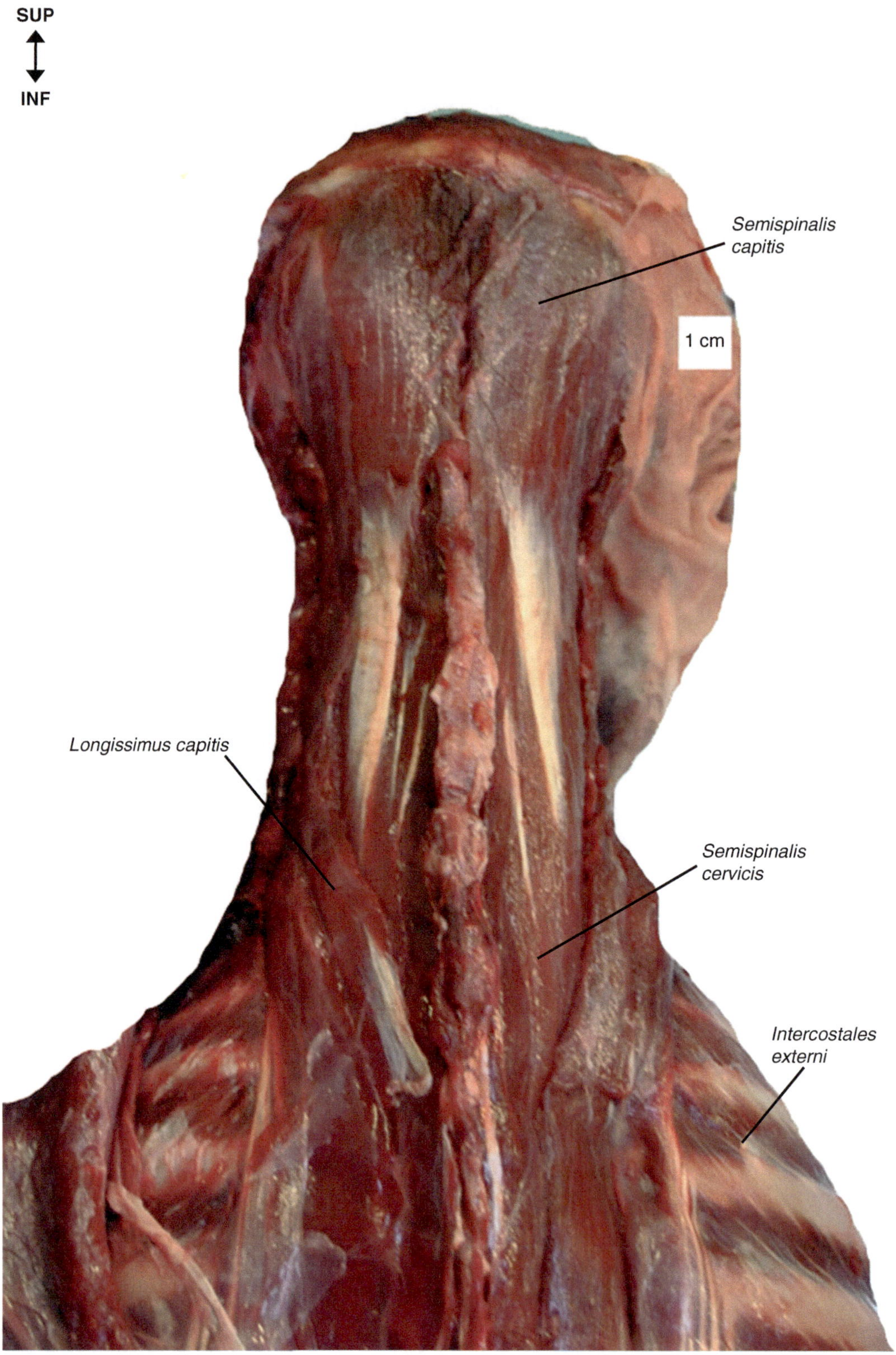


**Fig. 4.4** *Pan paniscus* (ZIMS 164031, Kidogo, adult male): dorsal view of the deeper back muscles

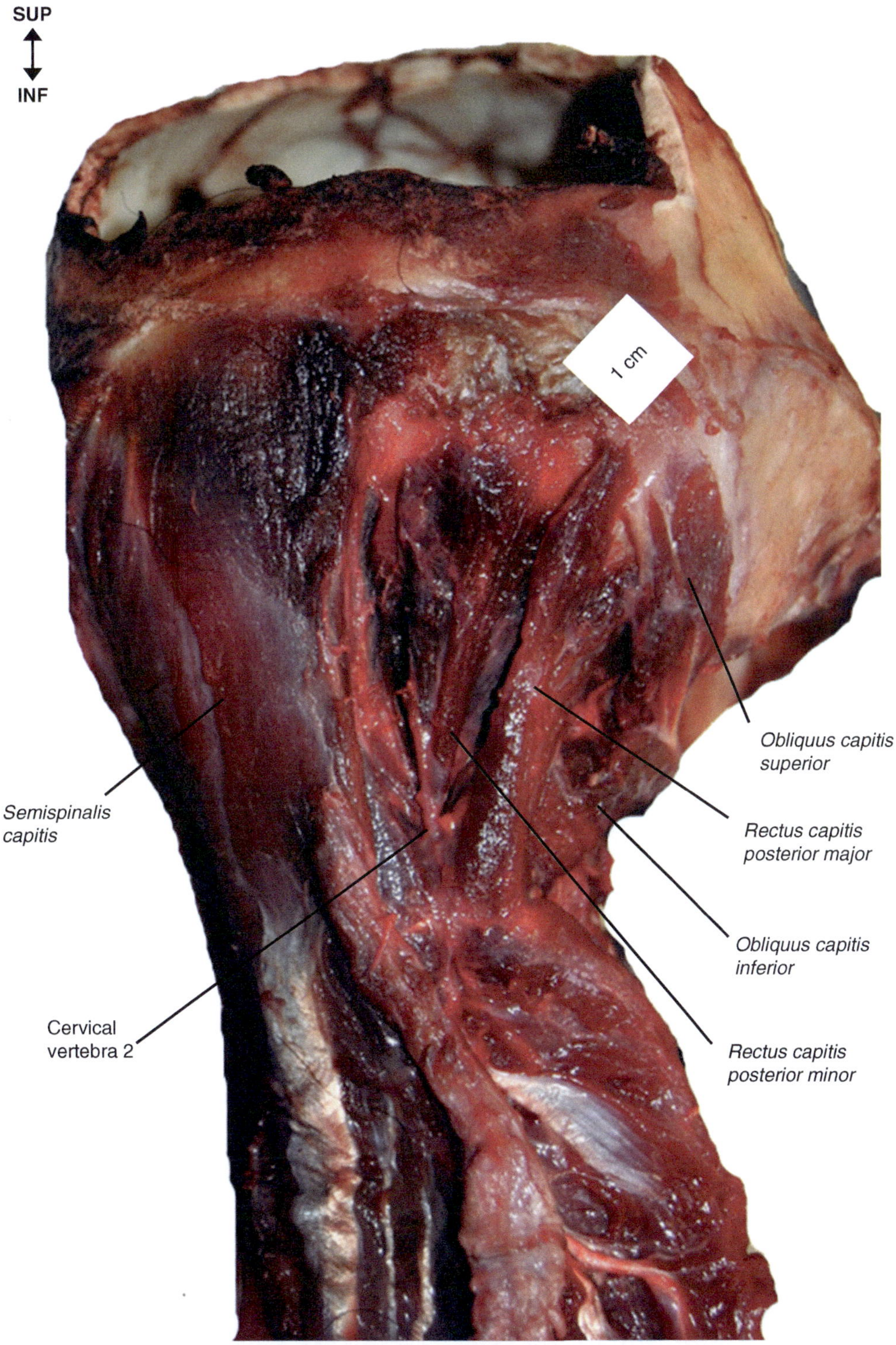


**Fig. 4.5** *Pan paniscus* (ZIMS 164031, Kidogo, adult male): dorsal view of the deeper back muscles

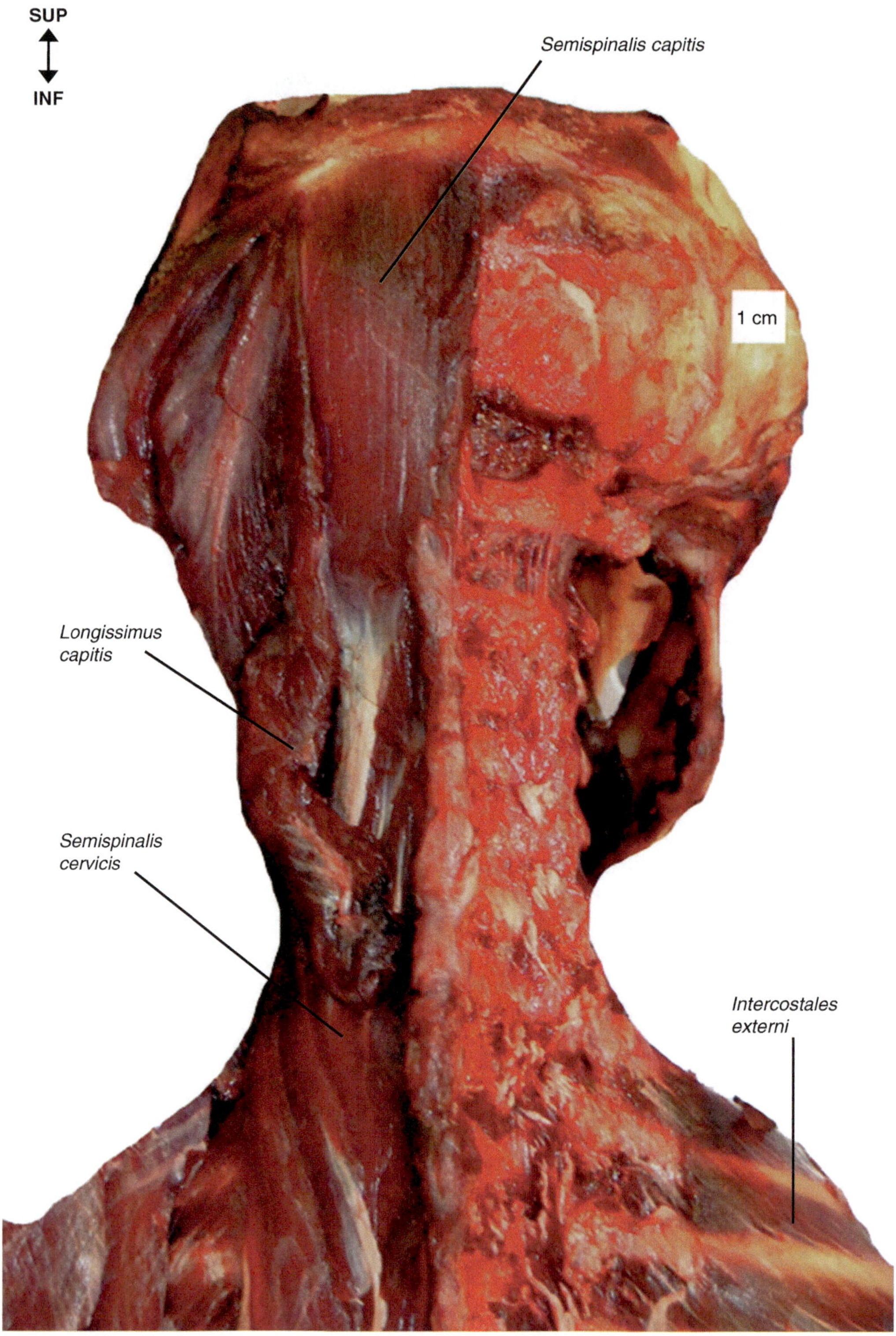


**Fig. 4.6** *Pan paniscus* (ZIMS 164031, Kidogo, adult male): dorsal view of the deeper back muscles

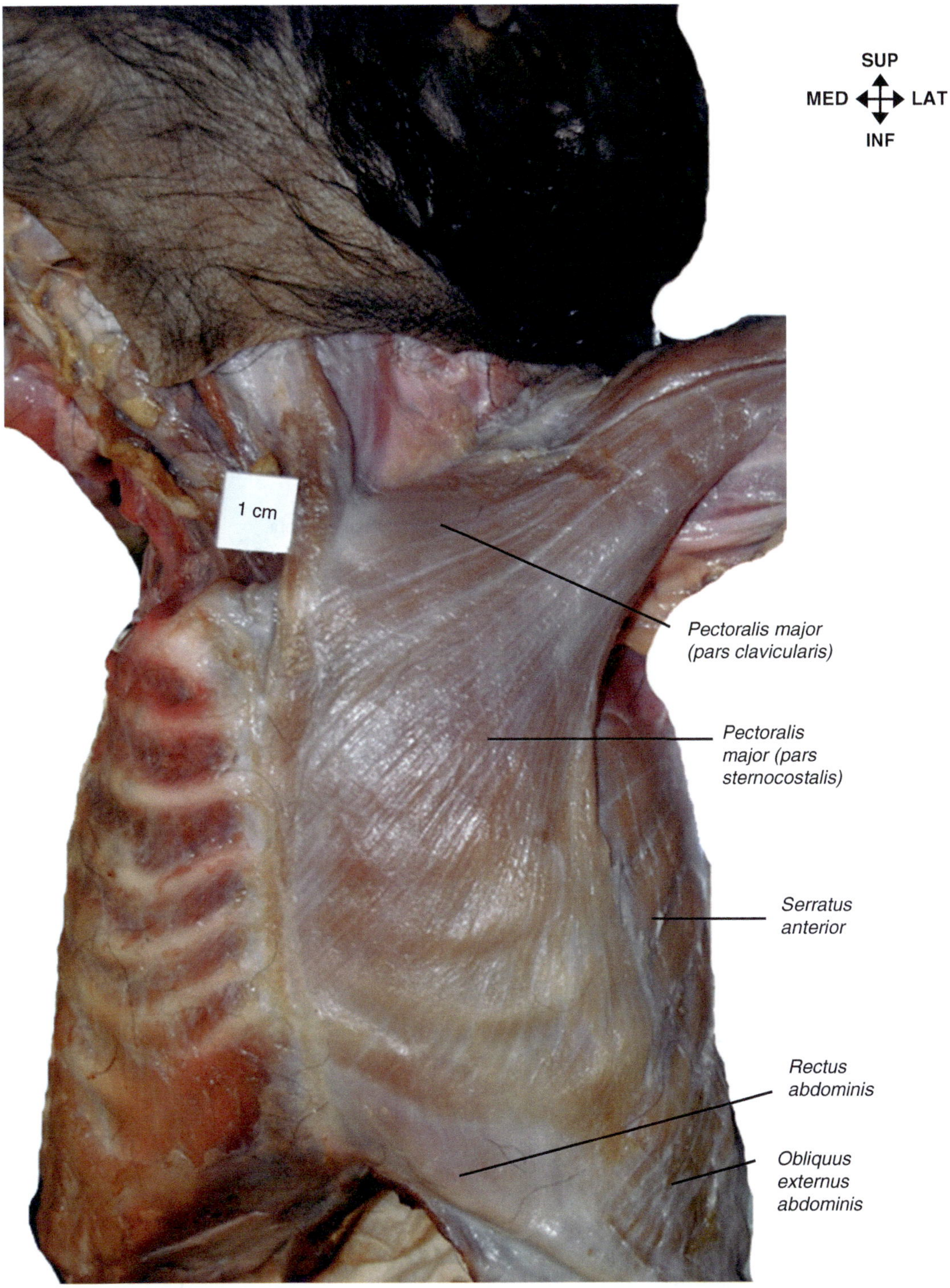


**Fig. 4.7** *Pan paniscus* (ZIMS 164041, Foyo, infant male): ventrolateral view of the shoulder and trunk muscles

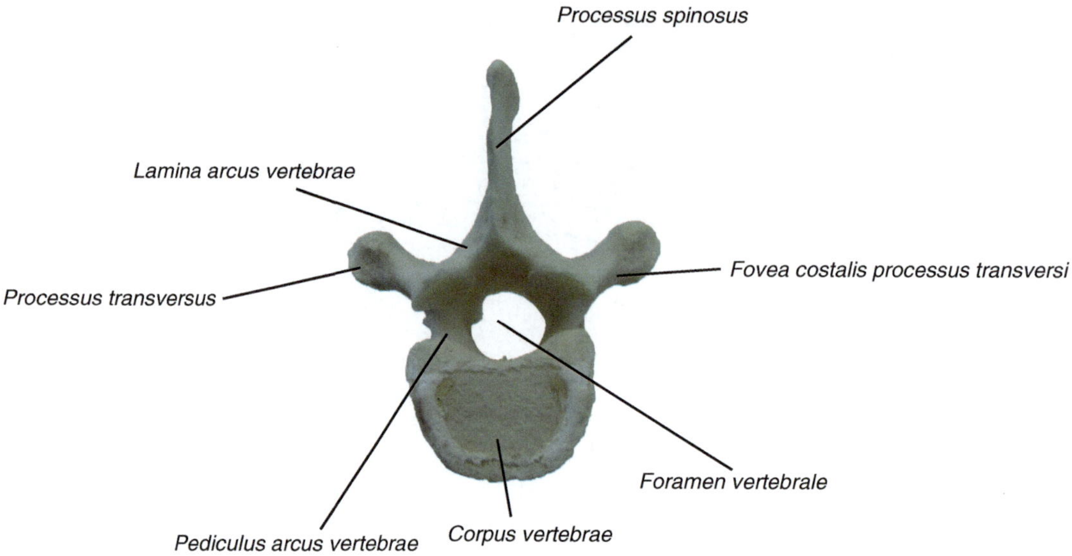


**Fig. 4.8** *Pan paniscus*: anterior (superior in human anatomy) view of trunk vertebra

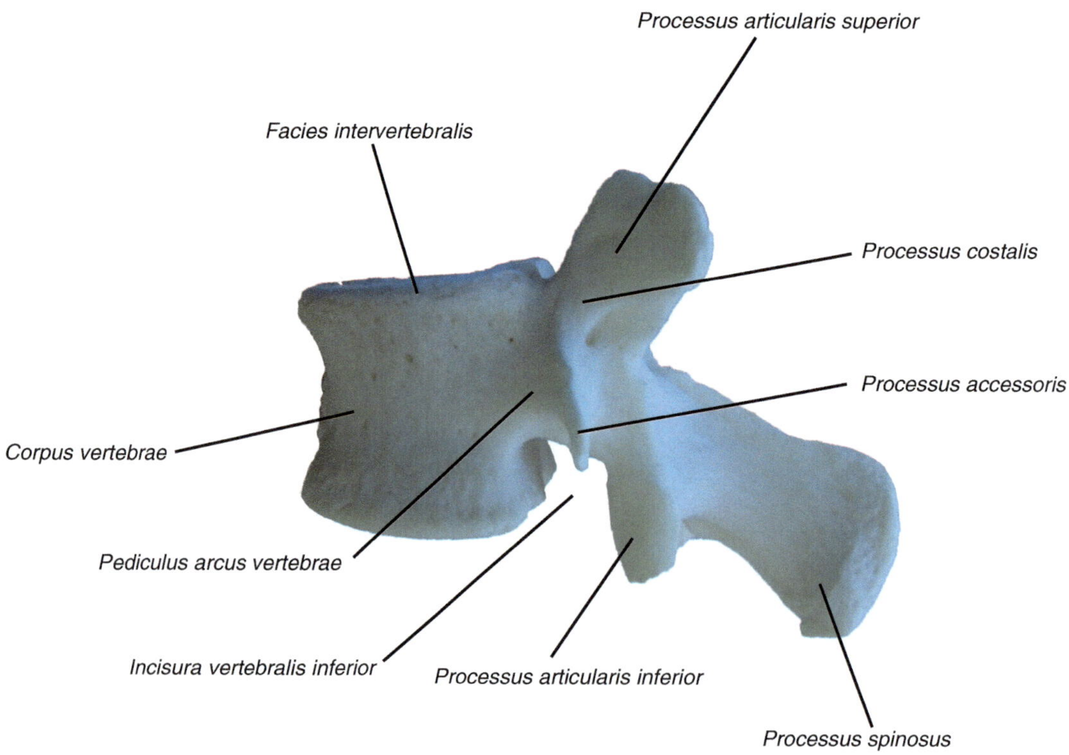


**Fig. 4.9** *Pan paniscus*: lateral view of trunk vertebra

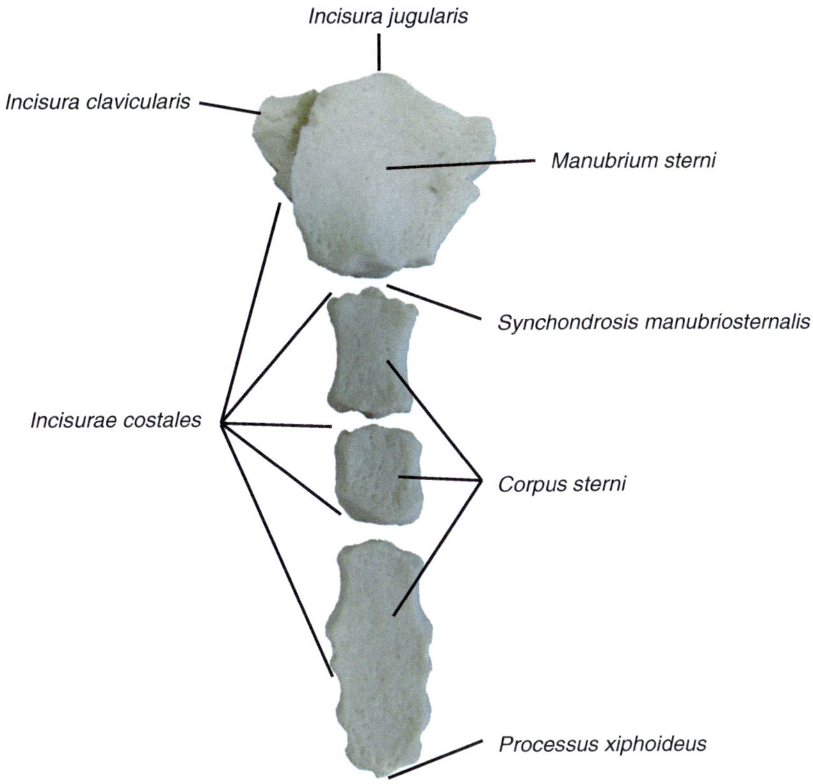


**Fig. 4.10** *Pan paniscus*: ventral/anterior aspect of sternum

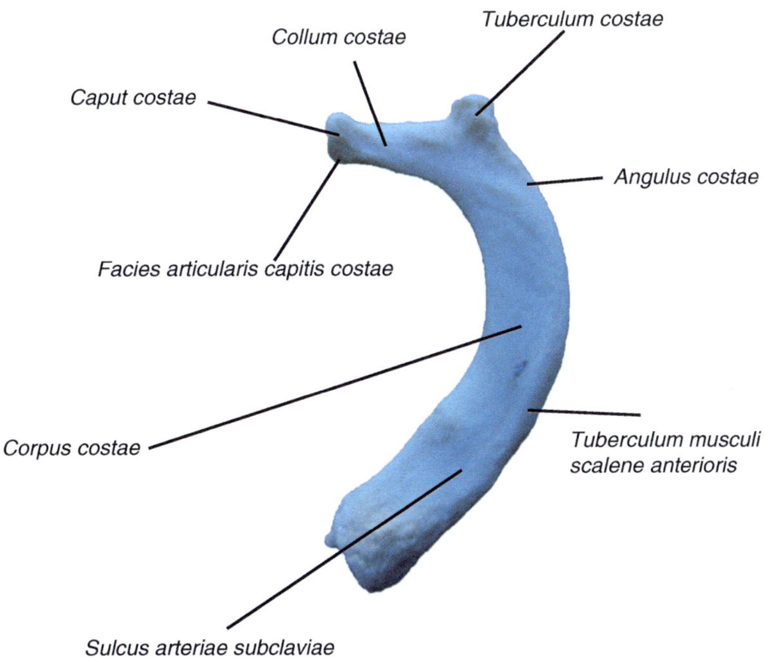


**Fig. 4.11** *Pan paniscus*: superior view of rib 1

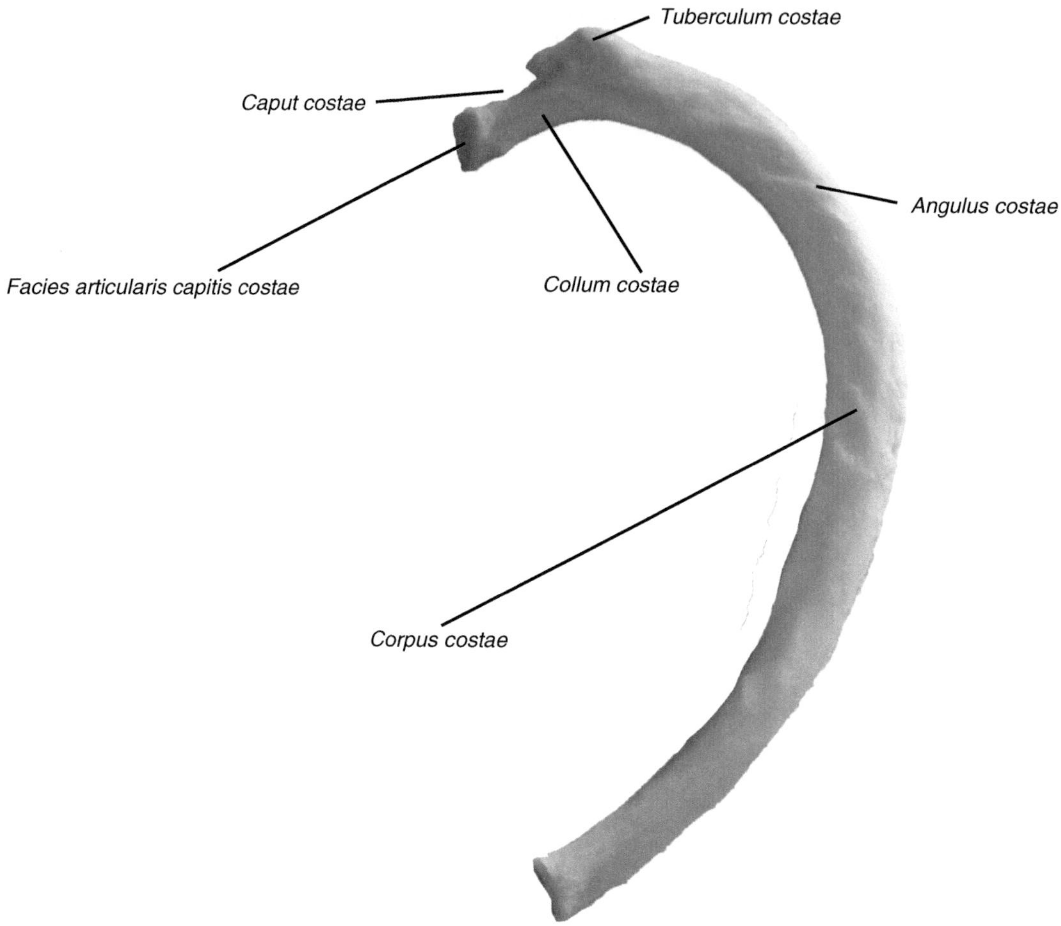


**Fig. 4.12** *Pan paniscus*: superior view of more posterior (inferior in human anatomy) rib

# Bibliography

## Appendix I: Literature Including Information about the Muscles of Bonobos and Common Chimpanzees

Blair DM (1923) A study of the central tendon of the diaphragm. J Anat Physiol 57:203–215

Champneys F (1872) The muscles and nerves of a Chimpanzee (*Troglodytes Niger*) and a *Cynocephalus Anubis*. J Anat Physiol 6:176–211

Dean MC (1985) Comparative myology of the hominoid cranial base, II, the muscles of the prevertebral and upper pharyngeal region. Folia Primatol 44:40–51

Diogo R, Wood BA (2012) Comparative anatomy and phylogeny of primate muscles and human evolution. Taylor & Francis, Oxford

Elftman HO (1932) The evolution of the pelvic floor of primates. Am J Anat 51:307–346

Gibbs S (1999) Comparative soft tissue morphology of the extant Hominoidea, including Man. Unpublished PhD Thesis, The University of Liverpool, Liverpool

Glidden EM, De Garis CF (1936) Arteries of the chimpanzee (*Pan* spec. ?), I, the aortic arch, II, arteries of the upper extremity, III, the descending aorta, IV, arteries of the lower extremity. Am J Anat 58:501–527

Gratiolet LP, Alix PHE (1866) Recherches sur l'anatomie du *Troglodytes aubryi*. Nouv Arch Mus Hist Nat Paris 2:1–264

Juraniec J (1972) The aortic and esophageral hiatus in the diaphragm of primates. Folia Morphol 31:197–207

Juraniec J, Szostakiewicz-Sawicka H (1968) The central tendon of the diaphragm in primates. Folia Morphol 27:183–194

Kohlbrügge JHF (1897) Muskeln und Periphere Nerven der Primaten, mit besonderer Berücksichtigung ihrer Anomalien. Verh K Akad Wet Amsterdam Sec 2(5):1–246

Körner F (1929) Vergleichend-anatomische Untersuchungen über den Faserverlauf der Pars lumbalis des Zwerchfells zur Begrenzung des Hiatus oesophageus bei Säugetieren. Morphol Jahrb 61:409–451

Low A (1907) A note on the crura of the diaphragm and the muscle of Treitz. J Anat 42:93–96

Lunn HF (1948) The comparative anatomy of the inguinal ligament. Anat Phys 82:58–67

Lunn HF (1949) Observations on the mammalian inguinal region. Proc Zool Soc 118:345–355

Macalister A (1871) On some points in the myology of the chimpanzee and others of the primates. Ann Mag Nat Hist 7:341–351

Mijsberg WA (1923) Über den Bau des Urogenitalapparates bei den männlichen Primaten. Verh K Akad Wet Amsterdam 23:1–92

Miller RA (1947) The inguinal canal of primates. Am J Anat 90:117–142

Miller RA (1952) The musculature of *Pan paniscus*. Am J Anat 91:182–232

Roberts JA, Seibold HR (1971) The histology of the primate urethra. Folia Primatol 14:59–69

Schultz AH (1936) Characters common to higher primates and characters specific for man. Q Rev Biol 11(259–283):425–455

Smith WC (1923) The levator ani muscle; its structure in man, and its comparative relationships. Anat Rec 26:175–204

Sonntag CF (1923) On the anatomy, physiology, and pathology of the chimpanzee. Proc Zool Soc Lond 23:323–429

Sonntag CF (1924) The morphology and evolution of the apes and man. John Bale Sons and Danielsson, Ltd., London

Stewart TD (1936) The musculature of the anthropoids, I, neck and trunk. Am J Phys Anthropol 21:141–204

Sutton JB (1883) On some points in the anatomy of the chimpanzee (*Anthropopithecus troglodytes*). J Anat Physiol 18:66–85

Swindler DR, Wood CD (1973) An atlas of primate gross anatomy: baboon, chimpanzee and men. University of Washington Press, Seattle

Thompson P (1901) On the arrangement of the fasciae of the pelvis and their relationship to the levator ani. J Anat Phys 35:127–141

Van den Broek AJP (1908) Ueber einige anatomische Merkmale von *Ateles* in Zusammenhang mit der Anatomie der Platyrrhinin. Anat Anz 33:111–124

Vrolik W (1841) Recherches d' anatornie comparé, sur le chimpanzé. Johannes Miller, Amsterdam

Waterman HC (1929) Studies on the evolution of the pelvis of man and other primates. Bull Am Mus Nat Hist 55:585–642

Winckler G (1947) Recherches sur le systéme musculaire du transverso-spinalis. Arch Anat Histol Embryol 30:151–231

Winckler G (1950) Contribution a l'étude des muscles larges de la paroi abdominale, Étude d'anatomie comparée. Arch Anat Histol Embryol 33:157–228

# Lower Limb Musculature and Lumbosacral Plexus

# 5

## 5.1 Lower Limb Musculature

***Quadratus lumborum*** (adult Kidogo 25.67 g; Figs. 5.1 and 5.25)

- Attachments in bonobos: From the last two ribs and lumbar vertebrae in Miller's 1952 adult bonobo or from the last rib and lumbar vertebrae in our adult Kidogo, to the medial lip of the iliac crest and adjacent iliolumbar and/or sacroiliac ligaments, its ventral fibres being continuous with those of the iliacus in Miller's 1952 adult bonobo.
- Usual innervation: Ventral rami of all of the lumbar nerves (Miller 1952: *P. paniscus*).
- Comparison with common chimpanzees: Similar to bonobos.

***Iliacus*** (infant Foyo iliacus + psoas major =7.79 g; fetus iliacus 1.19 g; Figs. 5.7, 5.19, 5.20, 5.25 and 5.40)

- Attachments in bonobos: From the iliac fossa (our fetus and adult Kidogo and adolescent Jasiri), iliac fossa and crest (infant Foyo) or entire inner surface of ilium (Miller's 1952 adult bonobo) to the lesser trochanter of the femur (our fetus and infant Etje) and sometimes extending distally, to the shaft of the femur distal to the lesser trochanter (Miller's 1952 adult bonobo and our adult Kidogo and adolescent Jasiri), in combination with the psoas major (see below). Fused with psoas major to form iliopsoas muscle.
- Usual innervation: Branches of femoral nerve (Miller 1952 and our dissections: *P. paniscus*) [same innervation in *P. troglodytes*: Sigmon 1974].
- Comparison with common chimpanzees: In *P. troglodytes*, it runs from the iliac fossa (Hepburn 1892) or the entire ventral surface of the ilium (Sigmon 1974) to the medial aspect of the lesser trochanter of the femur, in combination with the psoas major (Champneys 1872; Beddard 1893).

***Psoas major*** (adult Kidogo 291.85 g; infant Foyo iliacus + psoas major = 7.79 g; fetus psoas major 1.35 g; Figs. 5.7, 5.19, 5.20, 5.25, 5.39, 5.40, 5.41, 5.42, 5.43 and 5.44)

- Attachments in bonobos: Origin from vertebrae (last thoracic vertebra to L4 in bonobo fetus; lumbar vertebrae in Miller's 1952 adult bonobo; L1-L4 in our infant Foyo; last thoracic vertebra and various lumbar vertebrae in our adult Kidogo and adolescent Jasiri) and sometimes from iliac fossa (our adult Kidogo and adolescent Jasiri). Insertion onto lesser trochanter of femur (our fetus and infant Etje) and sometimes extending distally, to the shaft of the femur distal to this trochanter (Miller's 1952 adult bonobo and our adult Kidogo and adolescent Jasiri). Fused with iliacus to form *iliopsoas* muscle.


R. Diogo et al., *Photographic and Descriptive Musculoskeletal Atlas of Bonobos*,
DOI 10.1007/978-3-319-54106-8_5

- Usual innervation: Branches of the anterior divisions of the lumbar nerves via the lumbar plexus (Miller 1952: *P. paniscus*); small branches from the femoral nerve after ramification and nerve segments proximal to the femoral nerve from underside of muscle, nerve passing between psoas major and iliacus (our dissections: *P. paniscus*) [*P. troglodytes*: femoral nerve and first two or three lumbar nerves (Champneys 1872; Hepburn 1892; Sonntag 1923; Sigmon 1974)].
- Comparison with common chimpanzees: In *P. troglodytes*, it runs from the lateral surfaces of the bodies and the costal processes of the lumbar vertebrae (Champneys 1872; Hepburn 1892), extending proximally to T12 in some specimens (Champneys 1872; Sigmon 1974) and distally to S1 in most specimens (Hepburn 1892; Sigmon 1974). It may also take origin from the intervening intervertebral disc (Hepburn 1892; Sigmon 1974) and/or from the head of the twelfth rib (Champneys 1872).

***Psoas minor*** (adult Kidogo 10.79 g; infant Foyo 0.611 g; fetus 0.19 g; Figs. 5.40 and 5.42)

- Attachments in bonobos: Originates mainly from vertebrae (last thoracic to L3 in our fetus; last thoracic to L1 in Miller's 1952 adult bonobo and our adult Kidogo and adolescent Jasiri; L1 in our infant Foyo) to the iliopubic eminence (single insertion in our fetus) or the pectineal line of the pubis (single insertion in Miller's 1952 adult bonobo and in our infant Foyo) or the iliac pelvic brim (our adult Kidogo and adolescent Jasiri).
- Usual innervation: Branches of the anterior divisions of the lumbar nerves via the lumbar plexus (Miller 1952: *P. paniscus*) [*P. troglodytes*: first lumbar nerve (Champneys 1872; Hepburn 1892), but it may also be innervated by the twelfth thoracic nerve according to Hepburn (1892) and Sigmon (1974)].
- Comparison with common chimpanzees: In *P. troglodytes*, it mainly originates from the ventrolateral surface of L1 (Champneys 1872; Hepburn 1892; Swindler and Wood 1973). An origin from the last thoracic vertebra is present in half of the *P. troglodytes* specimens dissected by Champneys (1872), Hepburn (1892), Sigmon (1974) and Swindler and Wood (1973). An origin from L2 (Hepburn 1892) and/or from the intervertebral discs (Swindler and Wood 1973) may also be present in *P. troglodytes*. Insertion is onto the iliopubic eminence and the pectineal line in *P. troglodytes* (Champneys 1872; Hepburn 1892; Swindler and Wood 1973).

***Gluteus maximus*** (adult Kidogo 54.86 g; infant Foyo 4. 85 g; fetus 0.63 g; Figs. 5.2, 5.3, 5.21 and 5.29)

- Usual innervation: From surface of the sacrum and coccyx (our fetus and infant Foyo and our adult Kidogo and adolescent Jasiri and Miller's 1952 adult bonobo) and often also from the posterior spines of ilium (our fetus and infant Foyo and Miller's 1952 adult bonobo) and adjacent soft tissues such as the sacrotuberous ligament, posterior sacroiliac ligaments and erector spinae aponeurosis (our adult Kidogo and adolescent Jasiri), having also a firm attachment to the ischial tuberosity as it passes over the bone to the thigh (our fetus and infant Foyo and our adult Kidogo and adolescent Jasiri and Miller's 1952 adult bonobo). The gluteus maximus inserts onto the femur, extending from the base of the greater trochanter almost to the epicondyle, its superficial fibres being continuous with the iliotibial tract of the fascia lata in Miller's 1952 adult bonobo and our adult Kidogo and adolescent Jasiri, while the insertion was simply bony to the proximal 2/3 of the femur in our infant Foyo and on the right side of our fetus. The insertion on the left side of our fetus is similar, but there was no bony insertion, distally the muscle fused with the fascia lata and the tendon of origin of the ischiofemoralis muscle (i.e. it is fused with this latter muscle). A continuity between the gluteus maximus and ischiofemoralis is also seen in our adult Kidogo and adolescent Jasiri.

- Usual innervation: Inferior gluteal nerve from sacral plexus (Miller 1952: *P. paniscus*); inferior and superior gluteal nerves (our dissections: *P. paniscus*).
- Comparison with common chimpanzees: In *P. troglodytes*, it runs from the posterior iliac crest, thoracolumbar fascia, sacrum, coccyx, the sacrotuberous ligament and the fascia over gluteus medius (Hepburn 1892; Beddard 1893; Ranke 1897; MacDowell 1910; Van den Broek 1914; Mysberg 1917; Miller 1945; Robinson et al.1972; Sigmon 1974; Brown1983; Hamada 1985). There is often also an origin from the ischial tuberosity in *P. troglodytes* (Champneys 1872; Hepburn 1892; Beddard 1893; Dwight 1895; MacDowell 1910; Robinson et al.1972; Sigmon 1974; Zihlman and Brunker 1979; Brown1983; Hamada 1985). The muscle may share its origin with the long head of the biceps femoris (Champneys 1872) and may be subdivided into as many as three (Hamada 1985) or four (our *P. troglodytes* specimen GWUANT PT1) parts in *P. troglodytes*. The gluteus maximus inserts onto the iliotibial tract (when present) and the posterolateral aspect of the femur in the region of the gluteal tuberosity and occasionally onto the hypertrochanteric fossa on the lateral aspect of the femur in *P. troglodytes* (Humphry 1867; Champneys 1872; MacDowell 1910; Sigmon 1974; Brown1983; Hamada 1985). The insertion of the muscle generally extends more distally on the femur in apes than in modern humans, extending almost to the lateral condyle of the femur (Dwight 1895; Hepburn 1892;MacDowell 1910; Sigmon 1974; Zihlman and Brunker 1979; Hamada 1985).

***Gluteus medius*** (adult Kidogo 318.32 g; Infant Foyo = 6.15 g; fetus 1.94 g; Figs. 5.1, 5.2, 5.3, 5.4, 5.21 and 5.29)

- Attachments in bonobos: From the dorsal surface of the ilium and the gluteal and/or thoracolumbar fascia to the greater trochanter (Miller's 1952 adult bonobo and our adult Kidogo and adolescent Jasiri and fetus and infants Foyo and Etje), being often fused distally with the piriformis (our fetus and adult Kidogo and infants Foyo and Etje, and Miller's 1952 adult bonobo).
- Usual innervation: Superior gluteal nerve (Miller 1952, our dissections: *P. paniscus*) [same innervation in *P. troglodytes*: Sigmon 1974].
- Comparison with common chimpanzees: In *P. troglodytes*, it runs from the lateral surface of the ilium and the gluteal fascia (Champneys 1872) and also from the fascia lata (Beddard 1893), to the lateral aspect of the greater trochanter (Champneys 1872), its tendon of insertion being often split by the vastus lateralis (Champneys 1872; Hepburn 1892; Sonntag 1924) and fused with that of the piriformis (Champneys 1872; Hepburn 1892; Dwight 1895).

***Gluteus minimus*** (adult Kidogo 21.1 g; Infant Foyo = 6.15 g; fetus 0.08 g; Figs. 5.4 and 5.30)

- Attachments in bonobos: From the dorsal surface of the ilium body to the ventral surface of greater trochanter and/or intratrochanteric crest (our fetus and infant Foyo and Etje and adult Kidogo and adolescent Jasiri, and Miller's 1952 adult bonobo).
- Usual innervation: Superior gluteal nerve (Miller 1952, our dissections: *P. paniscus*) [same innervation in *P. troglodytes*: Sigmon 1974].
- Comparison with common chimpanzees: In *P. troglodytes*, it runs from the dorsolateral surface of the ilium, extending from just distal to the anterior superior iliac spine, towards the acetabulum (Champneys 1872; Sigmon 1974) and sometimes from the margin of the greater sciatic notch (Beddard 1893) and/or the ischial spine (Sigmon 1974), to the greater trochanter (Champneys 1872; Hepburn 1892; Sigmon 1974). The origin is in two parts according to Champneys's (1872) description of *P. troglodytes*. The muscle is triangular or fan-shaped (Sigmon 1974) and it is usually larger in African apes than in the Asian apes (Hepburn 1892).

***Scansorius*** (adult Kidogo 22.91 g; Infant Foyo = 1.34 g; fetus 0.43 g; Figs. 5.4 and 5.30)

- Attachments in bonobos: The scansorius runs from the ilium (infants Foyo and Etje from anterior superior iliac spine; ventral border in bonobo fetus; lateral iliac blade in our adult Kidogo and adolescent Jasiri) to greater trochanter, being completely separated from the gluteus minimus in our fetus and infant Etje, but not in Miller's 1952 adult bonobos (because she actually considered the scansorius to be part of the gluteus minimus) nor in our adult Kidogo and adolescent Jasiri, in which the two muscles were partially fused (see, e.g., Fig. 5.4).
- Usual innervation: Superior gluteal nerve (Miller 1952, our dissections: *P. paniscus*) [same innervation in *P. troglodytes*: Champneys 1872)].
- Comparison with common chimpanzees: We examined this region in detail in three common chimpanzees (specimens GWUANT PT1, GWUANT PT2 and VU PT1); in each the scansorius was not present as a distinct muscle. The scansorius is however said to be occasionally present in common chimpanzees (Wilder 1862; Champneys 1872; Hepburn 1892; Beddard 1893; Blake 1976), running from the anterolateral ilium adjacent to the acetabular rim (Beddard 1893;Dwight 1895; Sonntag 1923) and sometimes from the fascia lata (Beddard 1893), to the greater trochanter, distal to the insertion of the gluteus minimus (Wilder 1862;Champneys 1872; Hepburn 1892; Beddard 1893; Sonntag 1923). When present in *P. troglodytes*, it is a flat triangular muscle (Wilder 1862; Beddard 1893; Sigmon 1974) that is often blended with the gluteus minimus (Champneys 1872; Hepburn 1892; Dwight 1895).

***Ischiofemoralis*** (adult Kidogo 117.48 g; fetus 1.17 g; Figs. 5.2 and 5.29)

- Attachments in bonobos: From ischial tuberosity to proximal half of femur (our fetus) or to fascia lata and fascia of the vastus lateralis muscle (our adult Kidogo and adolescent Jasiri). It is likely that the ischiofemoralis was considered as part of the adductor magnus by Miller (1952). We could not discern if the ischiofemoralis was present, or not, in our infants Foyo and Etje.
- Usual innervation: Common gluteal nerve from sciatic nerve (our dissections: *P. paniscus*) [*P. troglodytes*: common fibular component of the sciatic nerve (Kaseda et al. 2008)].
- Comparison with common chimpanzees: Although initially described by some authorsas comprising the distal third of the gluteus maximus (Stern 1972; Sigmon 1974; Hamada 1985), the ischiofemoralis is now considered to be an independent muscle, particularly in light of the innervation studies of Kaseda et al. (2008). This muscle is usually missing in modern humans, but it is particularly developed in *Pongo*. In apes the muscle runs from the ischial tuberosity to the lateral aspect of the femoral shaft and the aponeurosis of the vastus lateralis. In the common chimpanzees dissected by us, the ischiofemoralis is blended with the gluteus maximus, running from the ischial tuberosity to the femoral diaphysis and the aponeurosis of the vastus lateralis.

***Gemellus superior*** (adult Kidogo 1.43 g; infant Foyo 0.10 g; fetus 0.01 g; Figs. 5.4, 5.5 and 5.30)

- Attachments in bonobos: From ischial spine (our fetus and infants Foyo and Etje and Miller's 1952 adult bonobo) or from sacrospinous ligament and ischium just below – and thus not directly from – the ischial spine (our adult Kidogo and adolescent Jasiri) to the intertrochanteric fossa (our fetus and infants Foyo and Etje and adult Kidogo and adolescent Jasiri), or just to greater trochanter in the less detailed description of Miller's 1952 adult bonobo, being fused with the obturator internus tendon (our adult Kidogo and adolescent Jasiri).
- Usual innervation: Branches of anterior divisions of the sacral nerves via the sacral plexus (Miller 1952: *P. paniscus*) [*P. troglodytes*:

branches of sacral plexus (Champneys 1872; Sigmon 1974)].
- Comparison with common chimpanzees: In *P. troglodytes*, it runs from the region of the ischial spine between the ischial spine and the ischial tuberosity (Sigmon 1974) to the intertrochanteric fossa together with the tendon of the obturator internus, with which it is often fused (Champneys 1872; Beddard 1893; Sigmon 1974).

***Gemellus inferior*** (adult Kidogo 1.70 g; Infant Foyo = 0.34 g; fetus 0.01 g; Figs. 5.4, 5.5 and 5.30)

- Attachments in bonobos: From inner surface of ischial tuberosity to the intertrochanteric fossa (bonobo fetus and infants Foyo and Etje and right side of adult Kidogo), or just to greater trochanter in the less detailed description of Miller's 1952 adult bonobo. In our fetus and infant Foyo, there was no fusion with the obturator internus, contrary to Miller's 1952 adult bonobo and our infant Etje and adult Kidogo and adolescent Jasiri, in which such a fusion was seen, to the point where a gemellus inferior cannot really be discerned, in the left side of our adult Kidogo and in our adolescent Jasiri (see Figs. 5.4 and 5.5).
- Usual innervation: Branches of anterior divisions of the sacral nerves via the sacral plexus (Miller 1952: *P. paniscus*) [*P. troglodytes*: branches of sacral plexus (Champneys 1872; Sigmon 1974)].
- Comparison with common chimpanzees: In *P. troglodytes*, it originates from the region of the ischial tuberosity (Champneys 1872; Hepburn 1892) to the intertrochanteric fossa of the femur (Champneys 1872; Beddard 1893; Sigmon 1974), being difficult to isolate from the quadratus femoris (Hepburn 1892) and/or from the obturator internus (our dissections).

***Obturatorius externus*** (Adult Kidogo 43.65 g; infant Foyo = 1.56 g; fetus 0.41 g; Figs. 5.5, 5.7, 5.25 and 5.28)

- Attachments in bonobos: From inner surfaces of obturator foramen (including pubic and ischial rami) and obturator membrane (our fetus, infant Foyo and adult Kidogo and adolescent Jasiri; Miller's 1952 adult bonobo), and sometimes also from hip joint capsule (adult Kidogo and adolescent Jasiri), to intertrochanteric fossa (our fetus, infant Foyo and adult Kidogo and adolescent Jasiri; Miller's 1952 adult bonobo).
- Usual innervation: Obturator nerve (Miller 1952, our dissections: *P. paniscus*) [same innervation in *P. troglodytes*: Sigmon 1974)].
- Comparison with common chimpanzees: In *P. troglodytes*, it runs from the external surface of the medial bony margin of the obturator foramen and from the obturator membrane (Sigmon 1974) and in our specimens also from the femoral neck, fusing with the obturator internus (Hepburn 1892) before it inserts with it onto the intertrochanteric fossa (Hepburn 1892; Sigmon 1974); occasionally it also inserts onto the lesser femoral trochanter (Beddard 1893).
- Synonymy: Obturator externus (Gibbs 1999).

***Obturatorius internus*** (Adult Kidogo 18.75 g; infant Foyo = 1.60 g; fetus 0.41 g; Figs. 5.4, 5.5 and 5.30)

- Attachments in bonobos: From inner surfaces of obturator foramen (including pubic and ischial rami) and obturator membrane, to intertrochanteric fossa (our fetus and infants Foyo and Etje and adult Kidogo and adolescent Jasiri, and Miller's 1952 adult bonobo).
- Usual innervation: Branches of anterior divisions of the sacral nerves via the sacral plexus (Miller 1952: *P. paniscus*) [*P. troglodytes*: sacral nerves (Sigmon 1974)].
- Comparison with common chimpanzees: In *P. troglodytes*, it originates from the margin of the obturator foramen and the obturator membrane, extending to the descending pubic ramus and the medial surface of the inferior ischial ramus (Champneys 1872; Sigmon 1974; Swindler and Wood 1973). The origin in *P. troglodytes* may extend to the ischial

tuberosity and pubic symphysis, and a tendinous arch running across the obturator foramen (Champneys 1872). The muscle in *P. troglodytes* runs through the lesser sciatic foramen (Hepburn 1892) and inserts together with the gemelli muscles onto the intertrochanteric fossa of the fermur (Champneys 1872; Beddard 1893; Sigmon 1974).

- Synonymy: Obturator internus (Gibbs 1999).

***Quadratus femoris*** (Adult Kidogo 2.13 g; infant Foyo = 0.45 g; fetus 0.04 g; Figs. 5.5 and 5.30)

- Attachments in bonobos: From ischial tuberosity (our fetus and infants Foyo and Etje and adult Kidogo and adolescent Jasiri and Miller's 1952 adult bonobo) and sometimes from hip joint capsule (adult Kidogo and adolescent Jasiri) to intertrochanteric crest (our fetus and infants Foyo and Etje and Miller's 1952 adult bonobo) or to intertrochanteric fossa (adult Kidogo and adolescent Jasiri). In our adult Kidogo and adolescent Jasiri, the tendon of the muscle blends with the obturator externus.
- Usual innervation: Branches of anterior divisions of the sacral nerves via the sacral plexus (Miller 1952: *P. paniscus*) [*P. troglodytes*: sacral nerves (Sigmon 1974)].
- Comparison with common chimpanzees: In *P. troglodytes*, it originates from the ventrolateral aspect of the ischial tuberosity (Beddard 1893; Dwight 1895; Miller 1945; Sigmon 1974), being a small, thick, fleshy muscle that is separated from the adductor magnus and closely related to the inferior gemellus (Hepburn 1892), and mainly inserting onto the intertrochanteric crest of the femur (Hepburn 1892; Dwight 1895; Sigmon 1974), sometimes being also attached to the dorsal surface of the greater trochanter (Hepburn 1892) and/or dorsal to the lesser trochanter (Champneys 1872; Dwight 1895; Hepburn 1892). Hepburn (1892) stated that the *articularis genu* is present in all apes, but the muscle is usually not reported in common chimpanzees (e.g. Sonntag 1923, 1924) and in the chimpanzees in which we examined this anatomical region in detail (*P. troglodytes* GWUANT PT1, GWUANT 2 and VU PT1 and our bonobo fetus, infant Foyo, adolescent Jasiri and adult Kidogo) the articularis genu is seemingly not present as a distinct muscle.

***Piriformis*** (Adult Kidogo 23.01 g; Infant Foyo = 0.86 g; fetus 0.19 g; Figs. 5.4 and 5.30)

- Attachments in bonobos: From sacrum (region of sacral vertebrae 1-3 in our fetus; region of sacral vertebrae 2–4 and upper border of greater sciatic notch in Miller's 1952 adult bonobo) to greater trochanter (our fetus and infants Foyo and Etje and adult Kidogo and adolescent Jasiri, and Miller's 1952 adult bonobo). In our adult Kidogo and adolescent Jasiri, the muscle also originates from the posterior inferior iliac spine.
- Usual innervation: Second and third sacral nerves via the sacral plexus (Miller 1952: *P. paniscus*) [*P. troglodytes*: branches of sacral plexus (Sigmon 1974)].
- Comparison with common chimpanzees: In *P. troglodytes*, it originates by slips (two well-differentiated heads of origin in our specimen GWUANT PT1) from the distal half of the sacrum (Champneys 1872; Hepburn 1892; Sigmon 1974), the posterior extent of the origin often being S5 (Champneys 1872; Hepburn 1892). An origin from the greater sciatic notch is often present in *P. troglodytes* (Chapman 1879; Sigmon 1974), and the muscle is often fused with the gluteus medius (Champneys 1872; Hepburn 1892; Dwight 1895; Sigmon 1974); it passes through the greater sciatic foramen to insert onto the ventromedial aspect of the tip of the greater trochanter of the femur (Champneys 1872; Hepburn 1892; Beddard 1893; Sigmon 1974). In the three common chimpanzees in which we examined this region in detail (GWUANT PT1, GWUANT PT2 and VU PT1), the piriformis is not fused with the gluteus medius.

***Rectus femoris*** (Adult Kidogo 85.13 g; infant Foyo quadriceps femoris as whole = 10.16 g; fetus rectus femoris 0.72 g; Figs. 5.6, 5.8, 5.20, 5.22, 5.23, 5.24, 5.26, 5.27, 5.28, 5.32, 5.40 and 5.42)

- Attachments in bonobos: From anterior inferior iliac spine and adjacent region (single head in our fetus and adult Kidogo and adolescent Jasiri) and often also from 'superior' (cranial sensu the present work) border of the acetabulum (additional head in Miller's 1952 adult bonobo and our infant Foyo), to patella, and then via patellar ligament to proximal tibia (Miller's 1952 adult bonobo and our fetus and infants Foyo and Etje and adult Kidogo and adolescent Jasiri). The rectus femoris, vastus intermedius, vastus lateralis and vastus medialis are all fused to form the *quadriceps femoris*.
- Usual innervation: Femoral nerve (Miller 1952, our dissections: *P. paniscus*).
- Comparison with common chimpanzees: In *P. troglodytes*, it possesses two heads of origin in 1/3 of the specimens dissected by Dwight (1895), Hepburn (1892) and Vrolik (1841). The short head, usually present in *P. troglodytes*, originates from the anterior inferior iliac spine (Champneys 1872); the reflected head, when present, originates from the ilium near the acetabulum (Hepburn 1892; Beddard 1893; Boyer 1935; Sigmon 1974), an origin that is similar to that of the single head of origin found in our *P. troglodytes* specimens VU PT1 and GWUANT PT1. In a single specimen of *P. troglodytes*, the rectus femoris was fused with the vastus lateralis (Beddard 1893). In *P. troglodytes*, the rectus femoris inserts together with the other muscles forming the 'quadriceps femoris' (i.e., vastus intermedius, vastus laterais and vastus medialis), mainly to the proximolateral border of the patella, and then via the patellar ligament to the tibial tuberosity (Beddard 1893; Hepburn 1892).

***Vastus intermedius*** (Adult Kidogo 73.76 g; infant Foyo quadriceps femoris as a whole =10.16 g; fetus vastus intermedius = 0.79 g; Figs. 5.6 and 5.33)

- Attachments in bonobos: From femoral shaft (distal 2/3 in adult Kidogo and adolescent Jasiri) to patella, and then via patellar ligament to proximal tibia (our infants Foyo and Etje and fetus and adult Kidogo and adolescent Jasiri, and Miller's 1952 adult bonobo).
- Usual innervation: Femoral nerve (Miller 1952, our dissections: *P. paniscus*).
- Comparison with common chimpanzees: In *P. troglodytes*, it originates from the dorsal side of the femoral shaft (Sigmon 1974) and inserts together with the other muscles forming the 'quadriceps femoris' (i.e. rectus femoris, vastus laterais and vastus medialis), mainly to the proximolateral border of the patella, and then via the patellar ligament to the tibial tuberosity (Beddard 1893; Hepburn 1892).

***Vastus lateralis*** (Adult Kidogo 110.96 g; Infant Foyo quadriceps femoris as a whole = 10.16 g; fetus vastus intermedius = 1.64 g; Figs. 5.21, 5.29, 5.30, 5.32 and 5.33)

- Attachments in bonobos: From femur, including greater trochanter in our fetus and infants Foyo and Etje and adult Kidogo and adolescent Jasiri but not in Miller's 1952 adult bonobo, to patella, and then via patellar ligament to proximal tibia.
- Usual innervation: Femoral nerve (Miller 1952, our dissections: *P. paniscus*).
- Comparison with common chimpanzees: In *P. troglodytes*, it originates from the lateral aspect of the greater femoral trocanter and the distal two-thirds of the lateral femoral shaft, in the region of the lateral lip of the linea aspera (Beddard 1893; Sigmon 1974), and occasionally from the iliofemoral ligament (Sigmon 1974). It inserts together with the other muscles forming the 'quadriceps femoris' (i.e. rectus femoris, vastus intermedius and vastus medialis) mainly to the proximolateral border of the patella, and then via the patellar ligament to the tibial tuberosity (Beddard 1893; Hepburn 1892).

***Vastus medialis*** (Adult Kidogo 323.4 g; infant Foyo quadriceps femoris as a whole = 10.16 g; fetus vastus intermedius = 0.73 g; Figs. 5.6, 5.8, 5.20, 5.22, 5.23, 5.24, 5.26, 5.27, 5.28, 5.32 and 5.33)

- Attachments in bonobos: From medial side of femur, including intertrochanteric line in our fetus and infants Foyo and Etje but not in Miller's 1952 adult bonobo, to patella, and then via patellar ligament to proximal tibia.
- Usual innervation: Femoral nerve (Miller 1952, our dissections: *P. paniscus*).
- Comparison with common chimpanzees: In *P. troglodytes*, it originates from the dorsomedial femoral shaft, in the region of the linea aspera (Champneys 1872; Beddard 1893; Sigmon 1974), and often also from the iliofemoral ligament (Beddard 1893;Sigmon 1974) and/or from the femoral neck (our specimens VU PT1 and GWUANT PT1). It inserts together with the other muscles forming the 'quadriceps femoris' (i.e., rectus femoris, vastus laterais and vastus intermedius) mainly to the proximolateral border of the patella, and then via the patellar ligament to the tibial tuberosity (Beddard 1893; Hepburn 1892).

***Sartorius*** (Adult Kidogo 63.84 g; infant Foyo 1.75 g; fetus 0.31 g; Figs. 5.6, 5.8, 5.19, 5.20, 5.26 and 5.42)

- Attachments in bonobos: From anterior superior iliac spine (our infants Etje and Foyo and fetus and Miller's 1952 adult bonobo), or instead from the lateral side of the lower iliac blade (adult Kidogo and adolescent Jasiri), to medial border of proximal 1/3 of tibia, its insertion being superficial to those of gracilis and semitendinosus in our fetus and infant Etje and adult Kidogo and adolescent Jasiri, contrary to the adult bonobo described by Miller 1952 and our infant Foyo, in which it inserted through a broad aponeurosis with the tendons of the gracilis and semitendinosus. In adult Kidogo and adolescent Jasiri, the sartorius also sends more tendinous fibres to the crural fascia and onto the medial collateral ligament.
- Usual innervation: Femoral nerve (Miller 1952, our dissections: *P. paniscus*) [same innervation in *P. troglodytes*: Sigmon 1974).
- Comparison with common chimpanzees: In *P. troglodytes*, it runs from the anterior superior iliac spine (Sigmon 1974; Humphry 1867; Champneys 1872; Hepburn 1892; Beddard 1893). The ribbon-shaped muscle courses obliquely over the thigh dorsal to the medial femoral condyle to insert onto the medial border of the tibial shaft, its insertion being often superficial to those of the gracilis and semitendinosus (Chapman 1879; Hepburn 1892; Beddard 1893; Sigmon 1974).

***Tensor fasciae latae*** (Adult Kidogo 32.5 g; fetus 0.20 g; Fig. 5.2)

- Attachments in bonobos: From ilium (crest adjacent to anterior superior iliac spine in Miller's 1952 adult bonobo and in our fetus and infant Etje; just below the anterior superior iliac spine in adolescent Jasiri or below this spine in adult Kidogo), to middle 1/3 (our fetus) or proximal 1/4 (our adult Kidogo and adolescent Jasiri) of lateral border of femur via the iliotibial tract, seemingly not extending directly to the tibia (in both Miller's 1952 adult bonobo and in our fetus and adult Kidogo and adolescent Jasiri). As a variation, the tensor fasciae latae was not present as a distinct muscle on both sides of our infant Foyo.
- Usual innervation: Inferior gluteal nerve from sacral plexus (Miller 1952, our dissections: *P. paniscus*) [*P. troglodytes*: superior gluteal nerve (Sigmon 1974), as is usually the case in humans, so this might be an interesting difference between bonobos versus common chimpanzees and humans].
- Comparison with common chimpanzees: In *P. troglodytes*, it runs from the region of the anterior superior iliac spine (Humphry 1867; Champneys 1872; Kaplan 1958a; Sigmon 1974) and often from the gluteal fascia (Kaplan 1958a; Sigmon 1974), to the iliotibial tract (Duvernoy 1855–1856; Hepburn 1892; Kaplan 1958a; Sigmon 1974).

***Adductor brevis*** (Adult Kidogo 17.78 g; infant Foyo adductor brevis + adductor minimis = 2.72 g; fetus adductor brevis 0.31 g; Figs. 5.7, 5.22, 5.23 and 5.27)

- Attachments in bonobos: It is possible that part of the adductor magnus *sensu* the present work was described as part of the 'adductor brevis' by Miller (1952), in her adult bonobo, in which she described the 'adductor brevis' as a large muscle with three parts, i.e., an outer part from the symphisis pubis together (fused with) the gracilis, a middle part from two heads from the region of the symphisis and crest of the pubis that then fuses with the adductor longus, and an inner part from the superior ramus of the pubis deep to the adductor longus, the outer and inner parts inserting together along the proximal half of the linea aspera of the femur as far as the lesser trochanter and the middle part inserting more distally upon the linea aspera with the adductor longus, according to her. In our fetus, there are two short muscle structures that were fused distally, the most superficial originates from the body and superior ramus of the pubis and inserts onto the femur just distal and posterior to the lesser trochanter, while the deeper structure originates from the iliopubic eminence and the capsule between the pelvic girdle and femur and inserts onto the femur together with the superficial structure. The deeper head might possible represent the adductor minimus of our adult Kidogo and adolescent Jasiri and infants Foyo and Etje, but in those bonobos there is actually an adductor brevis with two heads running from pubis to the proximal femur plus an adductor minimus running from the pubis to the proximal portion of the linea aspera just below the lesser trochanter. In our adult Kidogo and adolescent Jasiri, one head of the adductor brevis originates alongside the gracilis origin, lateral to the pubic symphysis, the second head originating from the superior pubic ramus; distally the two heads fuse with the adductor minimus and insert on the dorsomedial (= posteromedial in human anatomy) aspect of the femoral shaft.
- Usual innervation: Obturator nerve (Miller 1952, our dissections: *P. paniscus*) [*P. troglodytes*: ventral ('anterior') division of the obturator nerve (Champneys 1872; Sigmon 1974; Hamada 1985)].
- Comparison with common chimpanzees: In *P. troglodytes*, it originates from the body of the pubis (Hepburn 1892; Beddard 1893; Dwight 1895; Sigmon 1974), sometimes also from the inferior pubic ramus near the symphysis (Beddard 1893) and/or from the superior pubic ramus (Sigmon 1974) and occasionally from the ischium and intermuscular septa (Beddard 1893). In *P. troglodytes*, the muscle may be divided into two parts (Champneys 1872; Beddard 1893; Hepburn 1892) and is often partly fused with the short head of the adductor magnus (Hepburn 1892; Sigmon 1974). In a single *P. troglodytes* specimen, the adductor muscles were described as being indistinguishable from each other, forming an 'adductor mass' (Dwight 1895). In *P. troglodytes*, the insertion of the adductor brevis is distal to the lesser trochanter onto the proximal third of the medial lip of the linea aspera on the mid-dorsal femoral surface, onto the pectineal line (Champneys 1872; Hepburn 1892; Beddard 1893; Dwight 1895; Sigmon 1974).

***Adductor longus*** (Adult Kidogo 46.28 g; infant Foyo 1.23 g; fetus 0.27 g; Figs. 5.6, 5.7, 5.20, 5.22, 5.26, 5.27 and 5.42)

- Attachments in bonobos: Originates from body and anterior superior pubic ramus (our fetus and infants Foyo and Etje). It inserts onto (i.e. is fused with) adductor magnus, without any bony insertion in bonobo fetus, and with a bony insertion onto the lower 1/3 of the linea aspera of the femur in infant Etje, while in Miller's 1952 adult bonobo it inserts onto the lower half of the linea aspera of the femur, being fused at its insertion with the adductor magnus, and also more proximally with the adductor brevis, in that adult bonobo. In our adult Kidogo and adolescent Jasiri, the adductor longus fuses with a portion of adductor magnus to insert on the distal half of the dorsomedial (posteromedial of human anatomy)

aspect of the femoral shaft, where the adductor hiatus begins.
- Usual innervation: Obturator nerve (Miller 1952: *P. paniscus*); anterior branch of obturator nerve (our dissections: *P. paniscus*).
- Comparison with common chimpanzees: In *P. troglodytes*, it runs originates by a flat tendon from the anterior superior pubic ramus in the region of the pubic tubercle (Humphry 1867; Beddard 1893; Dwight 1895; Sigmon 1974) and occasionally from the intermuscular septum (Beddard 1893); it inserts onto the middle of the medial lip of the linea aspera or mediodorsal femoral shaft (Humphry 1867; Hepburn 1892; Beddard 1893; Dwight 1895; Sigmon 1974), besides and ventral to the proximal half of the insertion of the short head of the adductor magnus (Beddard 1893;Sigmon 1974). In *P. troglodytes*, there may be an expansion to the medial condyle (Humphry 1867), and the insertion of the adductor longus may be fused with the adductor magnus (Champneys 1872; Beddard 1893).

***Adductor magnus*** (Adult Kidogo 712.95 g; infant Foyo 9.01 g; fetus 3.23 g; Figs. 5.6, 5.7, 5.20, 5.26 and 5.27)

- Attachments in bonobos: With three heads that originate from pubis and ischial tuberosity (our fetus and infants Foyo and Etje; Miller 1952 just refers to an origin from the inferior pubic ramus, as a whole, in her adult bonobo) that join at its insertion onto the femur (distal fourth of the linea aspera of the femur together with the adductor longus, in Miller's 1952 adult bonobo). In the bonobo fetus and infant Foyo the most cranial head goes from the inferior pubic ramus to middle fifth of the femur, the next head originates from the ischiopubic ramus and inserts mainly onto the midlength of the other two heads, while the more caudal head originates from the ischipubic ramus and inserts mainly onto the medial epicondyle of the femur; all these three heads are deeply fused. In our adult Kidogo and adolescent Jasiri, there is a pubofemoral and ischiocondylar portion of the adductor magnus: the well-developed pubofemoral head begins just lateral to the pubic symphysis and inserts along with adductor longus into the dorsomedial aspect of the distal shaft; the ischiocondylar portion originates along the ventral aspect of the caudal portion of the ischial ramus, and inserts on the distal 2/3 of the linea aspera and the medial epicondyle of the femur. As noted above, it is possible that part of the adductor magnus sensu the present work was described as part of the 'adductor brevis' by Miller (1952).
- Usual innervation: Obturator nerve (Miller 1952: *P. paniscus*); posterior branch of obturator nerve (our dissections: *P. paniscus*) [*P. troglodytes*: short head by the obturator nerve (Champneys 1872; Hepburn 1892; Sigmon 1974); long head by flexores femoris nerve (Sigmon 1974) or by sciatic nerve (Champneys 1872; Hepburn 1892)]. According to Sigmon (1974) the flexor femoris nerve does not exist in modern humans, and the tibial nerve in apes (except in a few gorillas) does not supply the adductor magnus.
- Comparison with common chimpanzees: The configuration of the adductor magnus, with three heads and lying mainly superficial (anterior) to the adductor brevis, found in the bonobos dissected by us, is very similar to the one described and shown in common chimpanzees in our previous chimpanzee atlas. In *P. troglodytes* the adductor magnus runs from the ventral surface of the inferior pubic ramus, lateral to the symphysis, and from the inferior ischial ramus as far as the ischial tuberosity (Humphry 1867; Champneys 1872; Hepburn 1892; Beddard 1893; Dwight 1895; Sigmon 1974; Yirga 1987), and occasionally from the inferiomedial border of the semitendinosus and the long head of the biceps femoris (Sigmon 1974). Its insertion is onto the medial lip of the linea aspera or the dorsomedial surface of the femur, and to the adductor tubercle of the medial epicondyle (Humphry 1867; Champneys 1872; Hepburn 1892; Beddard 1893; Dwight 1895; Sigmon 1974), its short head often also attaching onto the lateral border of the common tendon of insertion of the adductor longus and adductor brevis (Champneys 1872; Beddard 1893).

***Adductor minimus*** (Adult Kidogo 20.56 g; infant Foyo adductor brevis + adductor minimis =2.72 g; Fig. 5.7)

- Attachments in bonobos: Note sure if this muscle is present or not as a distinct structure in our fetus and infant Foyo: see comments on adductor brevis, above. Regarding Miller's 1952 adult bonobo, she described a well-developed adductor minimus running from the symphysis pubis between the adductor brevis and obturator externus, to the most proximal portion of the linea aspera just below the lesser trochanter; this is also the condition found in our infants Foyo and Etje, in which the adductor minimus was partially fused with the adductor brevis. In our adult Kidogo and adolescent Jasiri, the adductor minimus originates on the inferior portion of the pubis, just lateral to the pubic symphysis, as well as on part of the ischiopubic ramus, blending with the adductor brevis to insert on the dorsomedial aspect of the femur.
- Usual innervation: Obturator nerve (Miller 1952, our dissections: *P. paniscus*).
- Comparison with common chimpanzees: The adductor minimus is absent from a third of great apes (Hepburn 1892; Sigmon 1974), being often described as a proximal subdivision of the adductor magnus or as a deep slip of the adductor brevis (Hepburn 1892). When present in it runs from the inferior pubic ramus (Hepburn 1892; Dwight 1895) to the linea aspera on the dorsal side of the femoral shaft (Hepburn 1892).

***Gracilis*** (Adult Kidogo 174.20 g; infant Foyo 4.91 g; fetus 0.77 g; Figs. 5.6, 5.8, 5.19, 5.20, 5.22, 5.26 and 5.42)

- Attachments in bonobos: From pubis (our fetus and infants Foyo and Etje), specifically from symphisis pubis and most of inferior pubic ramus (Miller's 1952 adult bonobo) or just lateral to pubic symphysis (our adult Kidogo and adolescent Jasiri), to ventromedial surface of proximal 1/3 of tibia, inserting deep to sartorius and superficial to semitendinosus in our bonobo fetus and infants Foyo and Etje, and in common (fused distally) with these two muscles in Miller's 1952 adult bonobo; she also described an additional fusion between the gracilis and the adductor brevis, in that adult bonobo. In our adult Kidogo and adolescent Jasiri, the gracilis inserts on the medial aspect of the proximal tibial shaft, just distal to the sartorius insertion and the medial collateral ligament; fibres from the semitendinosus merge with those of gracilis to create a conjoint tendon at that insertion.
- Usual innervation: Obturator nerve (Miller 1952: *P. paniscus*); anterior branch of obturator nerve (our dissections: *P. paniscus*)[*P. troglodytes*: ventral ('anterior') branch of obturator nerve (Sigmon 1974)].
- Comparison with common chimpanzees: In *P. troglodytes*, it runs from the inferior pubic ramus near to the pubic symphysis onto the ischial ramus (Champneys 1872; Hepburn 1892; Beddard 1893; Sigmon 1974) and sometimes from the whole pubic body (Hepburn 1892) or from the superior pubic ramus (Sonntag 1924). Insertion is to the ventromedial surface of the tibia (Humphry 1867; Champneys 1872; Hepburn 1892; Dwight 1895; Sigmon 1974), and often through an aponeurotic expansion to the fascia of the leg (Champneys 1872; Hepburn 1892; Dwight 1895).

***Pectineus*** (adult Kidogo 19.53 g; infant Foyo 0.56 g; fetus 0.08 g; Figs. 5.19, 5.20, 5.22 and 5.26)

- Attachments in bonobos: From superior or lateral pubic ramus to femur just distal to lesser trochanter (our fetus and infant Foyo and Etje and Miller's 1952 adult bonobo; specifically to proximal 1/4 of the humeral shaft in our adult Kidogo and adolescent Jasiri). Miller (1952) described a fusion with adductor longus, at its origin from the superior pubic ramus, in her adult bonobo.
- Usual innervation: Obturator and femoral nerves (Miller 1952: *P. paniscus*) [*P. troglodytes*: femoral nerve and occasionally also by

the ventral 'anterior' branch of the obturator nerve (Champneys 1872; Sigmon 1974)].

- Comparison with common chimpanzees: In *P. troglodytes*, it runs from the superior pubic ramus (Hepburn 1892; Sigmon 1974) to the femur just distal to the lesser trochanter (Hepburn 1892; Beddard 1893; Sigmon 1974). To our knowledge, the *iliocapsularis* has not been described as a distinct muscle in common chimpanzees nor in bonobos.

***Biceps femoris*** (adult Kidogo caput longum 113.06 g, caput breve 86.73 g, 'intermediate belly' 47.42 g; infant Foyo caput longum + caput breve = 3.59 g; fetus caput longum 0.57 g, caput breve 0.42 g; Figs. 5.2, 5.21, 5.29, 5.30 and 5.31)

- Attachments in bonobos: Long head from ischial tuberosity and short head from femur (midshaft in our fetus and infant Etje and Miller's 1952 adult bonobo; distal 2/3 in infant Foyo and adult Kidogo and adolescent Jasiri), the two heads fusing to attach onto the proximal fibula (our fetus and adult Kidogo, confirmed on the right side of this latter adult by BS) and often also onto the lateral condyle of the tibia (infant Foyo and adult Kidogo, Miller's 1952 adult bonobo). On the right side of our adult Kidogo, there is a variant 'intermediate' belly of the biceps femoris that originates with the short head of the biceps femoris from the ischial tuberosity and then fuses with the short head distally, inserting with it onto the lateral aspect of the proximal fibular head, also wrapping around to insert onto the tibia.
- Usual innervation: Tibial division of the sciatic nerve (Miller 1952: *P. paniscus*), although she might well have forgotten the innervation of the short head, which is usually by the fibular division of the sciatic nerve in humans and common chimpanzees; flexores femoris nerve for long and 'intermediate' heads of our adult Kidogo, and common fibular division of sciatic nerve for the short head of this specimen (our dissections: *P. paniscus*) [*P. troglodytes*: caput longum by flexores femoris nerve as is usually the case in other apes (and not by the tibial division of sciatic nerve, which is usually the case in modern humans); caput breve by common fibular nerve as is usually the case in other apes and in modern humans (Champneys 1872;Sigmon 1974)].
- Comparison with common chimpanzees: In *P. troglodytes*, the long head of the biceps femoris originates from the ischial tuberosity, often in common with the semitendinosus (Champneys 1872; Beddard 1893; Prejzner-Morawska and Urbanowicz 1971; Sigmon 1974; Hamada 1985; Kumakura 1989) and less often with the semimembranosus (Champneys 1872; Beddard 1893), the gluteus maximus (Champneys 1872; Sigmon 1974) and/or the posterior part of the gracilis (Champneys 1872). In *P. troglodytes*, the short head originates from the dorsolateral femur in the region of the lateral lip of the linea aspera (Beddard 1893; Prejzner-Morawska and Urbanowicz 1971; Sigmon 1974; Hamada 1985; Kumakura 1989), extending more distally in great apes than in *Hylobates* and modern humans (e.g. Sigmon 1974). The short head is absent from a single specimen of *P. troglodytes* (Sutton 1883), the two heads being fused in 21/26 common chimpanzees according to the literature review of Gibbs (1999). In our *P. troglodytes* specimen VU PT1, there is an additional belly of the short head of the biceps femoris inserting onto the tendon of insertion of the long head of this muscle. In *P. troglodytes*, the insertion of the long head of biceps femoris is usually to the tibial head (Hepburn 1892; Prejzner-Morawska and Urbanowicz 1971; Sigmon 1974) and to the tibial tuberosity, or condyle (Champneys 1872; Dwight 1895; Sonntag 1923; Kaplan 1958a, b; Prejzner-Morawska and Urbanowicz 1971; Sigmon 1974) and often to the fibular head and fascia of the leg (Champneys 1872; Beddard 1893;Dwight 1895; Prejzner-Morawska and Urbanowicz 1971; Sigmon 1974; Hamada 1985; Kumakura 1989) and/or to the iliotibial tract (Beddard 1893; Sigmon 1974; Hamada 1985). In *P. troglodytes*, the short head inserts onto the fibular head and fascia of the leg (Champneys 1872; Hepburn 1892; Beddard 1893; Dwight 1895; Prejzner-Morawska and Urbanowicz 1971; Sigmon 1974; Hamada 1985; Kumakura 1989) and occasionally onto

the lateral intermuscular septum (Prejzner-Morawska and Urbanowicz 1971).

***Semimembranosus*** (adult Kidogo 95.54 g; infant Foyo 1.39 g; fetus 0.53 g; Figs. 5.26, 5.27 and 5.28)

- Attachments in bonobos: From ischial tuberosity to medial tibial condyle (our fetus and infants Foyo and Etje and adult Kidogo and adolescent Jasiri), as well as adjacent structures such as the joint capsule and the medial collateral ligament (adult Kidogo and adolescent Jasiri). A variant was described in Miller's 1952 adult bonobo, in which there were two semimembranosus structures: the '*semimembranosus proprius*' running from ischial tuberosity to medial tibial condyle via an aponeurosis that extends down the leg with the crual fascia, and the '*semimembranosus accessorius*' running from the inferior ramus and adjacent tuberosity of the ischium to the medial epicondyle of the femur, fusing with the insertion of the adductor magnus onto the linea aspera of the femur.
- Usual innervation: Tibial division of the sciatic nerve (Miller 1952: *P. paniscus*); flexores femoris nerve (our dissections: *P. paniscus*) [*P. troglodytes*: by the flexores femoris nerve, as is usually the case in other apes (not by the tibial division of the sciatic nerve, as is usually the case in modern humans: Sigmon 1974)].
- Comparison with common chimpanzees: In *P. troglodytes*, it runs from the ischial tuberosity, distal and lateral to the semitendinosus (Champneys 1872; Sigmon 1974) and sometimes fused with this latter muscle (Beddard 1893; Champneys 1872), to the dorsal surface of the medial tibial condyle through a rounded tendon (Champneys 1872; Hepburn 1892; Beddard 1893; Sigmon 1974).

***Semitendinosus*** (adult Kidogo 138.94 g; infant Foyo 4.37 g; fetus 0.60 g; Figs. 5.2, 5.8, 5.21, 5.26, 5.27, 5.28 and 5.29)

- Attachments in bonobos: From ischial tuberosity to proximal tibia (fused with semimembranosus and long head of biceps femoris), deep to the gracilis and sartorius (our fetus and infants Foyo and Etje, Miller's 1952 adult bonobo). It does not seem to form a *pes anserinus* in our fetus and infant Edge, in the sense that it is not fused distally the gracilis and sartorius or closely associated with them distally through an aponeurosis, contrary to Miller's 1952 adult bonobo and our infant Foyo, in which the three muscles insert together via a common aponeurosis. Although the semitendinosus and gracilis blend their fibres in our adult Kidogo and adolescent Jasiri, there is not a true *pes anserinus* tendon because the sartorius does not contribute fibres to the insertion (see, e.g., Fig. 5.8).
- Usual innervation: Tibial division of the sciatic nerve (Miller 1952: *P. paniscus*); flexores femoris nerve (our dissections: *P. paniscus*) [*P. troglodytes*: by the flexores femoris nerve, as is usually the case in other apes (not by the tibial division of the sciatic nerve, as is usually the case in modern humans: Champneys 1872;Sigmon 1974)].
- Comparison with common chimpanzees: In *P. troglodytes*, it runs from the ischial tuberosity in common with the long head of the bíceps femoris (Champneys 1872; Hepburn 1892; Sigmon 1974) and sometimes also with the semimembranosus (Champneys 1872; Hepburn 1892; Beddard 1893; Sigmon 1974), to the medial tibial surface just distal to the tibial tuberosity through a narrow, flat tendon (Champneys 1872; Beddard 1893) extending further distally than in modern humans and *Gorilla* (Champneys 1872; Hepburn 1892). In some specimens of great apes, there is also an aponeurotic expansion to the fascia of the leg in this region (Humphry 1867; Champneys 1872). An oblique tendinous intersection in the muscle belly of the semitendinosus is occasionally present in common chimpanzees (MacAlister 1871; Hepburn 1892).

***Extensor digitorum longus*** (adult Kidogo 44.69 g; infant Foyo 1.76 g; fetus 0.34 g; Figs. 5.13, 5.14 and 5.34)

- Attachments in bonobos: The extensor digitorum longus is fused with the extensor hallucis longus proximally in Miller's 1952 adult bonobo. A single origin from the fibula (proximal 1/3), as well as from the interosseous membrane, was reported in the two adult bonobos described by Vereecke et al. (2005) and was also seen in our adult Kidogo and adolescent Jasiri. In the other specimens, the extensor digitorum longus runs from the tibia (lateral condyle in Miller's 1952 adult bonobo and our infants Foyo and Etje and fetus) and fibula (head and proximal 3/4 of the anterior surface in Miller's 1952 adult bonobo and our fetus; only fibular head in our infants Foyo and Etje). Insertion is onto digits 2–5 all the way to the distal phalanges via the extensor expansions (two adult bonobos of Vereecke et al. 2005; Miller's 1952 adult bonobo; our fetus and infants Foyo and Etje and adolescent Jasiri and adult Kidogo). The extensor digitorum longus has an extra tendon to the midshaft of MV in our fetus and infant Etje, very similar to the *fibularis tertius* of humans, or an extra tendon that partially attaches to MV but then continues to the proximal phalanx of digit 5, to attach onto the expansor expansion of this digit, in our adolescent Jasiri (Fig. 5.14). The configuration seen in Jasiri thus corresponds to that found in at least some non-human primates where the name '*fibularis digiti quinti*' has been used, thus showing that the use of this name in those primates was erroneous, because the extra tendon going to digit 5 is clearly part of the extensor digitorum longus, and also clearly homologous to the fibularis tertius present in our fetus and infant Etje (the main difference is that it extends a bit further distally to attach onto digit 5 itself, but its topology and group identity – i.e., from long extensor group – origin is the same). That is, it is not the short extensor of digit 5, which is the structure that gives rise to the true fibularis digiti quinti of other mammals: this is further shown by the fact that in, e.g., our fetus both the fibularis tertius (i.e. extra tendon of extensor digitorum longus to MV) and the tendon of the extensor digitorum brevis (i.e. the short extensor to digit 5) were both present, in the feet. These data thus strongly support the view that the fibularis tertius of modern humans is derived from the extensor digitorum longus and show this muscle is in fact present as a variant in bonobos (see comparison with common chimpanzees just below).

- Usual innervation: Branches of the deep fibular nerve (Miller 1952, our dissections: *P. paniscus*).
- Function: Dorsiflexion of ankle and toes, foot eversion (Vereecke et al. 2005: *P. paniscus*).
- Comparison with common chimpanzees: In *P. troglodytes*, it runs from the head and medial crest of the fibula, intermuscular septum and the lateral tibial condyle (Beddard 1893; MacDowell 1910; Kaplan 1958a; Lewis 1966, 1989) and sometimes also from the interosseus membrane (Beddard 1893; MacDowell 1910; Kaplan 1958a) and/or the crural fascia (Beddard 1893; Kaplan 1958a), to the extensor expansions of digits 2–5 - in our common chimpanzee photographic atlas (Diogo et al. 2013) we erroneously stated that it was inserted only onto digits 2, 3 and 4–(Hepburn 1892; Beddard 1893; MacDowell 1910; Lewis 1966), the slip to digit 2 being occasionally atrophied (Beddard 1893). According to the literature review of Gibbs (1999), the *fibularis tertius* ('peroneus tertius') is present in 5% of *P. troglodytes*; in the common chimpanzees in which we examined this region in detail (VU PT1, GWUANT 1 and GWUANT 2), the fibularis tertius is missing, but we did find a structure that seems to correspond to it in three of the bonobo specimens dissected by us, as noted above.

***Extensor hallucis longus*** (adult Kidogo 11.38 g; infant Foyo 0.78 g; fetus 0.12 g; Figs. 5.13, 5.14, 5.34 and 5.35)

- Attachments in bonobos: From middle 1/3 of fibula and interosseous membrane, passing along dorsal aponeurosis of hallux to insert onto proximal phalanx of hallux, extending to its distal phalanx via an extensor expansion (Miller's 1952 adult bonobo, Vereecke et al.'s 2005 two adult bonobos, our fetus and infants

Foyo and Etje), while it inserts only onto the distal phalanx of the big toe in our adult Kidogo and adolescent Jasiri.
- Usual innervation: Branches of the deep portion of the fibular nerve (Miller 1952: *P. paniscus*).
- Function: Extension and abduction of hallux, foot inversion (Vereecke et al. 2005: *P. paniscus*).
- Comparison with common chimpanzees: In *P. troglodytes*, it originates from the medial surface of the fibula and the interosseous membrane (Beddard 1893 Lewis 1966) and occasionally also from the lateral tibial condyle (Lewis 1966) and/or the crural fascia and intermuscular septa (Beddard 1893); the tendon of insertion passes along the shaft of MI (Hepburn 1892; Beddard 1893) and inserts onto the distal phalanx of the hallux (Beddard 1893; Lewis 1966).

***Tibialis anterior*** (adult Kidogo 65.01 g; infant Foyo 3.32 g; fetus 0.51 g; Figs. 5.13 and 5.35)

- Attachments in bonobos: It runs from tibia (our adult Kidogo and adolescent Jasiri; specifically from lateral condyle and proximal 1/2 in Miller's 1952 adult bonobo, our fetus, and Vereecke et al.'s 2005 two adult bonobos; proximal 2/3 in our infant Foyo) and often also from the interosseous membrane (our adult Kidogo and adolescent Jasiri), splitting into two bellies relatively close to its origin and then inserting onto the plantar surface of the medial cuneiform and onto the proximal end of the MI (Miller's 1952 adult bonobo, our fetus and our adult Kidogo and adolescent Jasiri, and Vereecke et al.'s 2005 two adult bonobos). Therefore, the naming of the belly with the metatarsal insertion as a separate muscle, the '*abductor hallucis longus*' (see Fig. 5.36) in common chimpanzees (see below) would make some sense although this structure is considered here as a second head of the muscle tibialis anterior. We could not see – but this does not necessarily mean it was not present – an insertion onto the medial cuneiform in our infant Foyo.
- Usual innervation: Branch of the deep fibular nerve (Miller 1952, our dissections: *P. paniscus*).
- Function: Foot dorsiflexion and inversion (Vereecke et al. 2005: *P. paniscus*).
- Comparison with common chimpanzees: In *P. troglodytes*, it runs from the lateral aspect of the tibia (Sutton 1883; Beddard 1893; MacDowell 1910;Lewis 1966), splitting into two bellies relatively close to its origin and then inserting onto the plantar surface of the medial cuneiform and onto the proximal end of the MI (Humphry 1867; Sutton 1883; Hepburn 1892; Beddard 1893; MacDowell 1910; Lewis 1966). The metatarsal insertion has sometimes been described as part of a separate muscle, the '*abductor hallucis longus*' (Humphry 1867; Sutton 1883): see just above.

***Extensor digitorum brevis*** (adult Kidogo 7.59 g; infant Foyo 0.57 g; fetus 0.12 g; Figs. 5.14 and 5.35)

- Attachments in bonobos: From calcaneus to proximal phalanges of digits 2, 3 and 4 (Vereecke et al.'s 2005 two adult bonobos and our infant Etje and adult Kidogo and adolescent Jasiri), and often also 5 (Miller's 1952 adult bonobo, our fetus and infant Foyo) (Fig. 5.35), extending to distal phalanges via extensor expansions; we could not see – but it does not mean it was not present – an insertion onto digit 2 in the infant Foyo (it seemingly inserted onto digits 3-5 only).
- Usual innervation: Deep branch of the fibular nerve (Miller 1952, our dissections: *P. paniscus*).
- Function: Extension of digits 2-4 (Vereecke et al. 2005: *P. paniscus*).
- Comparison with common chimpanzees: A calcaneal origin and division into three tendons for the dorsal aponeurosis of digits 2, 3 and 4 is the configuration usually present in common chimpanzees (Champneys 1872; Ruge 1878a; Hepburn 1892; Beddard 1893; Lewis 1966), but sometimes there is an additional tendon for digit 5 that merges with the tendon of the fibularis brevis (Gibbs 1999).

***Extensor hallucis brevis*** (adult Kidogo 4.34 g; infant Foyo 0.21 g; fetus 0.05 g; Figs. 5.34 and 5.35)

- Attachments in bonobos: From calcaneus to base of proximal phalanx of hallux (Miller's 1952 adult bonobo, our fetus and infants Foyo and Etje and adolescent Jasiri, Vereecke et al.'s 2005 two adult bonobos); as a variant, in our adult Kidogo it inserted onto the head of MI.
- Usual innervation: Deep branch of the fibular nerve (Miller 1952: *P. paniscus*).
- Function: Extension of big toe, foot supination (Vereecke et al. 2005: *P. paniscus*).
- Comparison with common chimpanzees: In *P. troglodytes*, it is a separate muscle that originates from the extensor digitorum brevis according to Champneys (1872), Hepburn (1892), Beddard (1893) and Lewis (1966) and which according to Hepburn (1892), and Lewis (1966) inserts onto the base of the proximal phalanx of the hallux, while Dwight (1895) reported an insertion onto the distal phalanx of this digit in one chimpanzee specimen. In our *P. troglodytes* specimens GWUANT PT2 and VU PT1, the extensor hallucis brevis is a distinct muscle running from the calcaneus to the proximal phalanx of the hallux.

***Fibularis brevis*** (adult Kidogo 27.74 g; infant Foyo 0.93 g; fetus 0.19 g; Figs. 5.13 and 5.37)

- Attachments in bonobos: From distal 1/2 of fibula – and often from intermuscular septa (Miller's 1952 adult bonobo) – to base of MV (Miller's 1952 adult bonobo; our fetus and infants Foyo and Etje and adult Kidogo and adolescent Jasiri; Vereecke's et al. 2005 two adult bonobos).
- Usual innervation: Branches of the superficial fibular nerve (Miller 1952, our dissections: *P. paniscus*).
- Function: Foot eversion and plantar flexion (Vereecke et al. 2005: *P. paniscus*).
- Comparison with common chimpanzees: In *P. troglodytes*, it runs from the distolateral fibula and from the intermuscular septa (Beddard 1893; MacDowell 1910) to the tuberosity at the base of MV and the extensor digitorum tendon (Champneys 1872; Hepburn 1892; Beddard 1893; Dwight 1895; MacDowell 1910; Lewis 1966), the muscle having sometimes a double insertion (Champneys 1872) and often having a small tendon to the proximal and/or middle phalanges of digit 5 (Vrolik 1841; Champneys 1872; MacDowell 1910; Prejzner-Morawska and Urbanowicz 1971). In our *P. troglodytes* VU PT1 specimen, the muscle has an insertion onto the base of the proximal phalanx of digit 5, whereas in our *P. troglodytes* GWUANT PT2 specimen it inserts onto MV.
- Synonymy: Peroneus brevis (Gibbs 1999).

***Fibularis longus*** (adult Kidogo 42.16 g; infant Foyo 1.87 g; fetus 0.38 g; Figs. 5.13, 5.14, 5.34, 5.35 and 5.37)

- Attachments in bonobos: From proximal 1/2 of fibula, including fibular head (Miller's 1952 adult bonobo, our fetus and infant Etje and adult Kidogo and adolescent Jasiri; Vereecke et al.'s 2005 two adult bonobos), and sometimes also from lateral tibial condyle (Miller's 1952 adult bonobo) and/or the intermuscular septum (our adult Kidogo and adolescent Jasiri), passing through a groove on cuboid bone (Miller's 1952 adult bonobo) to insert onto base of MI (Miller's 1952 adult bonobo, Vereecke et al.'s 2005 two adult bonobos, our fetus and infants Foyo and Etje and adult Kidogo and adolescent Jasiri).
- Usual innervation: Branches of the superficial fibular nerve (Miller 1952, our dissections: *P. paniscus*).
- Function: Foot eversion and plantar flexion; flexion and adduction of hallux (Vereecke et al. 2005: *P. paniscus*).
- Comparison with common chimpanzees: In *P. troglodytes*, it runs from the fibular head and adjacent proximal fibula (Beddard 1893; MacDowell 1910; Lewis 1966) and sometimes also from the lateral tibial condyle (Lewis 1966) and/or the intermuscular septa (Beddard 1893; Boyer 1935), to the tuberosity

of MI (Hepburn 1892; Beddard 1893; Dwight 1895; MacDowell 1910; Lewis 1966) and occasionally via a fibrous attachment to MV (Lewis 1966; our *P. troglodytes* specimen GWUANT PT2).
- Synonymy: Peroneus longus (Gibbs 1999; Miller 1952).

***Gastrocnemius*** (adult Kidogo 167.81 g; infant Foyo 5.43 g; fetus 0.69 g; Figs. 5.6, 5.8, 5.13, 5.31, 5.36, 5.37 and 5.38)

- Attachments in bonobos: Two heads, running from medial and lateral femoral condyles, and sometimes the capsule of knee joint (Miller's 1952 adult bonobo and our adult Kidogo and adolescent Jasiri), to calcaneal tuberosity, fusing with soleus and plantaris (when present) distally to form the calcaneal (Achilles) tendon (Miller's 1952 adult bonobo, our fetus and infants Foyo and Etje; Vereecke et al.'s 2005 two adult bonobos).
- Usual innervation: Tibial nerve (Miller 1952: *P. paniscus*).
- Function: Foot plantar flexion (Vereecke et al. 2005: *P. paniscus*).
- Comparison with common chimpanzees: In *P. troglodytes*, the gastrocnemius has two heads and joins with the soleus, originating from the medial and lateral femoral condyles and the capsule of the knee joint (Beddard 1893; Hepburn 1892; Frey 1913), inserting onto the calcaneal tuberosity (MacAlister 1871; Beddard 1893). Within the primates dissected by us, the tendo calcaneus is better defined and more substantial in *Hylobates* and *Pongo* than it is in *Gorilla* and *Pan*, but in no extant apes is it as well defined as is usually the case in modern humans.

***Plantaris*** (adult Kidogo 5.86 g; infant Foyo 0.17 g)

- Attachments in bonobos: From lateral head of gastrocnemius (i.e. fused with that muscle proximally in both sides of our infants Foyo and Etje and one side of Miller's 1952 adult bonobo) or from lateral epicondyle of femur and adjacent knee joint (on the other side of Miller's 1952 adult bonobo, so only fused with the gastrocnemius and soleus distally on that side; the configuration of this latter side is found on both sides of our adult Kidogo and adolescent Jasiri) or just the lateral femoral condyle (Vereecke et al.'s 2005 two adult bonobos) to calcaneal tuberosity, together with soleus and gastrocnemius (both sides of Miller's 1952 adult bonobo and of our infants Foyo and Etje and adult Kidogo and adolescent Jasiri, and Vereecke et al.'s 2005 two adult bonobos). The muscle was absent, or completely fused/undifferentiated with the gastrocnemius, in the both lower limbs of our bonobo fetus.
- Usual innervation: Tibial nerve (Miller 1952: *P. paniscus*).
- Function: Foot plantar flexion (Vereecke et al. 2005: *P. paniscus*).
- Comparison with common chimpanzees: According to the literature review of Gibbs (1999), the plantaris in common chimpanzees is absent in 40%, is bilaterally present in 40% and unilaterally present in 20%. The muscle was unilaterally present in the two common chimpanzee specimens in which we analysed this region in detail (GWUANT PT2 and VU PT1), running from the lateral femoral supracondylar line, the lateral femoral epicondyle and the oblique popliteal ligament to the calcaneal tuberosity.

***Soleus*** (adult Kidogo 125.37 g; infant Foyo 0.67 g; fetus 0.61 g; Figs. 5.9, 5.37 and 5.38)

- Attachments in bonobos: From fibula (head and proximal 1/3 of fibula, plus capsule of knee joint, in Miller's 1952 adult bonobo; just fibular head in our fetus and infants Foyo and Etje; fibular head and interosseus membrane in Vereecke's et al.2005 two adult bonobos) and sometimes from proximal tibia (adult Kidogo and adolescent Jasiri) to calcaneal tuberosity (Miller's 1952 adult bonobo and our fetus and infants Foyo and Etje and adult Kidogo and adolescent Jasiri, and Vereecke's et al.2005 two adult bonobos). In our adult

Kidogo and adolescent Jasiri, the soleus was partially blended proximally with the flexor hallucis longus.

- Usual innervation: Tibial nerve (Miller 1952: *P. paniscus*).
- Function: Foot plantar flexion (Vereecke et al. 2005: *P. paniscus*).
- Comparison with common chimpanzees: In *P. troglodytes*, it has an origin from the head and dorsal aspect of the fibular shaft (Humphry 1867; Sutton 1883; Hepburn 1892; Beddard 1893; Lewis 1962), although this origin is sometimes reported as absent (Vrolik 1841); the tibial origin is as often absent (Champneys 1872; MacAlister 1871; MacDowell 1910) as present (Vrolik 1841; Humphry 1867; Champneys 1872; MacAlister 1871; Hepburn 1892; MacDowell 1910; Frey 1913; Lewis 1962), but it was present in the two common chimpanzees (GWUANT PT2 and VU PT1) in which we analysed this region in detail. In *P. troglodytes* the tibial head may be included with the lateral part of the gastrocnemius in those cases where it is reported as absent (Gibbs 1999). In common chimpanzees, the insertion of the soleus is onto the calcaneal tuberosity (MacAlister 1871; Beddard 1893).

***Popliteus*** (adult Kidogo 21.66 g; infant Foyo 0.82 g; fetus 0.16 g; Fig. 5.12)

- Attachments in bonobos: From lateral femoral condylar head and often also from knee capsule (Miller's 1952 adult bonobo, our fetus and infants Foyo and Etje), as well as from fibular head by a distinct tendon (our fetus and infant Etje and adolescent Jasiri and adult Kidogo) (Fig. 5.12) and sometimes also from medial femoral condyle via a tendon that seems to resemble the menisco-femoral ligament of modern humans (our fetus). It inserts onto tibia (Miller 1952; proximal 1/3 in our fetus and infants Foyo and Etje and adult Kidogo and adolescent Jasiri), along the edge of the medial collateral ligament in our adult Kidogo and adolescent Jasiri, where the muscle is fused with the flexor digitorum longus. It was not possible to discern if the infant Foyo had the distinct tendon of origin from the fibular head.
- Usual innervation: Tibial nerve (Miller 1952: *P. paniscus*).
- Comparison with common chimpanzees: In *P. troglodytes*, it runs from the lateral femoral condylar head (Hepburn 1892; Beddard 1893) and often also from the knee capsule (Hepburn 1892), as well as from the tibial collateral ligament in our *P. troglodytes* specimen VU PT1, to the dorsal tibial surface (Beddard 1893) as well as to the lateral surface of this latter bone in this specimen VU PT1.

***Flexor digitorum longus*** (infant Foyo 2.21 g; fetus 0.35 g; Figs.5.9, 5.10, 5.11, 5.15, 5.18, 5.36 and 5.38)

- Attachments in bonobos: From tibia (midshaft in our fetus; proximal 2/3 in infant Foyo; proximal 1/3 in Miller's 1952 adult bonobo and Vereecke et al.'s 2005 two adult bonobos) to distal phalanges of digits 2 and 5 (our fetus and infants Foyo and Etje and adult Kidogo and adolescent Jasiri, and Miller's 1952 adult bonobo and Vereecke et al.'s 2005 two adult bonobos). Fused with flexor hallucis longus and flexor digitorum brevis, as well as with lumbricales (Miller's 1952 adult bonobo, Vereecke et al.'s 2005 two adult bonobos, and our fetus and infants Etje and Foyo and adult Kidogo and adolescent Jasiri).
- Usual innervation: Tibial nerve (Miller 1952: *P. paniscus*).
- Function: Flexion of digits; foot plantar flexion (Vereecke et al. 2005: *P. paniscus*).
- Comparison with common chimpanzees: In *P. troglodytes*, it runs from the dorsal aspect of the tibial shaft (Sutton 1883; Beddard 1893; MacDowell 1910) and sometimes from the intermuscular septum (Beddard 1893), the tendon distribution to the digits varies among the apes. According to the literature review of Gibbs (1999), digit 5 is supplied in all apes, and in common chimpanzees digit 4 is supplied in 4/23, digit 3 in 2/23 specimens and digit 2 in 19/23 specimens (Beddard 1893; Champneys 1872; Dwight 1895; Hepburn 1892; Humphry

1867; Keith1894a, b; Lewis 1962; MacDowell 1910; Sutton 1883). When present in *P. troglodytes*, the tendon to digit 2 is often fused with the flexor hallucis longus (Champneys 1872), and the muscle may be fused to a lesser or greater degree with the tendon of flexor digitorum brevis (Sutton 1883). However, at least some of these reports about *P. troglodytes* are probably erroneous due to a confusion between the tendons of the flexor digitorum longus and flexor hallucis longus (see below), because in our *P. troglodytes* specimens GWUANT PT2 and VU PT1 the insertion is onto digits 2, 3, 4 and 5, as it is in the bonobos dissected by us, so it seems likely that an insertion onto these four digits is in fact the common condition for both common chimpanzees and bonobos.
- Synonymy: Flexor digitorum medialis or tibialis (Gibbs 1999; Vereecke et al. 2005).

***Flexor hallucis longus*** (infant Foyo 4.69 g; fetus 0.94 g; Figs.5.9, 5.10, 5.11, 5.12, 5.15, 5.17, and 5.37)

- Attachments in bonobos: From interosseous membrane and posterior crural intermuscular septum and often also from fibula (Miller's 1952 adult bonobo; Vereecke et al.'s 2005 two adult bonobos from proximal 1/2 of fibula and interosseous membrane), or simply from the whole dorsal side of fibula (our fetus) or proximal 2/3 of fibula (our infant Foyo) or proximal 1/4 of fibula (adult Kidogo and adolescent Jasiri) to distal phalanges of digits 1, 3 and 4 (Miller's 1952 adult bonobo, our fetus and infants Foyo and Etje and adult Kidogo and adolescent Jasiri, and Vereecke et al.'s 2005 two adult bonobos), seemingly also sending a tendon to digit 2 in and adolescent Jasiri), as a variant.
- Usual innervation: Tibial nerve (Miller 1952: *P. paniscus*).
- Function: Flexion of digits; foot plantar flexion and inversion (Vereecke et al. 2005: *P. paniscus*).
- Comparison with common chimpanzees: In *P. troglodytes*, it runs from the interosseous membrane and dorsal side of crural intermuscular septum and often also from the distodorsal aspect of the fibula (Sutton 1883; Champneys 1872; Beddard 1893); in one case the fibular origin was reported as the proximal two-thirds (Beddard 1893). In *P. troglodytes* the distal attachment is to the base of the distal phalanx of the hallux and sometimes to digits 2, 3 and/or 4 (Humphry 1867; Champneys 1872; Sutton 1883; Hepburn 1892; MacDowell 1910; Lewis 1962). In our *P. troglodytes* specimen VU PT1, the muscle inserted onto digits 1, 3 and 4, while in our *P. troglodytes* specimen GWUANT PT2 it inserted onto digits 1, 2, 3 and 4.
- Synonymy: Flexor digitorum lateralis or fibularis (Gibbs 1999; Vereecke et al. 2005).

***Tibialis posterior*** (adult Kidogo 78.27 g; infant Foyo 1.82 g; fetus 0.43 g; Figs. 5.9 and 5.36)

- Attachments in bonobos: From interosseous membrane and adjoining surfaces of tibia (proximal 2/3 in our fetus; proximal 1/2 in Miller's 1952 adult bonobo and in Vereecke et al.'s 2005 two bonobo adults; proximal 4/5 in our adult Kidogo and adolescent Jasiri) and fibula (proximal 1/2 in our fetus and in Miller's 1952 adult bonobo; proximal 2/3 in our adult Kidogo and adolescent Jasiri) to navicular, MII, MIII and MIV, and medial, intermediate and lateral cuneiforms (Miller's 1952 adult bonobo, our fetus and adult Kidogo and adolescent Jasiri, and in Vereecke et al.'s 2005 two bonobo fetus), but simply to medial cuneiform and navicular in Vereecke et al.'s 2005 two bonobo adults. The tendon of insertion of the tibialis posterior was fused with the flexor hallucis brevis. Attachments onto the interosseous membrane and navicular were also seen in our infants Foyo and Etje, but we could not discern the other attachments.
- Usual innervation: Tibial nerve (Miller 1952: *P. paniscus*).
- Function: Foot plantar flexion and inversion (Vereecke et al. 2005: *P. paniscus*).
- Comparison with common chimpanzees: In *P. troglodytes*, it runs from the interosseous membrane and the adjoining surfaces of the

tibia and fibula (Beddard 1893) and occasionally from the intermuscular septa (Beddard 1893), the medial part of the muscle inserting onto the navicular bone (Hepburn 1892; Lewis 1964) and the lateral part onto MII, MIII and MIV (Lewis 1964), onto the plantar ligaments (Hepburn 1892; Beddard 1893), and/or onto the sheath of the tendon of fibularis longus (Humphry 1867; Hepburn 1892; Lewis 1964). The cuneiform insertion in *P. troglodytes* can be to the medial (Humphry 1867; Lewis 1964), medial and lateral (Hepburn 1892) or lateral (Champneys 1872) cuneiforms, but never to all three.

***Abductor digiti minimi*** (adult Kidogo 6.81 g; infant Foyo 0.38 g; fetus 0.10 g; Figs. 5.15 and 5.18)

- Attachments in bonobos: From calcaneus to base of proximal phalanx of digit 5 (our fetus and infants Foyo and Etje and adolescent Jasiri and adult Kidogo, Miller's 1952 adult bonobo, and Vereecke et al.'s 2005 two adult bonobos: in these two latter bonobos, the abductor digiti minimi sent two tendons to the proximal phalanx of digit 5 and was fused distally to the flexor digiti minimi brevis) and sometimes also to base of MV (Miller's 1952 adult bonobo and our adult Kidogo and adolescent Jasiri), forming a well-developed '*abductor ossis metatarsi quinti*' bundle *sensu* Miller (1952). It should be noted that this so-called 'abductor ossis metatarsi quinti' is just a bundle of the abductor digiti minimi, not a separate muscle and that it does not correspond to the opponens digiti minimi *sensu* the present work, which was also described by Miller (1952) in her adult bonobo.
- Usual innervation: Lateral plantar nerve (Miller 1952: *P. paniscus*).
- Function: Flexion and abduction of digit 5(Vereecke et al. 2005: *P. paniscus*).
- Comparison with common chimpanzees: In *P. troglodytes*, it runs from the medial and lateral calcaneus (Champneys 1872; Beddard 1893) and sometimes from the plantar aponeurosis (Champneys 1872; Beddard 1893) and/or the calcaneal tuberosity (our specimen VU PT1), to the proximal phalanx of digit 5 (Champneys 1872; Beddard 1893) and often to MV Champneys (1872) and Beddard (1893), and occasionally to the distal middle phalanx of digit 5 (Vrolik 1841). It should be noted that the part of the 'abductor digiti minimi' reported by authors such as Champneys (1872) and Beddard (1893) to insert onto the MV in common chimpanzees corresponds to the '*abductor ossis metatarsi quinti*' bundle *sensu* Miller (1952), i.e., it is just a bundle of the abductor digiti minimi, as explained just above.
- Synonymy: Abductor digiti quinti (Miller 1952; Gibbs 1999).

***Abductor hallucis*** (adult Kidogo opponens hallucis + abductor hallucis =26.85 g; infant Foyo opponens hallucis + abductor hallucis =0.80 g; fetus adductor hallucis 0.13 g; Figs. 5.15 and 5.17)

- Attachments in bonobos: Our dissections indicate that the '*abductor ossis metatarsi hallucis*' *sensu* Miller 1952 (and *sensu* Chapman 1879 in *P. troglodytes*) is in fact simply an *opponens hallucis*, which can be more separated from (e.g. our fetus), or more fused with (e.g. our infants Foyo and Etje and adult Kidogo and adolescent Jasiri), the abductor hallucis, because it is deep to the abductor hallucis and runs from the medial cuneiform to the distal part of MI (our fetus and infants Foyo and Etje and adult Kidogo and adolescent Jasiri). The abductor hallucis proper runs from calcaneal tuberosity to base of proximal phalanx of digit 1 and/or adjacent structures such as the first metatarsophalangeal joint (our fetus and infants Foyo and Etje and adolescent Jasiri and adult Kidogo, and Miller's 1952 adult bonobo). The muscle 'abductor hallucis' described by Vereecke et al. (2005) thus probably corresponds to the opponens hallucis + abductor hallucis *sensu* the present work, because it is a pennate structure that has two heads and runs from the calcaneal tuberosity and navicular tuberosity to the proximal phalanx of the big toe and the adjacent medial sesamoid bone.

- Usual innervation: Medial plantar nerve (Miller 1952: *P. paniscus*).
- Function: Abductor hallucis + opponens hallucis do abduction and flexion of big toe (Vereecke et al. 2005: *P. paniscus*).
- Comparison with common chimpanzees: In *P. troglodytes*, the abductor hallucis runs from the medial and plantar surfaces of the calcaneus (Brooks 1887; Hepburn 1892; Beddard 1893) and sometimes also from the medial part of the plantar aponeurosis (Hepburn 1892; Beddard 1893), to the base of the proximal phalanx of the hallux (Vrolik 1841; Brooks 1887; Hepburn 1892; Beddard 1893) and occasionally also to the medial and/or lateral cuneiforms (Vrolik 1841). According to Chapman (1879), some common chimpanzees have a separate slip that goes to the base of MI, which has been named the '*abductor ossis metatarsi hallucis*', but corresponds very likely to the opponens hallucis *sensu* the present work, as noted just above. In fact, an *opponens hallucis* has been described as a variant in *P. troglodytes* (Vrolik 1841; Champneys 1872). In our *P. troglodytes* specimen GWANT PT2, there is a slip deep to the main body of the abductor hallucis that runs from the medial surface of the calcaneus and from the navicular bone to insert together with the main tendon of the abductor hallucis onto the proximal phalanx of the hallux; this slip could to correspond to the 'abductor ossis metatarsi hallucis' and thus to the opponens hallucis, but it does not insert onto the MI and originates more proximally to the opponens hallucis we found in our bonobos.
- Synonymy: Part of 'abductor digiti minimi' sensu Miller (1952) and Vereecke et al. (2005), the other part actually corresponding to the opponens digiti minimi sensu the present work (see below).

***Opponens hallucis*** (adult Kidogo opponens hallucis + abductor hallucis = 26.85 g; infant Foyo opponens hallucis + abductor hallucis 0.80 g; fetus opponens hallucis 0.02 g)

- Attachments in bonobos: As noted just above, the '*abductor ossis metatarsi hallucis*' *sensu* Miller 1952 (and *sensu* Chapman 1879 in *P. troglodytes*) is in fact simply an *opponens hallucis*, which can be more separated from (e.g. our fetus) or more fused with (e.g. our infants Foyo and Etje and adult Kidogo and adolescent Jasiri) the abductor hallucis, as it is deep to the abductor hallucis and runs from the medial cuneiform to the distal part of MI (our infants Foyo and Etje and adult Kidogo and adolescent Jasiri and fetus). The abductor hallucis proper runs from the calcaneal tuberosity to the base of the proximal phalanx of digit 1 (our fetus and infants Foyo and Etje and adolescent Jasiri, and Miller's 1952 adult bonobo). The muscle 'abductor hallucis' described by Vereecke et al. (2005) thus probably corresponds to the opponens hallucis + abductor hallucis *sensu* the present work, because it is a pennate and has two heads and runs from the calcaneal tuberosity and navicular tuberosity to the proximal phalanx of the big toe and the adjacent medial sesamoid bone.
- Usual innervation: Medial plantar nerve (Miller 1952: *P. paniscus*), because it was described by this author as part of the abductor hallucis.
- Function: Abductor hallucis + opponens hallucis abduct and flex the big toe (Vereecke et al. 2005: *P. paniscus*).
- Comparison with common chimpanzees: As noted just above, according to Chapman (1879), some common chimpanzees have a separate slip that goes to the base of MI, which has been named the '*abductor ossis metatarsi hallucis*', which corresponds very likely to the opponens hallucis *sensu* the present work. An *opponens hallucis* has been actually described as a variant in *P. troglodytes* (Vrolik 1841; Champneys 1872). In our *P. troglodytes* specimen GWANT PT2, there is a slip deep to the main body of the abductor hallucis that runs from the medial surface of the calcaneus and from the navicular bone to insert together with the main tendon of the abductor hallucis onto the proximal phalanx of the hallux; this slip could correspond to the 'abductor ossis

metatarsi hallucis' and thus to the opponens hallucis, but it does not insert onto the MI and originates more proximally to the opponens hallucis we found in our bonobos (see above).

- Synonymy: Part of 'abductor digiti minimi' sensu Miller (1952) and Vereecke et al. (2005), the other part actually corresponding to the abductor hallucis sensu the present work (see above).

***Flexor digitorum brevis*** (adult Kidogo 12.16 g; infant Foyo 0.51 g; fetus 0.11 g; Figs. 5.17 and 5.18)

- Attachments in bonobos: From calcaneus to digits 2 (Miller's 1952 adult bonobo and our fetus and infants Foyo and Etje and adult Kidogo and adolescent Jasiri and Vereecke et al.'s 2005 two adult bonobos), 3 (Miller's 1952 adult bonobo and our fetus and infants Foyo and Etje and adult Kidogo and adolescent Jasiri and Vereecke et al.'s 2005 two adult bonobos), 4 (Miller's 1952 adult bonobo and our fetus and infants Foyo and Etje and Vereecke et al.'s 2005 two adult bonobos; there was no tendon to digit 4 in our adult Kidogo, this being thus a variant; we could not discern if a tendon to digit 4 was present or not in our adolescent Jasiri) and/or 5 (left side of our fetus). Fused with flexor digitorum longus and flexor hallucis longus in Miller's 1952 adult bonobo and Vereecke et al.'s 2005 two adult bonobos and in our adult Kidogo and adolescent Jasiri and infant Etje and Foyo, but not in our fetus; in all specimens, the flexor digitorum brevis was partially fused with the four lumbricales (see Fig. 5.17). The insertion onto digit 5 seen on the left side of the fetus might be seen as a variation within bonobos, but actually it is possible that this is a trait that might represent an early developmental condition for bonobos, the insertion onto digit 5 being then possibly lost in later developmental stages.
- Usual innervation: Medial plantar nerve (Miller 1952: *P. paniscus*).
- Function: Foot plantar flexion and flexion of digits (Vereecke et al. 2005: *P. paniscus*).
- Comparison with common chimpanzees: In *P. troglodytes*, it runs from the calcaneus, usually the medial and ventral surfaces (Champneys 1872; Sutton 1883; Hepburn 1892; Beddard 1893; Dwight 1895; MacDowell 1910; Sarmiento 1983), and sometimes also from the proximal plantar aponeurosis (Champneys 1872) and/or the plantar aspect of the flexor digitorum longus (Sutton 1883; Hepburn 1892; MacDowell 1910), to digits 2 and 3 and often to digits 4 and 5, the tendon to digit 5 being derived from a deep head that occasionally supplies digit 4 (Hepburn 1892; Sarmiento 1983). The muscle goes to digits 2, 3 and 4 in our *P. troglodytes* specimen VU PT1 and to digits 2 and 3 in our *P. troglodytes* specimen GWUANT PT2.

***Quadratus plantae*** (adult Kidogo 0.56 g; infant Foyo 0.12 g; Figs. 5.10 and 5.15)

- Attachments in bonobos: When present, it is a relatively poorly developed muscle running from the lateral margin of the plantar surface of the calcaneus to the tendon of the flexor digitorum longus (our infant Foyo and adult Kidogo and adolescent Jasiri); absent in Miller's 1952 adult bonobo and in three of the four feet of Vereecke et al.'s 2005 two adult bonobos and seemingly in our fetus); we could not discern if the muscle is present or not in our infant Etje (it seemed to be absent, but we are not completely sure).
- Usual innervation: Seemingly by lateral plantar nerve, as in modern humans (see, e.g., *P. paniscus* specimen shown in Fig. 5.15).
- Comparison with common chimpanzees: In *P. troglodytes*, it runs from the lateral margin of the plantar surface of the calcaneus (lateral head, because the medial head is absent in *Pan*; Champneys 1872; Hepburn 1892) to the common tendon of the flexor digitorum longus before the differentiation of the flexor hallucis longus (Chapman 1879; Hepburn 1892), with the exception of some specimens in which the quadratus plantae was present unilaterally and did not reach the flexor digitorum longus (Humphry 1867; Champneys 1872).

Attachment to the flexor digitorum longus tendons to digits 3 and 4 was reported in a specimen of *P. troglodytes* (Dwight 1895).

- Synonymy: Flexor accessorius (Vereecke et al. 2005; Gibbs 1999)

***Lumbricales*** (adult Kidogo lumbricalis 1 = 1.56 g, lumbricalis 2 = 1.40 g, lumbricalis 3 = 1.05 g, lumbricalis 4 = 0.72 g; infant Foyo lumbricales 1 + 2 + 3 + 4 = 0.42 g; fetus lumbricalis 1 = 0.01 g, lumbricalis 2 = 0.01 g, lumbricalis 3 = 0.01 g, lumbricalis 4 = 0.01 g; Figs. 5.15 and 5.18)

- Attachments in bonobos: Lumbrical 1 from tendon of flexor digitorum profundus to digit 2 (our fetus and infants Foyo and Etje); lumbrical 2 from tendons of flexor digitorum profundus to digits 2 and 3 (our fetus and infants Foyo and Etje; Miller's 1952 adult bonobo and Vereecke et al.'s 2005 two adult bonobos reported to originate only from tendon to digit 3); lumbrical 3 from tendons of flexor digitorum profundus to digits 3 and 4 (our fetus and infants Foyo and Etje; Miller's 1952 adult bonobo and Vereecke et al.'s 2005 two adult bonobos reported to originate only from tendon to digit 4); lumbrical 4 from tendons of flexor digitorum profundus to digits 4 and 5 (our fetus and infants Foyo and Etje; Miller's 1952 adult bonobo and Vereecke et al.'s 2005 two adult bonobos reported to originate only from tendon to digit 5). They insert onto the proximal phalanx and then to the extensor expansions of each digit. The four lumbricals were not only fused with the flexor digitorum longus, but also with the flexor digitorum brevis and flexor hallucis longus, as noted above.
- Usual innervation: Medial (to lumbricales 1 and 2) and lateral (to lumbricales 3 and 4) plantar nerves (Miller 1952: *P. paniscus*). Therefore, the innervation of the lumbricales seems to indicate that the axis of the bonobo foot seems to be digit number 3, as seen in the bonobo and common chimpanzee and modern human hand but contrary to the modern human foot, in which it is digit 2 instead.
- Function: Assists in flexion of proximal phalanges (Vereecke et al. 2005: *P. paniscus*).
- Comparison with common chimpanzees: In *P. troglodytes*, there are usually four lumbricales (Sutton 1883; Hepburn 1892; Beddard 1893; Dwight 1895; our specimens GWUANT PT2 and VU PT1) although occasionally only three muscles are present (Beddard 1893); Humphry (1867) reported seven lumbricales in *P. troglodytes*, but this is because he included two muscle structures going to the hallux. In *P. troglodytes*, the second, third and fourth lumbricals have a double origin and insert respectively onto digits 3, 4 and 5, while the first lumbrical usually has a single head from the flexor digitorum longus tendon to digit 2 (Champneys 1872; Sutton 1883; Hepburn 1892; Beddard 1893; Dwight 1895), although in Beddard's (1893) specimen with three lumbricales the first one was reported as having a double origin, with the additional origin arising from the tendon of the flexor hallucis longus to digit 2. In *P. troglodytes*, the tendons of the lumbricales radiate into the extensor aponeuroses of their respective digits (Sutton 1883).

***Adductor hallucis + Adductor hallucis accessorius*** (adult Kidogo adductor hallucis + adductor hallucis accessorius =13.33 g; infant Foyo adductor hallucis caput transversum 0.42 g and caput obliquum 0.63 g; adolescent Jasiri adductor hallucis accessorius 1.20 g; fetus adductor hallucis caput transversum 0.10 g and caput obliquum 0.07 g; fetus adductor hallucis accessorius = 0.08 g; Figs. 5.15, 5.16, 5.17 and 5.18)

- Attachments in bonobos: Oblique head from MII and MIII (our fetus and infants Foyo and Etje and adolescent Jasiri and adult Kidogo) or just MIII (Vereecke et al.'s two adult bonobos) or just MII (Miller's 1952 adult bonobo) and contrahentes raphe (Miller's 1952 adult bonobo and our fetus and infants Foyo and Etje and adolescent Jasiri and adult Kidogo), transverse head from MIII (our fetus and infant Etje) or from both MII and MIII (our infant Foyo) or even from MII, MIII and MIV together (Miller's 1952 adult bonobo and Vereecke et al.'s two adult bonobos) and contrahens raphe (our fetus and infants Foyo and Etje and adolescent Jasiri and adult Kidogo). The *adductor hallucis accessorius* – which is very well developed, and separate proximally in our fetus and infant Etje and adolescent Jasiri – originates from bases of MI and MII in our fetus and from medial cuneiform and base of MI in our adolescent Jasiri (Fig. 5.17). The oblique and transverse head of the adductor hallucis and the adductor hallucis accessorius fuse distally and insert onto the fibular side of the proximal phalanx of digit 1, of metatarsophalangeal capsule and/or of the lateral sesamoid bone present in this region of the big toe (our fetus and infants Foyo and Etje and adolescent Jasiri, Vereecke et al.'s two adult bonobos). We could not discern if an adductor hallucis accessorius is present or not in our adult Kidogo and fetus Foyo, and almost no other authors have specifically analysed if this muscle was present in the *Pan* specimens examined by them, making it difficult to quantify the frequency of its presence in chimpanzees (see comparison with common chimpanzees just below).

- Usual innervation: Adductor hallucis by lateral plantar nerve (Miller 1952: *P. paniscus*) [*P. troglodytes*: adductor hallucis by lateral plantar nerve and also by medial plantar nerve (Brooks 1887; Hepburn 1892; Sonntag 1923].
- Function: Adductor hallucis flexes and adducts the big toe (Vereecke et al. 2005: *P. paniscus*).
- Comparison with common chimpanzees: The two heads of the adductor hallucis are variably united in *P. troglodytes*, the oblique head originating from MII and MIII and the sheath of the tendon of the fibularis longus (Sutton 1883; Champneys 1872; Brooks 1887; Beddard 1893) and sometimes also from MIV (Sutton 1883; Brooks 1887) and/or from the cuboid (Vrolik 1841), and the transverse head originating from MII and MIII and often also from MIV (Vrolik 1841; Dwight 1895), the third and fourth metatarsophalangeal joints and associated ligaments and/or the second metatarsophalangeal joint and interosseous fascia (Brooks 1887; Beddard 1893). The oblique head inserts onto the base of the proximal phalanx of the hallux (Sutton 1883; Brooks 1887) and often also to MI (Champneys 1872; Brooks 1887; Beddard 1893) and occasionally also to the distal phalanx (Brooks 1887). The insertion of the transverse head is to the base of the proximal phalanx (Sutton 1883; Brooks 1887) and sometimes also to MI (Champneys 1872; Brooks 1887; Beddard 1893). In modern humans, the '*transversalis pedis*' (not listed in Terminologia Anatomica 1998) corresponds to the caput transversum of the adductor hallucis. Regarding the *adductor hallucis accessorius*, this is a muscle that is often present in modern humans, at least at early developmental stages (Cihak 1972), and that topologically corresponds to adductor pollicis accessorius of the hand (see upper limb muscles, above), being very likely derived from the oblique head of the adductor hallucis (as the adductor pollicis accessorius is derived from the oblique head of the adductor pollicis: see notes on adductor pollicis above). So far, the adductor hallucis accessorius had almost never been described in non-human primates, but it was clearly present in the two bonobos in which we specifically searched for it (i.e. adolescent Jasiri and fetus), as well as in an adult chimpanzee specimen that we dissected together with the bonobos, to directly compare bonobos with common chimpanzees, so it is very likely that this muscle is often, and perhaps even usually, present as a separate structure in *Pan* (and probably other non-human primates: we plan to undertake a future work to address this issue).

***Contrahentes pedis*** (infant Foyo contrahens to fifth digit 0.06 g; adolescent Jasiri contrahens to fifth digit 0.28 g; and to contrahens to fourth digit 0.12 g; fetus contrahens to digit 4 + contrahens to digit 5 = 0.01 g; Figs. 5.16 and 5.18)

- Attachments in bonobos: Contrahens to digit 4 is often thin (our fetus and infant Etje and adolescent Jasiri and adult Kidogo and

Miller's 1952 adult bonobo) or even reduced to connective/fascial tissue or completely missing (e.g. our infant Foyo), while contrahens to digit 5 is usually more developed (our fetus and adolescent Jasiri and adult Kidogo and infants Foyo and Etje and Miller's 1952 adult bonobo); these contrahentes run from the contrahens raphe to digits 4 and 5, respectively. Vereecke et al. (2005) stated that they could not find contrahentes muscles other than the adductor hallucis in their two adult bonobos, but this might well be because these are very thin muscles in general, because at least some contrahentes going to digits other than digit 1 were found in all the bonobos dissected for this book, as well as by Miller (1952).

- Usual innervation: Lateral plantar nerve (Miller 1952: *P. paniscus*).
- Comparison with common chimpanzees: The contrahentes of the foot have been mainly neglected in the literature of *P. troglodytes*, but it is likely that the configuration is similar to those of bonobos, as contrahentes to digits 4 and 5 are also often seen in the hand of common chimpanzees, but many authors did not report them (see upper limb muscles, above).

***Flexor digiti minimi brevis*** (adult Kidogo 1.63 g; infant Foyo 0.15 g; fetus 0.02 g; Fig. 5.18)

- Attachments in bonobos: From base of MV to base of proximal phalanx of digit 5 and/or adjacent metatarso-phalangeal joint (our fetus and infants Foyo and Etje and adolescent Jasiri and adult kidogo; Vereecke et al.'s 2005 two adult bonobos, Miller's 1952 adult bonobo) being sometimes proximally fused with the abductor digiti minimi belly (Vereecke et al.'s 2005 two adult bonobos).
- Usual innervation: Lateral plantar nerve (Miller 1952: *P. paniscus*).
- Function: Flexion of digit 5(Vereecke et al. 2005: *P. paniscus*).
- Comparison with common chimpanzees: In *P. troglodytes*, it runs from the base of MV to the base of the proximal phalanx of digit 5, the origin being sometimes shared with the opponens digiti minimi and the plantar interossei (Champneys 1872). The flexor digiti minimi brevis may be absent in some *P. troglodytes* according to Champneys (1872) and is seemed to be absent in our *P. troglodytes* specimen GWUANT PT2, but was is present in our *P. troglodytes* specimen VU PT1.
- Synonymy: Flexor digiti quinti brevis or flexor digiti minimi or flexor digiti quinti (Miller 1952; Gibbs 1999).

***Flexor hallucis brevis and flexor brevis profundus 2***(infant Foyo 0.44 g; fetus caput mediale 0.05 g, caput laterale 0.03 g; Figs.5.15 and 5.17)

- Attachments in bonobos: The flexor hallucis brevis ('medial head of 'flexor hallucis brevis' of human anatomy) originates from medial cuneiform and/or ligaments associated to it in our infants Foyo and Etje and fetus and adolescent Jasiri and Vereecke et al.'s 2005 two adult bonobos and also from MI in Miller's 1952 adult bonobo, and also from intermediate cuneiform in our adult Kidogo and adolescent Jasiri. It inserts onto medial side of base of proximal phalanx of hallux and/or adjacent structures such as the first metatarsophalangeal joint in our fetus and infants Foyo and Etje and adolescent Jasiri and adult Kidogo and Miller's 1952 adult bonobo, and simply to the sesamoid bone associated with it Vereecke et al.'s 2005 two adult bonobos. The flexor brevis profundus 2 ('lateral head of flexor hallucis brevis' of human anatomy) runs from base of MI in our fetus or from medial cuneiform in our infant Foyo or from the lateral cuneiform plus MI in Miller's adult bonobo, or from medial and intermediate cuneiforms in our adolescent Jasiri and adult kidogo, to lateral side of base of proximal phalanx of hallux and/or adjacent structures such as the first metatarsophalangeal joint in our fetus and infant Foyo and adolescent Jasiri and adult Kidogo and Miller's 1952 adult bonobo, or simply to the sesamoid bone associated with it in Vereecke et al.'s 2005 two adult bonobos. One of these structures, or both, are fused with the abductor hallucis and with the adductor hallucis in Miller's 1952 adult bonobo and in our fetus, and with the abductor hallucis in our adult Kidogo and adolescent Jasiri.

- Usual innervation: Medial plantar nerve (Miller 1952: *P. paniscus*) [same innervation in *P. troglodytes*: Champneys 1872;Brooks 1887).
- Function: Flexion of MI (Vereecke et al. 2005: *P. paniscus*).
- Comparison with common chimpanzees: In *P. troglodytes*, the origin of the 'flexor hallucis brevis' of human anatomy is usually (Brooks 1887; Hepburn 1892; Beddard 1893), but according to Lewis (1964) not always, double-headed, originating from the intermediate cuneiform and the tendon of the tibialis posterior (Vrolik 1841; Lewis 1964), the medial and lateral cuneiforms (Champneys 1872; Brooks 1887) and/or the metatarsophalangeal joint (Brooks 1887) and occasionally from the cuboid (Beddard 1893). Its so-called 'two heads', which correspond to the flexor hallucis brevis and flexor brevis profundus 2 *sensu* the present work, are subequal in size in *P. troglodytes* (Brooks 1887), the entire complexed formed by them being often fused with the abductor hallucis and the lateral head being often fused with the adductor hallucis (Champneys 1872; Ruge 1878b; Beddard 1893) and being usually inserted onto the proximal phalanx of the hallux and sometimes also onto MI (Brooks 1887; Beddard 1893).

***Opponens digiti minimi*** (adolescent Jasiri 1.02 g; Fig. 5.18)

- Attachments in bonobos: Not present in our fetus, but present in our adolescent Jasiri and adult Kidogo and infant Etje (we did not analyse this feature in detail in our infant Foyo), running from the flexor retinaculum to MV, being partially blended with the flexor digiti minimi brevis. Opponens digiti minimi also described in Miller's 1952 adult bonobo, from base of MV to fibular side and head of MV. It is likely that the opponens digiti minimi was considered to be a deep fascicle of the flexor digiti minimi – as it often happens in common chimpanzees (see just below) – in Vereecke et al.'s 2005 two adult bonobos, because they referred to a two-headed flexor digiti minimi brevis in their descriptions, although they did state that a structure that could correspond to the opponens digiti minimi was present in one of these four feet, running from the joint between the cuboid and MV, to MV. Therefore, the opponens digiti minimi seems too often – and perhaps usually – present in bonobos.
- Usual innervation: Lateral plantar nerve (Miller 1952: *P. paniscus*).
- Comparison with common chimpanzees: In orangutans and modern humans, the opponens digiti minimi is usually considered to be a deep fascicle of the flexor digiti minimi (for instance, it was not listed in Terminologia Anatomica 1998), but it has been described as a distinct muscle in *Pan* and *Gorilla* (e.g., Gibbs 1999). When present in *P. troglodytes*, it has a common origin from the flexor digiti minimi brevis and the fourth plantar interosseous muscle according to Champneys (1872), and inserts onto the lateral surface of MV (Champneys 1872), also extending onto the plantar surface of this latter bone according to Champneys (1872). We also found an insertion of the opponens digit minimi onto the proximal phalanx of digit 5 in our common chimpanzee specimens.
- Synonymy: Opponens digiti quinti (Miller 1952; Gibbs 1999).

***Interossei dorsales*** (adult Kidogo interosseus dorsalis 1 = 4.15 g; interosseus dorsalis 2 = 2.97 g, interosseus dorsalis 3 = 2.39 g, interosseus dorsalis 4 = 1.63 g; infant Foyo interossei dorsales 1 + 2 + 3 + 4 = 0.70 g; fetus interosseus dorsalis 1 = 0.06 g, interosseus dorsalis 2 = 0.04 g, interosseus dorsalis 3 = 0.08 g, interosseus dorsalis 4 = 0.07 g; Fig. 5.18)

- Attachments in bonobos: In all bonobos dissected by us and described by others, the intermetatarsales are mainly fused with four flexores breves profundi to form the dorsal interossei of the foot (see Fig. 5.18). The dorsal interosseus 1 runs from MII and MI to medial side of proximal phalanx of digit 2, dorsal interosseus 2 running from MII and MIII to medial side of proximal phalanx of digit 3, dorsal interosseus 3 running from MIII and MIV to lateral side of proximal phalanx of digit 3 and dorsal interosseus 4 running from MIV and MV to lateral side of proximal phalanx of digit 4 (our fetus and infants Foyo and Etje and adolescent Jasiri and adult Kidogo and Miller's 1952 adult bonobo and Vereecke et al.'s 2005 two adult bonobos, even if these authors prefer to use a different nomenclature). Therefore, the configuration of the dorsal interossei – as the innervation of the lumbricales (see above) – seems to indicate that the axis of the bonobo foot seems to be digit number 3, as seen in the bonobo and common chimpanzees and modern human hand but contrary to the modern human foot, in which the axis is digit 2 instead.

- Usual innervation: Deep ramus of the lateral plantar nerve (Miller 1952: *P. paniscus*).
- Comparison with common chimpanzees: In *P. troglodytes*, there are usually four dorsal interossei muscles (Humphry 1867), with the exception of the single specimen dissected by Dwight (1895) that had six, according to the author, a description that is questionable or that might indicate that at least some intermetatarsales were not completely fused with the flexores breves profundi to form the dorsal interossei. Each dorsal interosseus usually has two heads of origin, the reference line for action being digit 3 in *P. troglodytes* (Champneys 1872), as is the case in bonobos (see above). In *P. troglodytes*, the first dorsal interosseous originates from the lateral surface of MI and the medial surface of MII (Champneys 1872; Brooks 1887; Dwight 1895; Sonntag 1924), although the hallucial head may be rudimentary or absent (Brooks 1887). The second dorsal interosseous originates from the lateral surface of MII and the medial surface of MIII (Dwight 1895). The third dorsal interosseous originates from the lateral surface of MIII and the medial surface of MIV (Dwight 1895). The fourth dorsal interosseous originates from the lateral surface of MIV and the medial surface of MV. In the specimen of *P. troglodytes* reported by Dwight (1895), there was a first dorsal accessory inerosseous muscle lying between the third and fifth dorsal interossei muscles, originating from the dorsal surface of MIV, the second accessory dorsal interosseous muscle arose with the fourth dorsal interosseous from the lateral surface of MIV and the medial surface of MV; the fourth dorsal and second accessory dorsal interossei muscles might be considered as a single muscle with two separate insertions according to Dwight (1895). The first dorsal interosseous inserts onto the medial surface of the base of the proximal phalanx of digit 2, the second dorsal interosseous inserting onto the medial surface of digit 3, the third dorsal interosseous inserting onto the lateral surface of digit 3, and the fourth dorsal interosseous inserting onto the lateral aspect of the proximal phalanx of digit 4; the first and second accessory dorsal interossei described above insert onto the medial surface of digit 4 and onto the medial surface of digit 5, respectively (Humphry 1867; Dwight 1895; Miller 1952).

***Interossei plantares*** (adult Kidogo interosseus palmaris 1 = 2.28 g, interosseus palmaris 2 = 1.90 g, interosseus palmaris 3 = 1.29 g; infant Foyo interossei plantares 1 + 2 + 3 + 4 = 0.70 g; fetus interosseus palmaris 1 = 0.01 g, interosseus palmaris 2 = 0.02 g, interosseus palmaris 3 = 0.01 g; Fig. 5.18)

- Attachments in bonobos: Plantar interosseus 1 from MII to lateral side of proximal phalanx of digit 2; plantar interosseus 2 from MIV to medial side of proximal phalanx of digit 4; plantar interosseus 3 from MV to medial side of proximal phalanx of digit 5 (our fetus and infants Foyo and Etje and adolescent Jasiri and adult Kidogo and Miller's 1952 adult bonobo and Vereecke et al.'s 2005 two adult bonobos, even if these latter authors prefer to use a different nomenclature).
- Usual innervation: Lateral plantar nerve (Miller 1952: *P. paniscus*).

- Comparison with common chimpanzees: In *P. troglodytes*, there are usually three plantar interossei muscles, each with a single head of origin (Humphry 1867), with the exception of a specimen of *Pan* stated to possess five muscles, which might indicate that at least some of the intermetatarsales were not fused with the flexores breves profundi to form dorsal interossei (Dwight 1895).In *P. troglodytes*, the first plantar interosseous usually originates from MII, the second from the medial surface of MIV and the third from the lateral side of MIV (Dwight 1895). The two accessory plantar interossei reported in *P. troglodytes* by Dwight (1895) originated from the medial and lateral sides of MIII, respectively. The first plantar interosseous inserts onto the lateral side of the proximal phalanx of digit 2, the second onto medial side of digit 4 and the third onto the medial side of digit 5, the two accessory plantar interossei described above inserting onto the medial and lateral sides of the proximal phalanx of digit 3, respectively (Humphry 1867; Dwight 1895; Humphry 1867).

## 5.2 Lumbosacral Plexus

### 5.2.1 Description of the *plexus lumbosacralis* of Adolescent Female Jasiri (Fig. 5.39)

The left *plexus lumbosacralis* begins just distal to the crural insertions of the *m. diaphragm* with nerve contributions from the first lumbar nerve. The *plexus lumbalis* proper is comprised of nerve roots L1, L2, L3 and L4 and provides a distal branch to the *plexus sacralis* via the *truncus lumbosacralis*. The *truncus lumbosacralis* is comprised of nerves L3-4, while the *plexus sacralis* provides additional innervation from sacral nerves S1-2.The combined *plexus lumbosacralis* does not receive a direct contribution from the *n. subcostalis*, as is occasionally seen in modern humans. The overall morphology of this nerve complex is similar to that seen in modern humans, although different root levels contribute to the distal segments of nerves. As with the *plexus brachialis* observed in the specimens described here, no histological studies of nerve fibre contributions were conducted of the *plexus lumbosacralis* for this atlas. All assignations of root contributions to nerves, designated by the nerve number in brackets, e.g., (L1-2), are hypotheses that remain to be tested. The discrete nerves found and followed to their terminal points in the hindlimb of Jasiri are as follows.

The *plexus lumbalis* is formed by a series of contributions from L1-4 and S1-2, as follows. A distinct *iliohypogastricus* was not observed in Jasiri. The *n. ilioinguinalis* forms from a caudally descending branch of L1 connecting with a laterally outbranching of L2, though it is unlikely that this nerve receives any axons from L2 given its branching position superior to the junction point. A distinct *n. genitofemoralis* was not observed in Jasiri. The *n. femoris* forms mainly as a junction of contributions from L1 and L2, but also receives a small branch from L3. The nerve runs inferiolaterally, deep to the belly of *m. psoas major* and superficial to the *m.* iliacus, both of which it gives off short nerve branches to for innervation. The *n. femoris* passes deep to the *ligamentum inguinale* into the thigh, where it divides into several branches that innervate the muscles of the anterior compartment from their deep surfaces. A single branch goes on to become *n. saphenous* in the leg. This branch does not pass under or through the aponeurotic covering of the *canalis adductorius* as in modern humans, but runs superficial to the muscles and directly enters the leg on the medial aspect of the knee. In Jasiri, the *n. obturatorius* begins with a small branch from L1, which descends to join with larger branches of L2 and L3. When formed, it travels inferiorly along the medial border of the *m. psoas major* and *m.* iliacus (deep to but not within the muscle bellies), over the pelvic brim and into the pelvic bowl on the ventral surface of the *os sacrum*. This nerve runs lateral to (but parallel with) the *truncus lumbosacralis* and makes no connections with distal nerves. The contributions from L1-L3 all travel superficial to the rest of the *plexus lumbalis*. The *n. obturatorius* then travels distally through the *canalis obturatorius*, where it then enters the thigh to innervate the adductor compartment. The anterior branch innervates and pierces adductor brevis to terminate in the inferior belly of pectineus. The posterior branch provides innervation to the anterior portion of *m obturatorius externus*. In other specimens (Kidogo, Ano, Etje), the nerve forms mainly as a combination of contributions from L2-L3, with a small contribution from L4.

The *truncus lumbosacralis* is a medium sized nerve comprised of L3-L4 that connects the *plexus lumbalis* with the *plexus sacralis*, providing nerve fibres from the superior part of this neural complex to the inferior portion. It runs along the medial border of the *m. psoas major*, over the pelvic brim and onto the ventral surface of the *os sacrum* (to which it is closely adherent), where it connects with a branch of the first sacral nerve to form the beginnings of the *n. ishiadicus*. The *n. cutaneous femoris lateralis* branches directly from the lateral aspect of the femoral nerve roots (L1-2) prior to the contribution from L3. It receives branches from both dorsal and ventral aspects of the *plexus lumbalis*, and passes deep to the *m. psoas major*, travelling laterally and exiting deep to the *ligamentum inguinalae* to innervate the skin of the lateral thigh. The *n. ischiadicus* is a large nerve that forms from a combination of the *truncus lumbosacralis* (L3-L4) joining with the sacral nerves (S1-2). Unlike the condition seen in modern humans (S1-S3 or S4) or *P. troglodytes* (S1-S3), the *plexus sacralis* only contributes two nerve roots to the *n. ischiadicus* in the majority of our specimens, perhaps correlating to the decreased overall muscle-mass in *P. paniscus*, though this hypothesis remains to be tested. This nerve courses deep to the hamstrings muscles toward the popliteal region where it bifurcates into the *n. tibialis* and *n. fibularis comminus*. Along the way, the *pars tibialis* of the *n. ischiadicus* innervates the hamstrings muscles, excluding the *m. biceps femoris caput brevis*, which receives innervation from the *n. fibularis comminus*. The *n. gluteus superioris* forms from a short branch of the *truncus lumbosacralis* arising from the combination of L4-S1. It exists through the cranial aspect of the *foramen ischiadicum majus*, superior to the *m. piriformis*. The *n. gluteus inferoris* forms from a branch distal to the *n. gluteus superioris* (also arising from L4-S1), exits through the caudal aspect of the *foramen ischiadicum majus* and travels with the *A.* and *V. gluteae inferior* just inferior to the *m. piriformis*. The *n. musculi piriformis* is a direct branch off of the dorsal division of the ventral rami of the sacral plexus, arising from roots L4-S1. The *n. gemellus superior* originates from L3-4 and does not appear to receive any functional fibres from the sacral plexus, although without further study it cannot be ruled out that a small contribution of S1 also travels with this nerve. The *n. musculi obturatorii interni* originates from L3-4 and short direct branches from S1. The *n. gemellus inferior* originates from L3-4 and does not appear to receive any functional fibres from the sacral plexus, although without further study it cannot be ruled out that a small contribution of S1 also travels with this nerve. The *n. quadratus femoris* originates from L3-4, and short direct branches from S1 and possibly S2. Lastly, *n. puboischiofemoralis* is a direct nerve from the *plexus lumbosacralis* that innervates the *m. puboischofemoralis* by the roots of L3-S1. This nerve also sends short contributions to that muscle. The pudendal nerves were not investigated in this study.

### 5.2.2 Intraspecific Variations or Polymorphisms of the Lumbosacral Plexus

The following are descriptions of the intraspecific variation in our study sample. For consistency, we used the left side on all specimens as a template and here expand upon these notes with observations from the right side where dissected. Most variations observed were in the proximal branching patterns seen in the formation of the *plexus lumbalis*. For example, in Jasiri, the *n. femoralis* sends a superficial contribution to the *n. obturatorius*.

Figure 5.40, adult male Kidogo: This figure demonstrates the typical positioning of the *n. femoris* as it emerges from deep to the *m. psoas major*. The split in the distal branches of the *n. femoris* occurs more proximally in this specimen than in others, but is not outside the normal range of variations observed in our sample. The muscular branches of the *n. femoris* can be observed entering the inferior aspect of the *m. vastii.*

Figure 5.41, adult male Kidogo: This figure depicts the form of the *plexus lumbosacralis* without the surrounding musculature. No *n. iliohypogastricus* was observed in this specimen. No branch of a combined *n. iliohypogastricus/n. ilioinguinalis* was observed, as is sometimes the case in *Homo sapiens*. The *truncuslumbosacralis* is relatively small in this specimen and is derived mostly from L4, with minor potential inputs from L1-3.

Figure 5.42, female fetus Ano: Ano displays the typical configuration seen in the other *P. paniscus* specimens observed in this study, with the exception of a visible *n. iliohypogastricus* that derives purely from L1. It can be observed emerging dorsolaterally from behind *p. major*.

Figure 5.43, female fetus Ano: This figure depicts the deeper morphology of the *plexus lumbosacralis* of Ano with the overlaying musculature removed. The *n. cutaneus femoris lateralis* is derived from L2-3, with a potential contribution from L1, a condition not typically observed. The *n. cutaneus femoris lateralis* also does not arise directly from the *n. femoris* as seen in Jasiri and Kidogo.

Figure 5.44, infant male Etje: Etje displays what is possibly a*n. iliohypogastricus*, which travels deep to the path of the *n. ilioingualis*, and is derived from the first pre-lumbar vertebrae.

### 5.2.3 Summary

Based on the overview of the different specimens dissected for this atlas, we can surmise that the following description of the *plexus lumbosacralis* summarizes the morphology most common to *P. paniscus*:

- The roots of the *plexus lumbosacralis* are consistently L1-4 and S1-2. There were no visible contributions from more distal branches of the *plexus sacralis*, nor any visible contributions from the *n. subcostalis*.
- The *plexus lumbalis* forms both within and behind the *m. psoas major*.
- Then *plexus sacralis* only receives motor contributions to the limbs from S1-2.
- No *n. genitofemoralis* was observed in a distinct form in any specimen, which has been proposed to have been a result of the shortening of the lumbar column in hominoids.
- The *n. cutaneus femoris lateralis* was the least consistent nerve in terms of its formation and root contributions, though its distribution pattern was consistent distally.
- The distal distribution patterns of the *plexus lumbosacralis* are consistent in their routes with that seen in *Homo sapiens*.

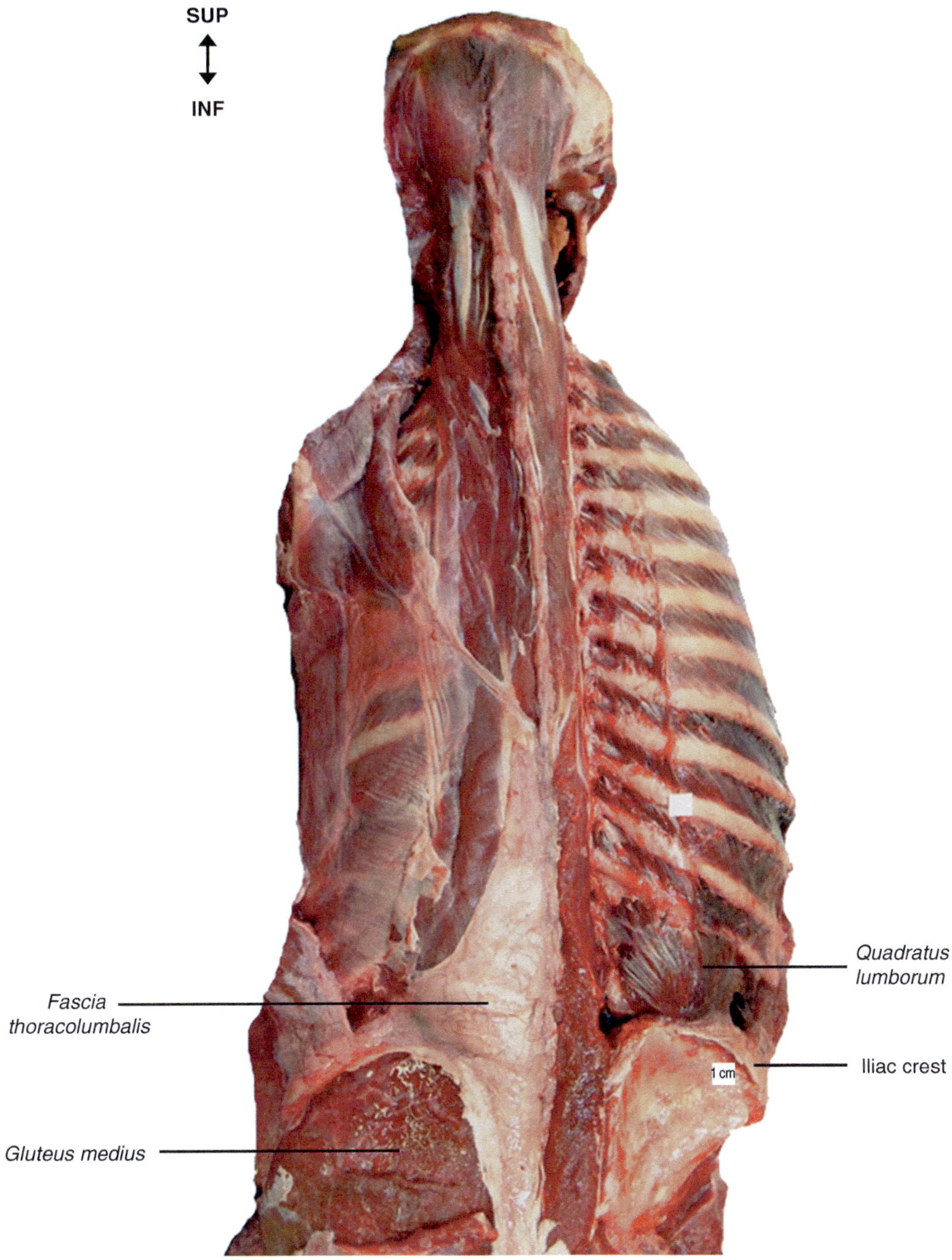


**Fig. 5.1** *Pan paniscus* (ZIMS 164031, Kidogo, adult male): dorsal view of the back, showing the lower limb muscles quadratus lumborum and gluteus medius

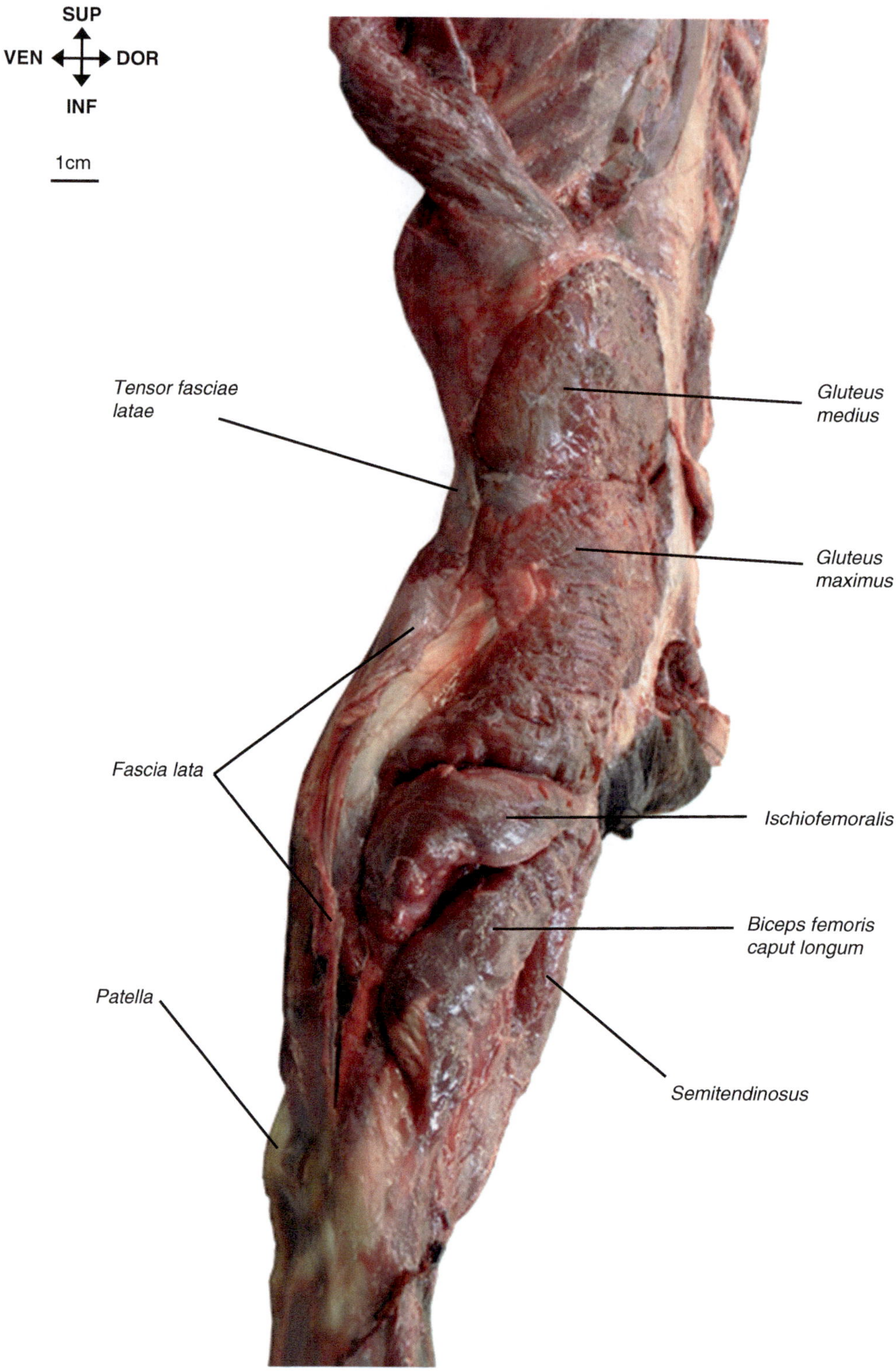


**Fig. 5.2** *Pan paniscus* (ZIMS 164031, Kidogo, adult male): lateral view of the left thigh and lower torso

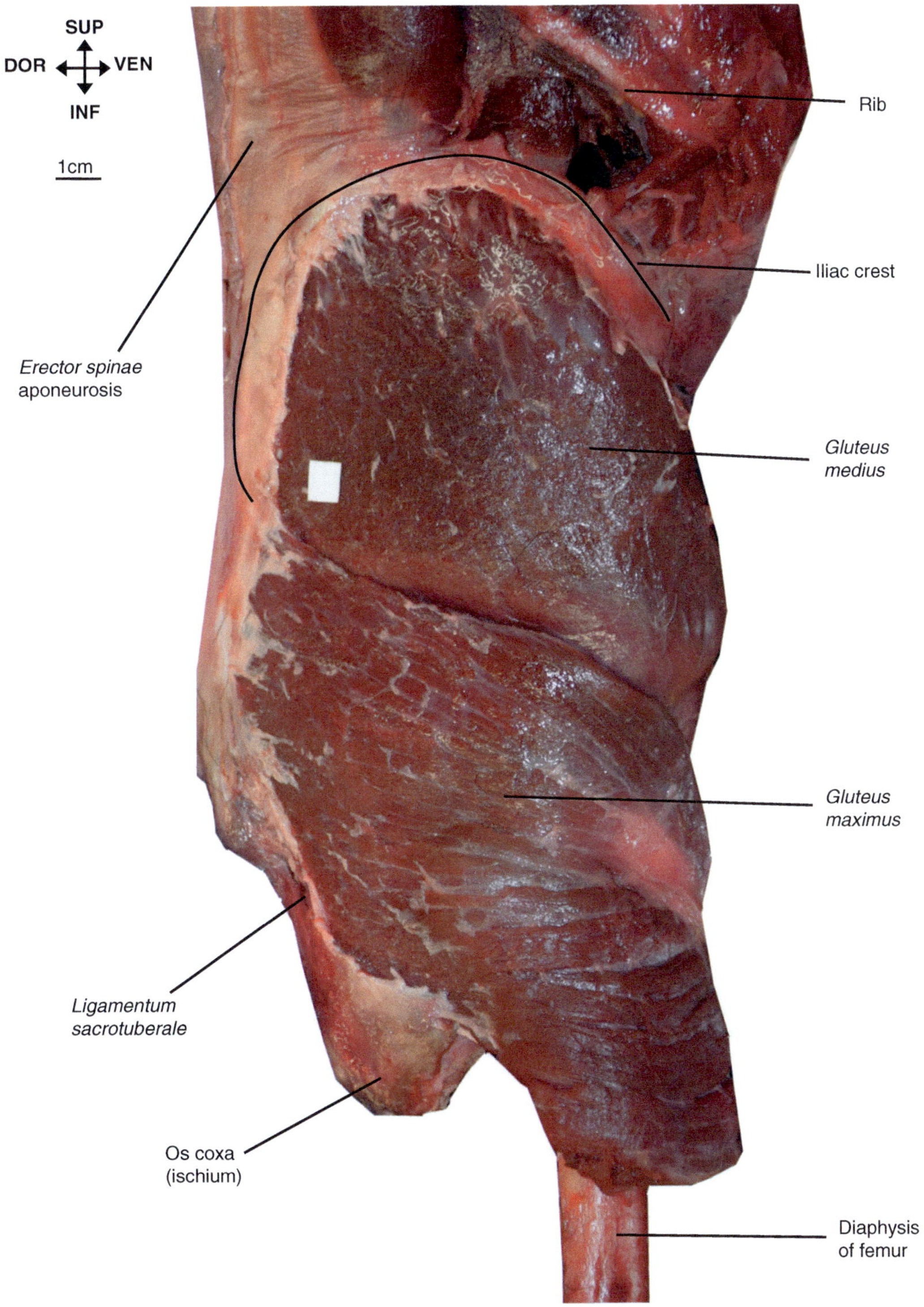


**Fig. 5.3** *Pan paniscus* (ZIMS 164031, Kidogo, adult male): lateral view of the right gluteal region after removal of thigh musculature. Note that the gluteus maximus does not originate from the iliac crest

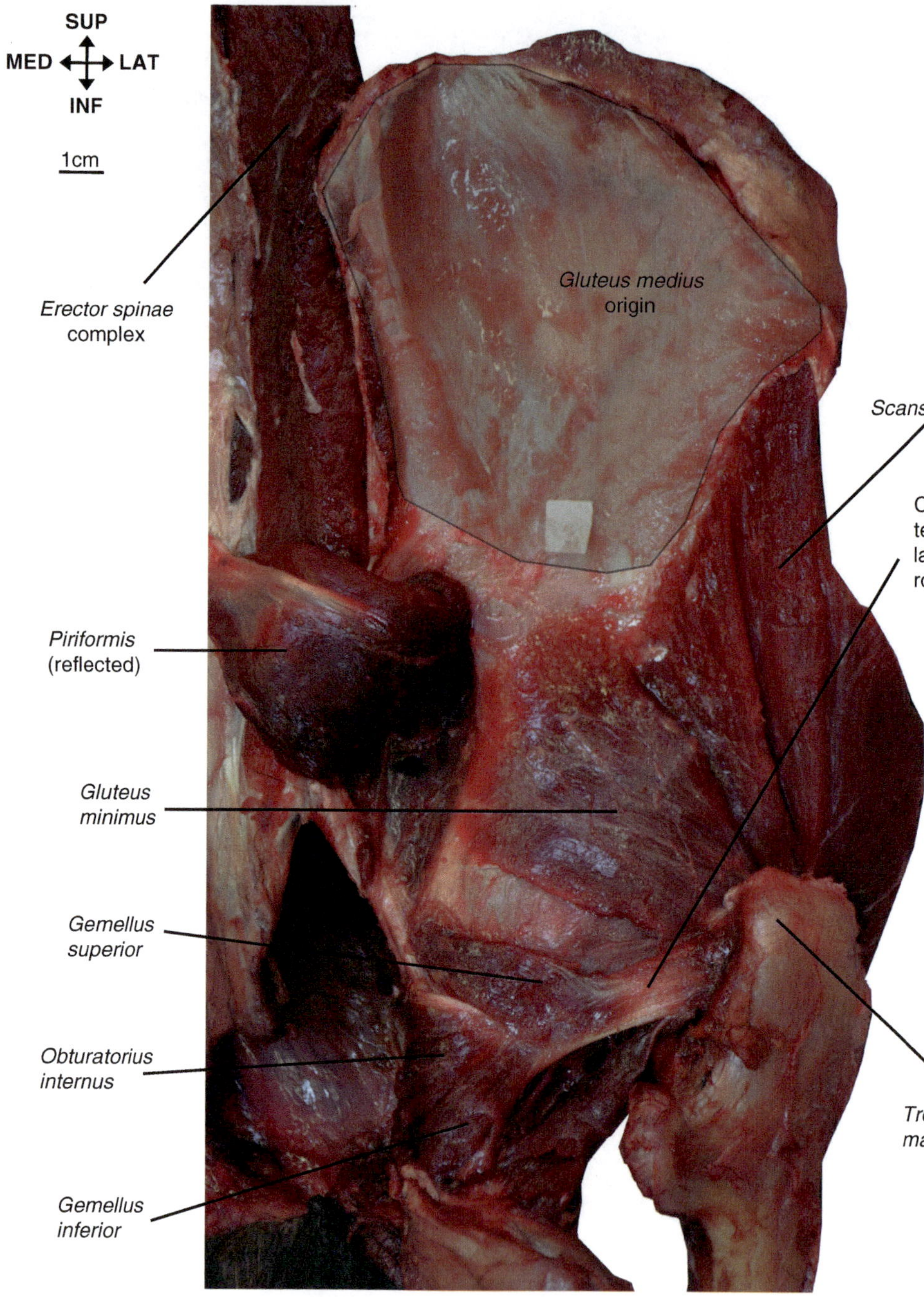


**Fig. 5.4** *Pan paniscus* (ZIMS 164031, Kidogo, adult male): dorsal view of the gluteal muscles after removal of gluteus maximus and gluteus medius, showing in detail the origin of the gluteus medius

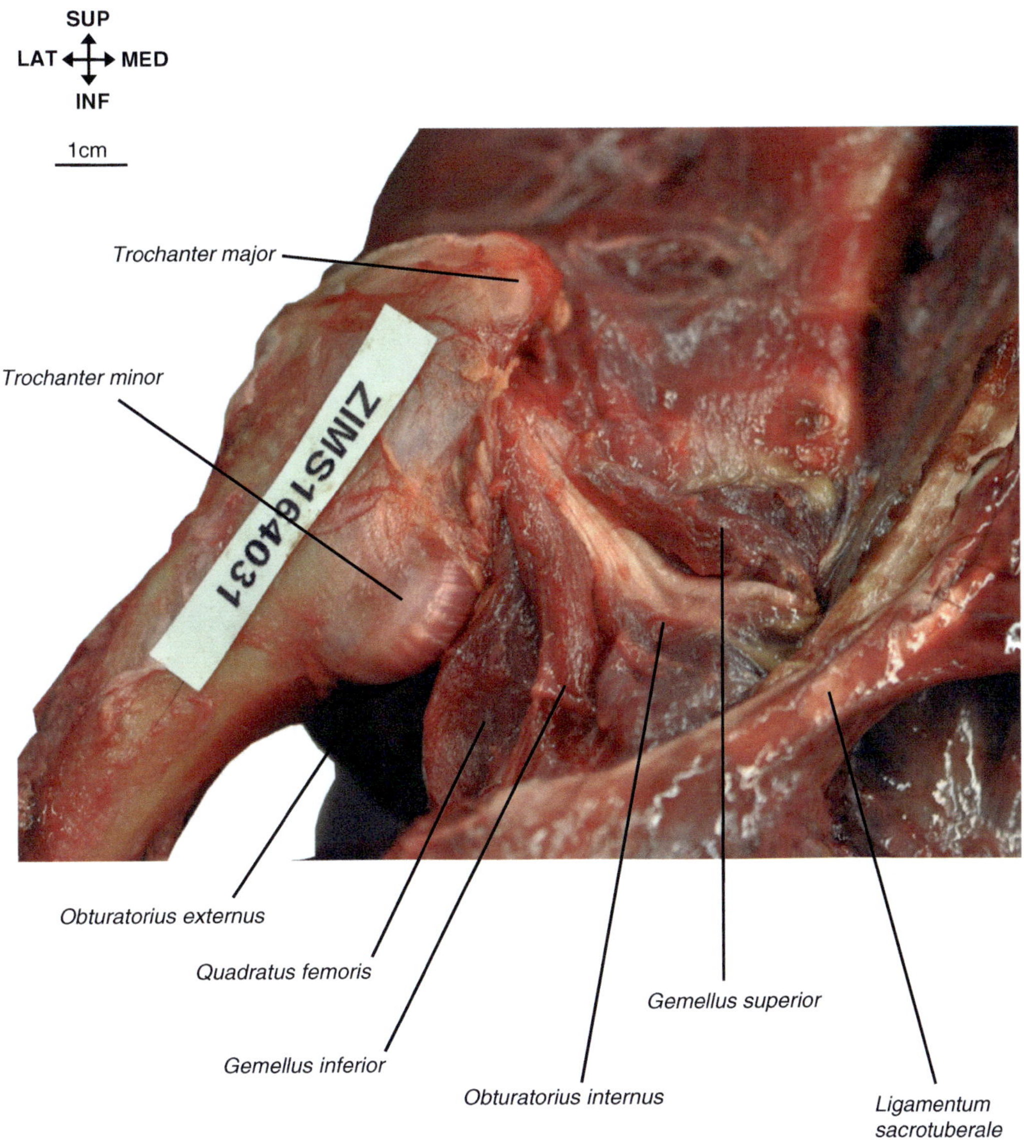


**Fig. 5.5** *Pan paniscus* (ZIMS 164031, Kidogo, adult male): dorso-inferior view of the small gluteal muscles after removal of gluteus maximus and gluteus medius

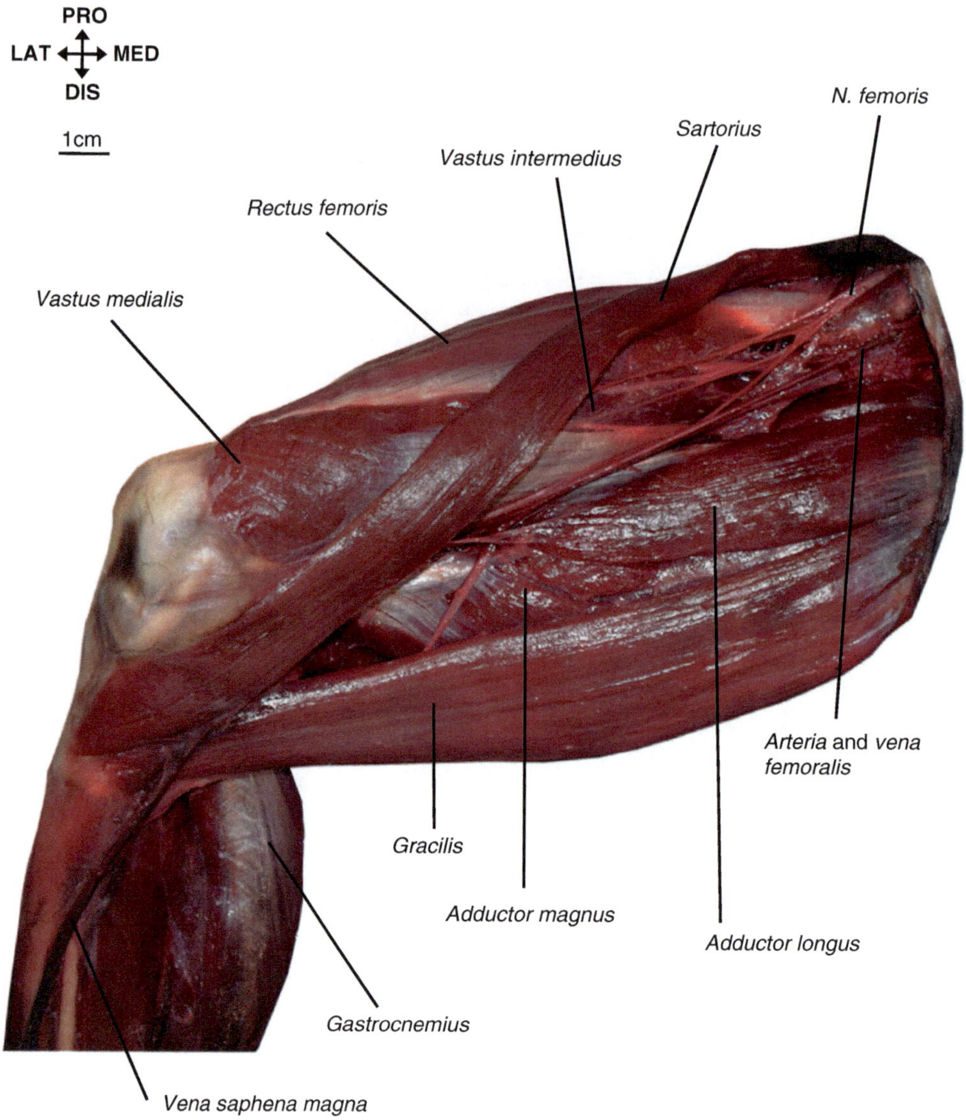


**Fig. 5.6** *Pan paniscus* (ZIMS 164031, Kidogo, adult male): medial view of the right thigh muscles

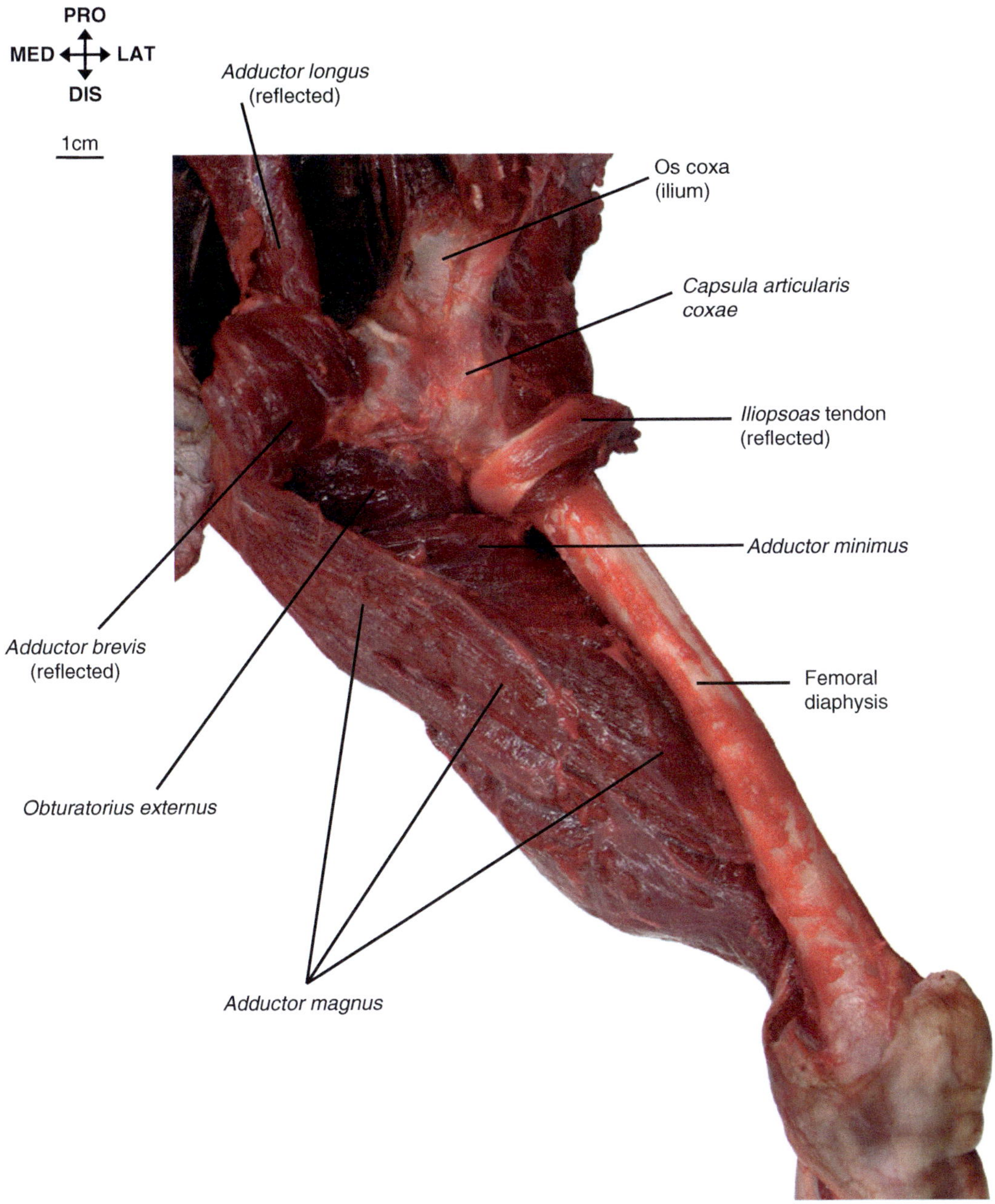


**Fig. 5.7** *Pan paniscus* (ZIMS 16047, Jasiri, adolescent female): ventro-medial view of the left thigh muscles after removal of the quadriceps muscles

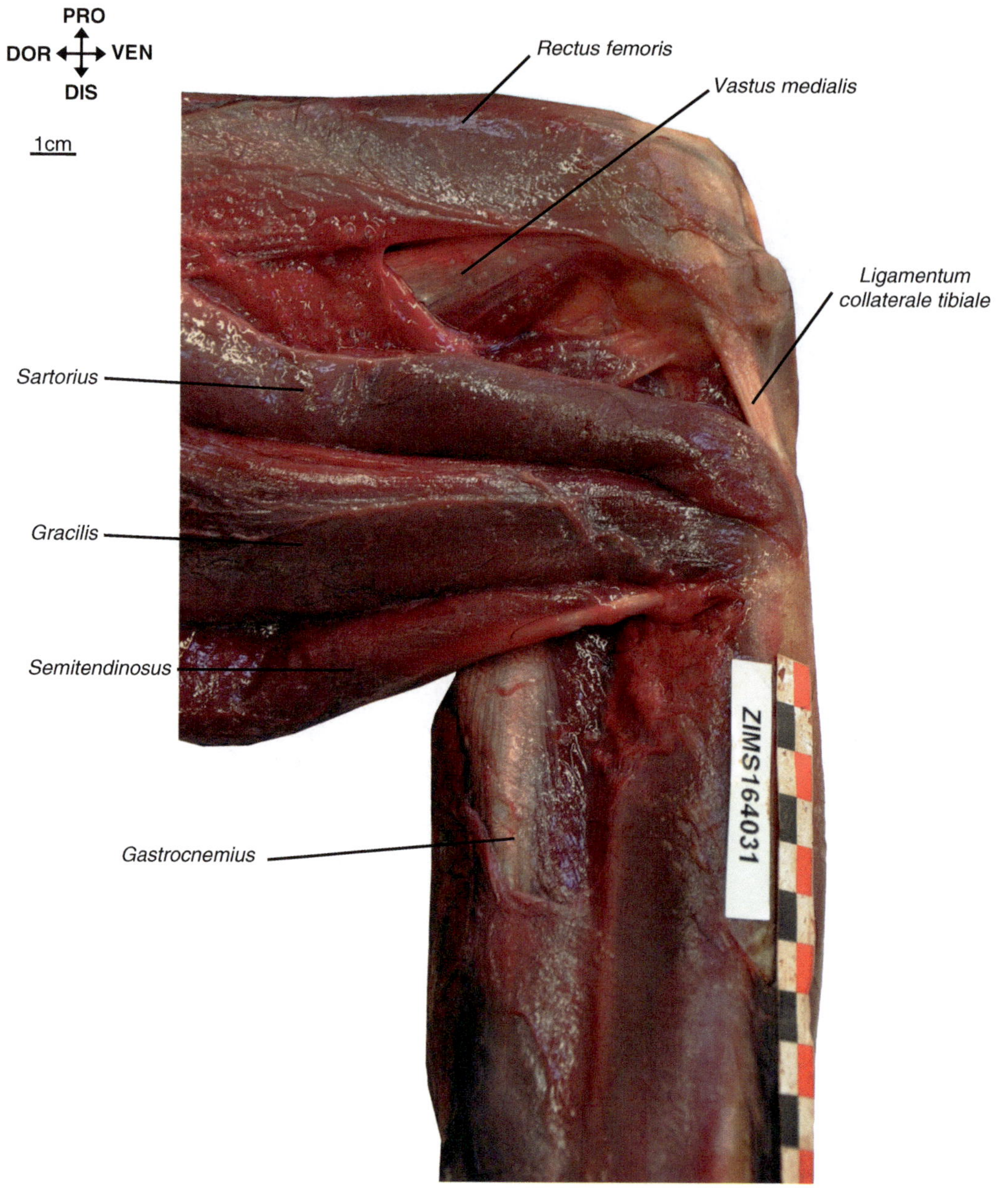


**Fig. 5.8** *Pan paniscus* (ZIMS 164031, Kidogo, adult male): medial view of the left knee. Note that the insertion of the sartorius, gracilis, and semitendinosus does not form a conjoint tendon (i.e. a *pes anserinus*)

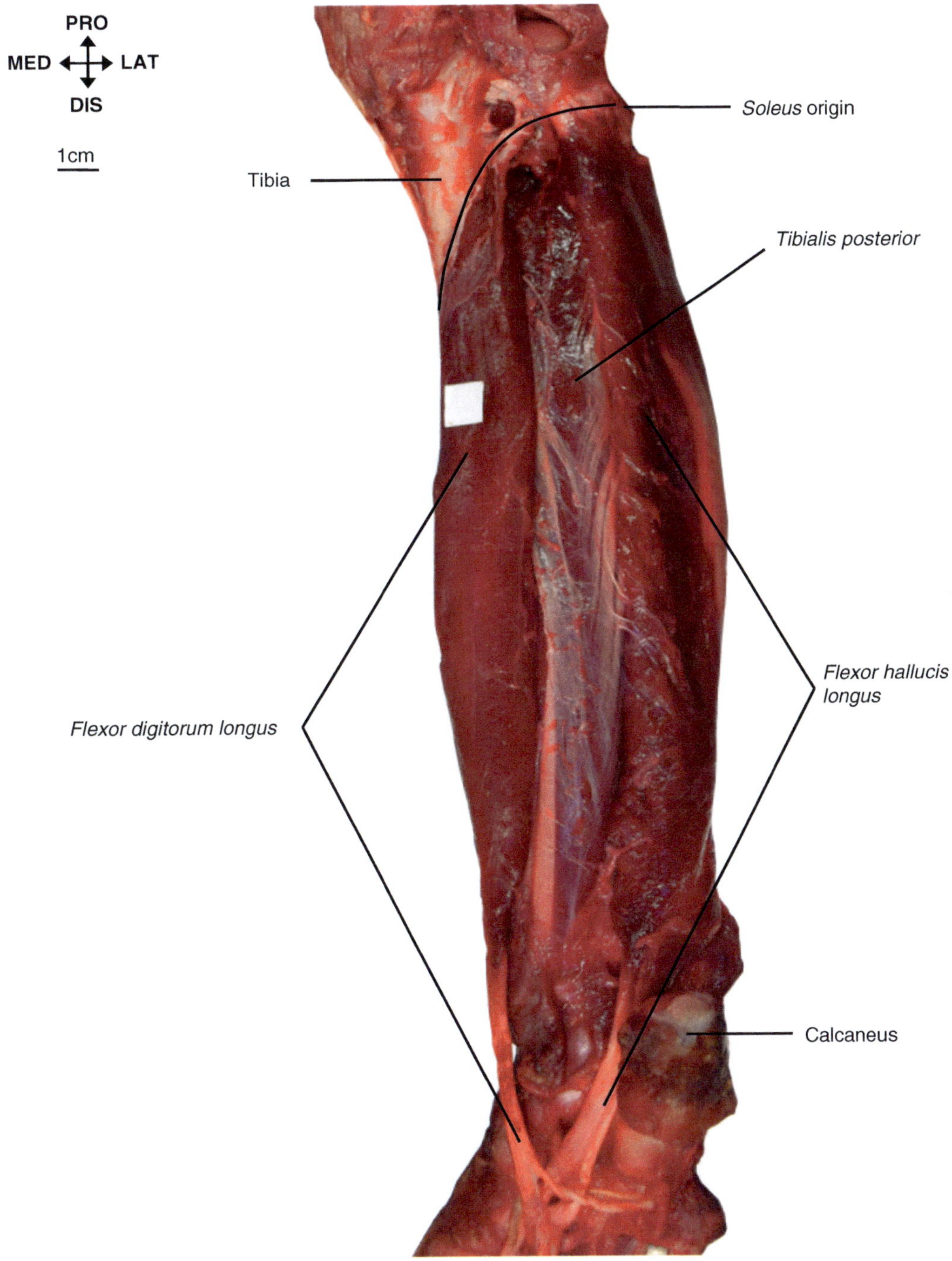


**Fig. 5.9** *Pan paniscus* (ZIMS 164031, Kidogo, adult male): dorsal view of the deep leg muscles of the right leg after removal of gastrocnemius and soleus

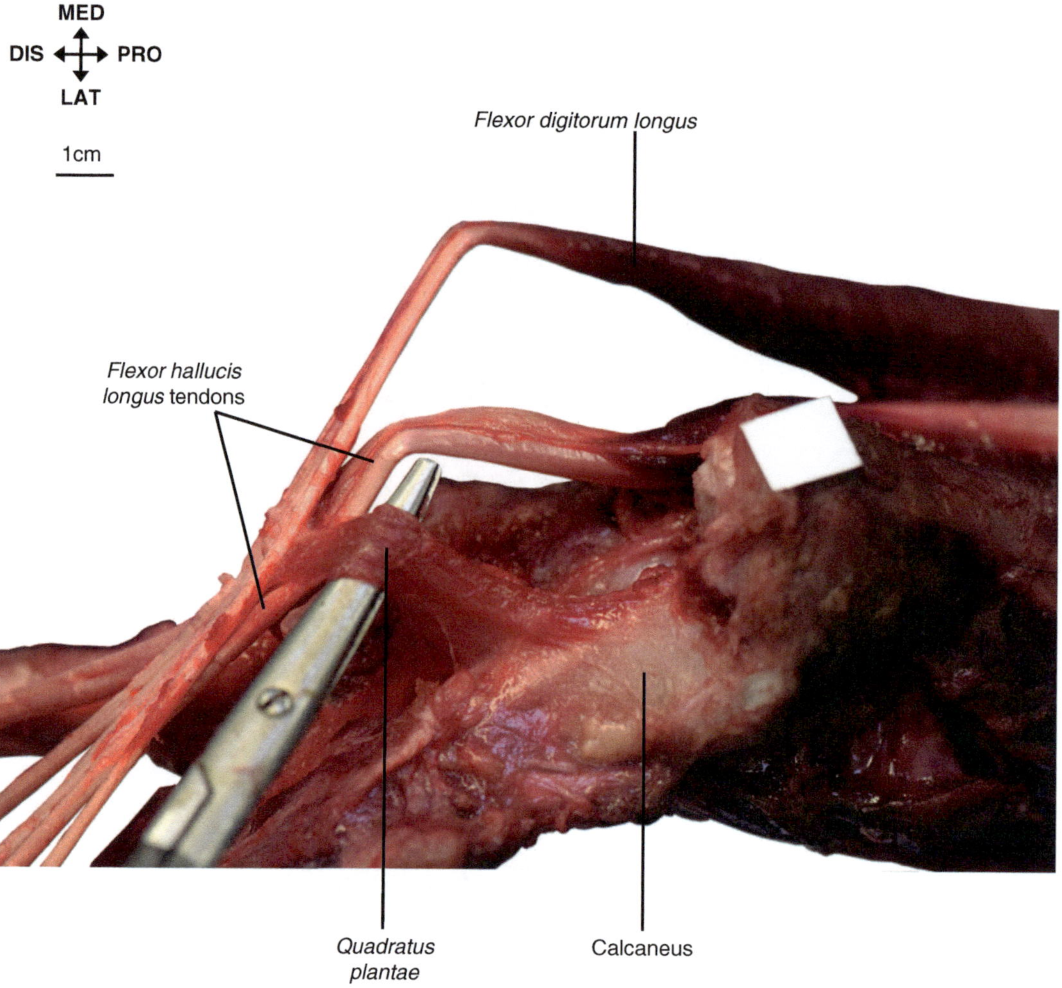


**Fig. 5.10** *Pan paniscus* (ZIMS 164047, Jasiri, adolescent female): latero-plantar view of the right foot showing the contributions of flexor hallucis longus tendons to lateral digits, i.e., contrary to modern humans it does not insert exclusively onto the most medial digit (big toe)

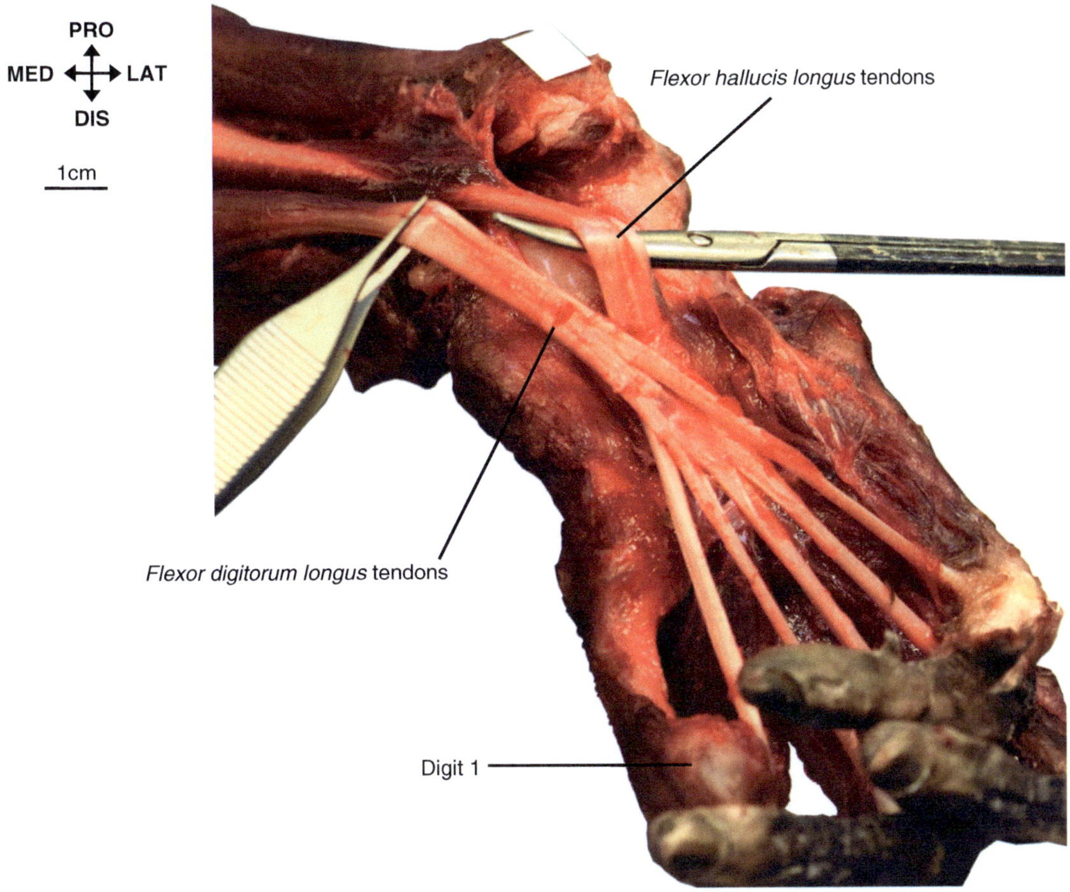


**Fig. 5.11** *Pan paniscus* (ZIMS 164047, Jasiri, adolescent female): plantar view of the right foot after removal of the skin, plantar fascia and more superficial muscles

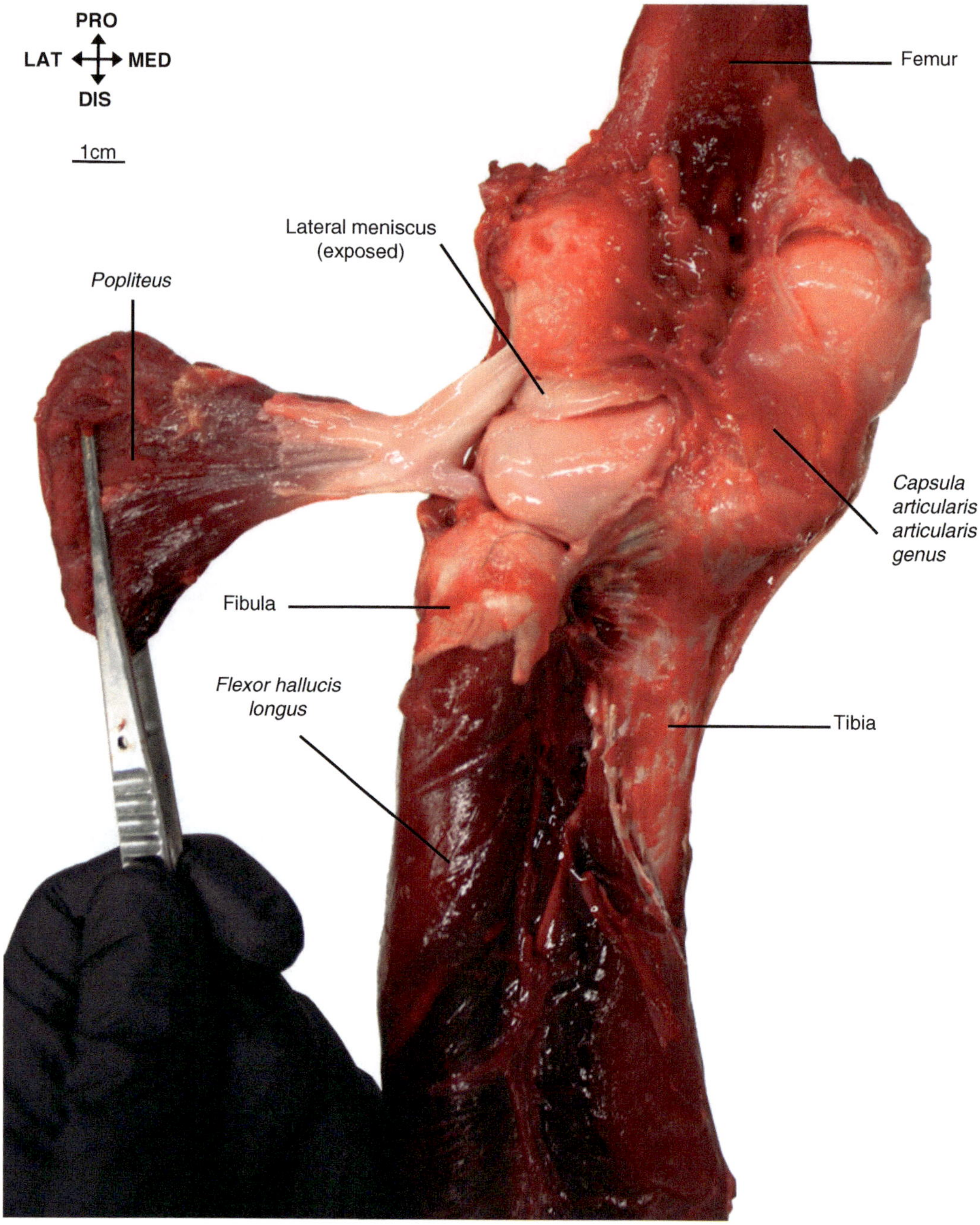


**Fig. 5.12** *Pan paniscus* (ZIMS 164047, Jasiri, adolescent female): dorsal view of the left popliteal region after reflection of popliteus, showing its additional tendon to the proximal fibula

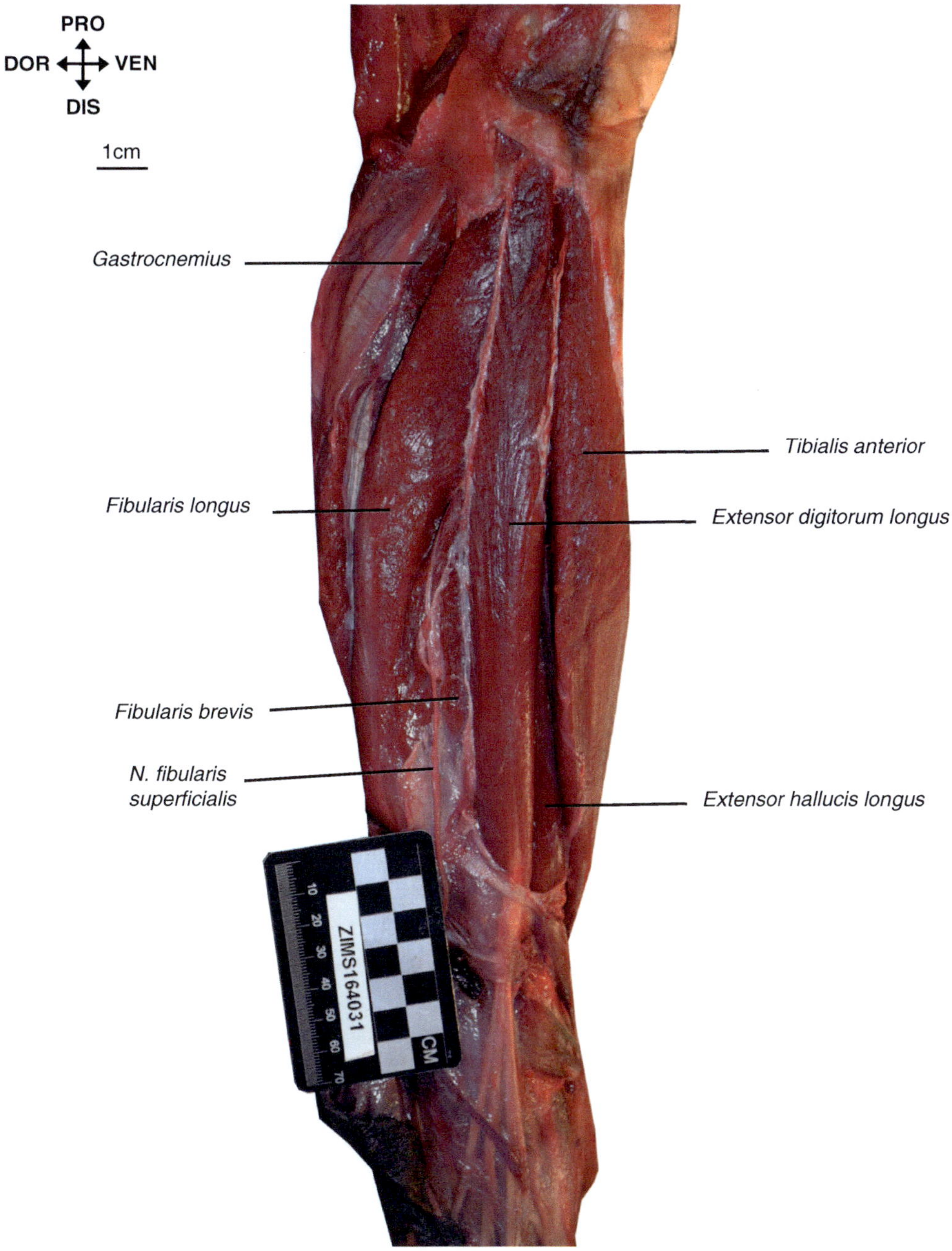


**Fig. 5.13** *Pan paniscus* (ZIMS 164031, Kidogo, adult male): lateral view of the right leg muscles after removal of the crural fascia

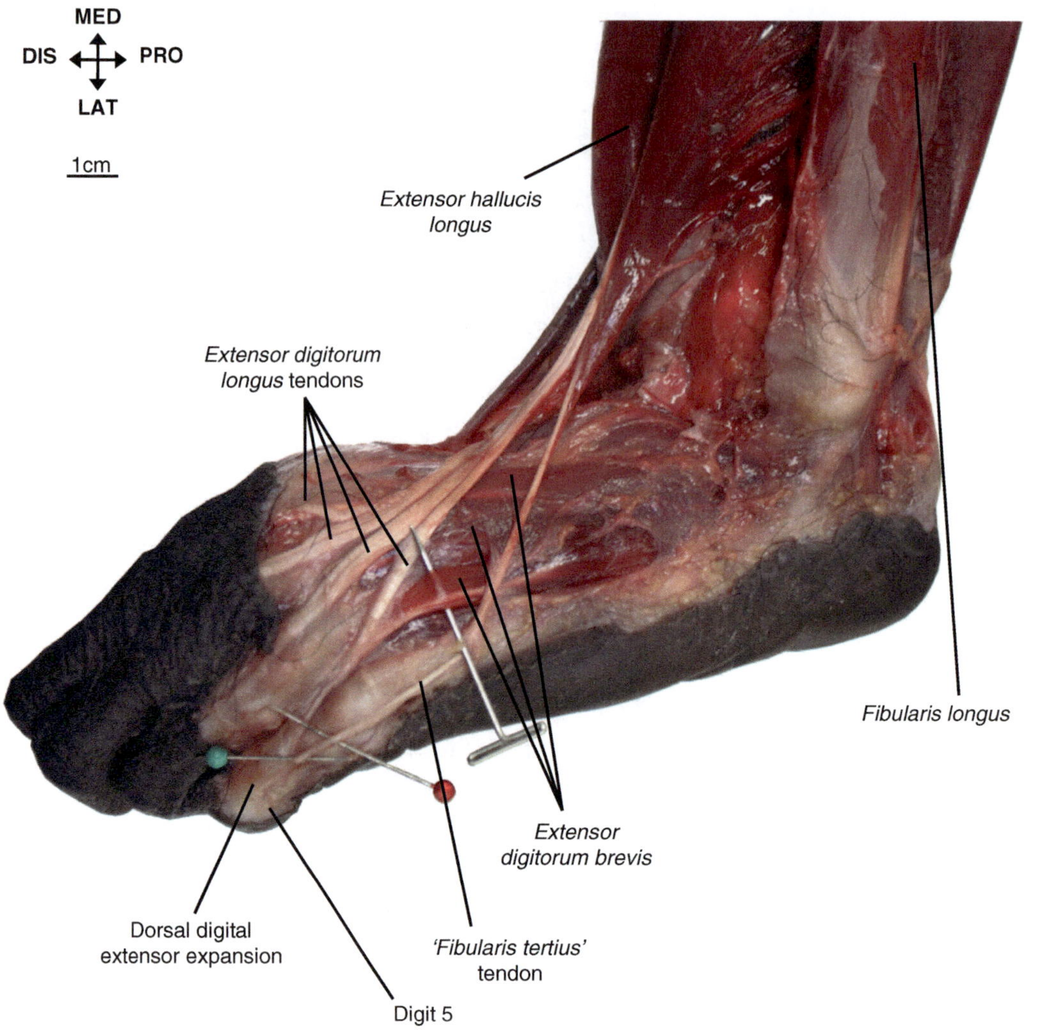


**Fig. 5.14** *Pan paniscus* (ZIMS 164047, Jasiri, adolescent female): lateral view of the left foot, showing a structure that is consistent with the fibularis tertius of modern humans (see text)

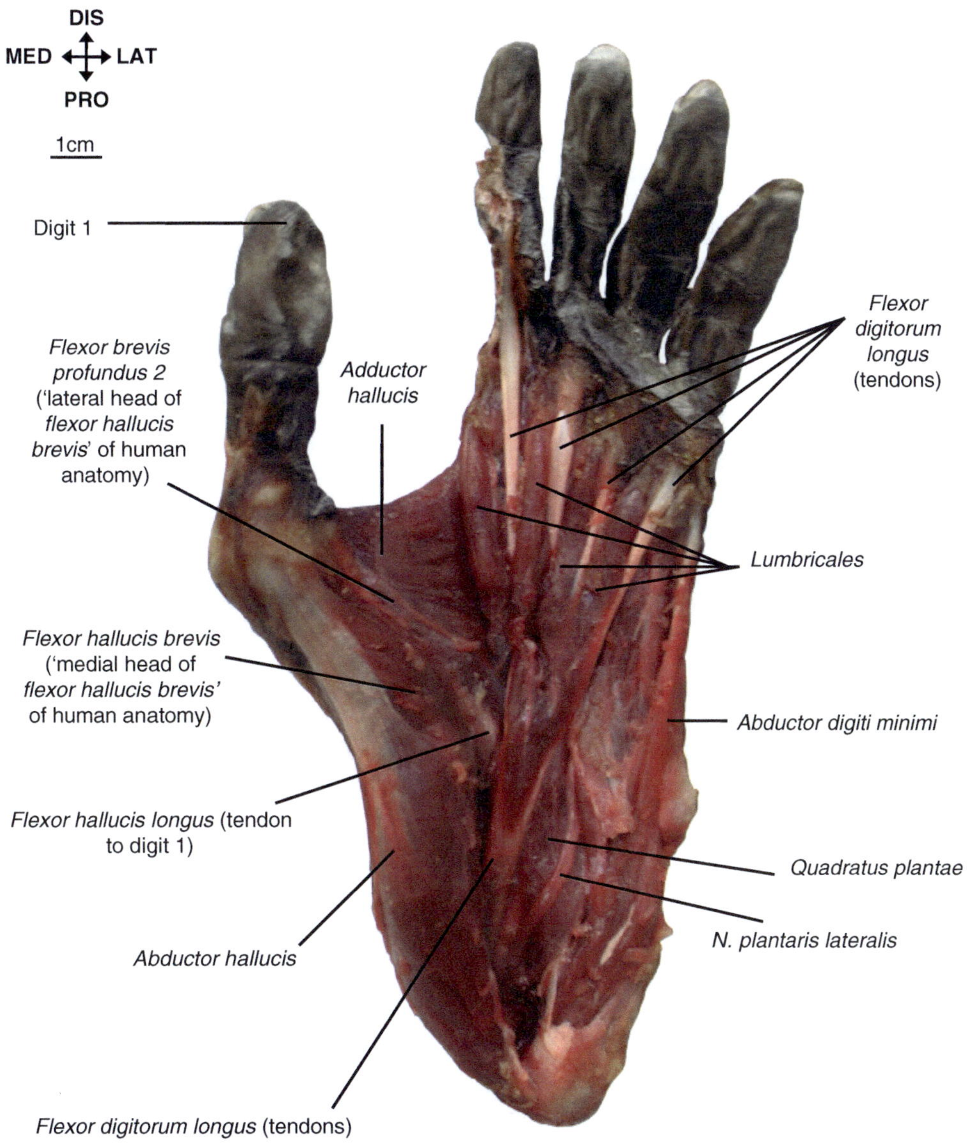


**Fig. 5.15** *Pan paniscus* (ZIMS 164031, Kidogo, adult male): plantar view of foot muscles after removal of flexor digitorum brevis

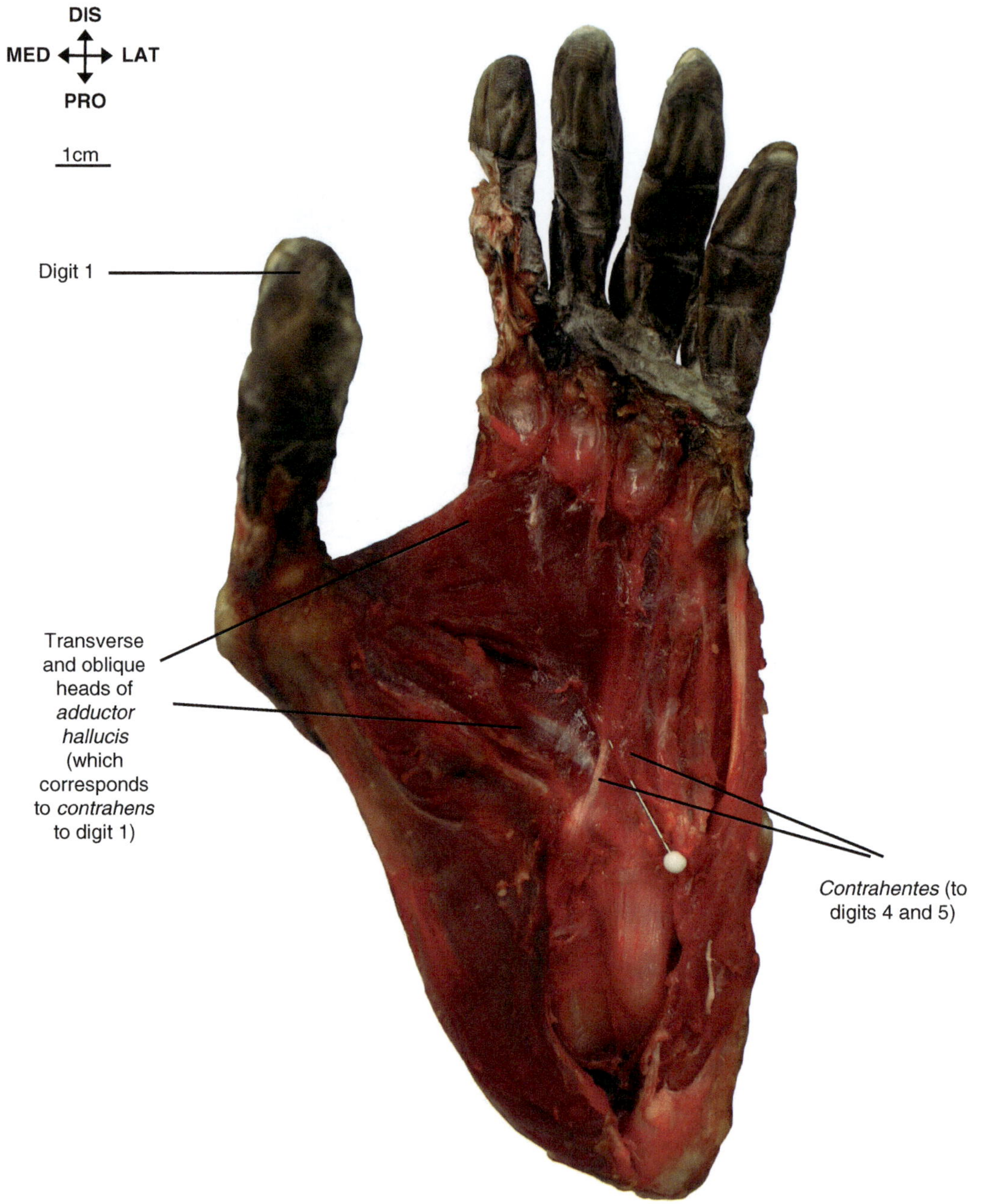


**Fig. 5.16** *Pan paniscus* (ZIMS 164031, Kidogo, adult male): plantar view of contrahentes layer

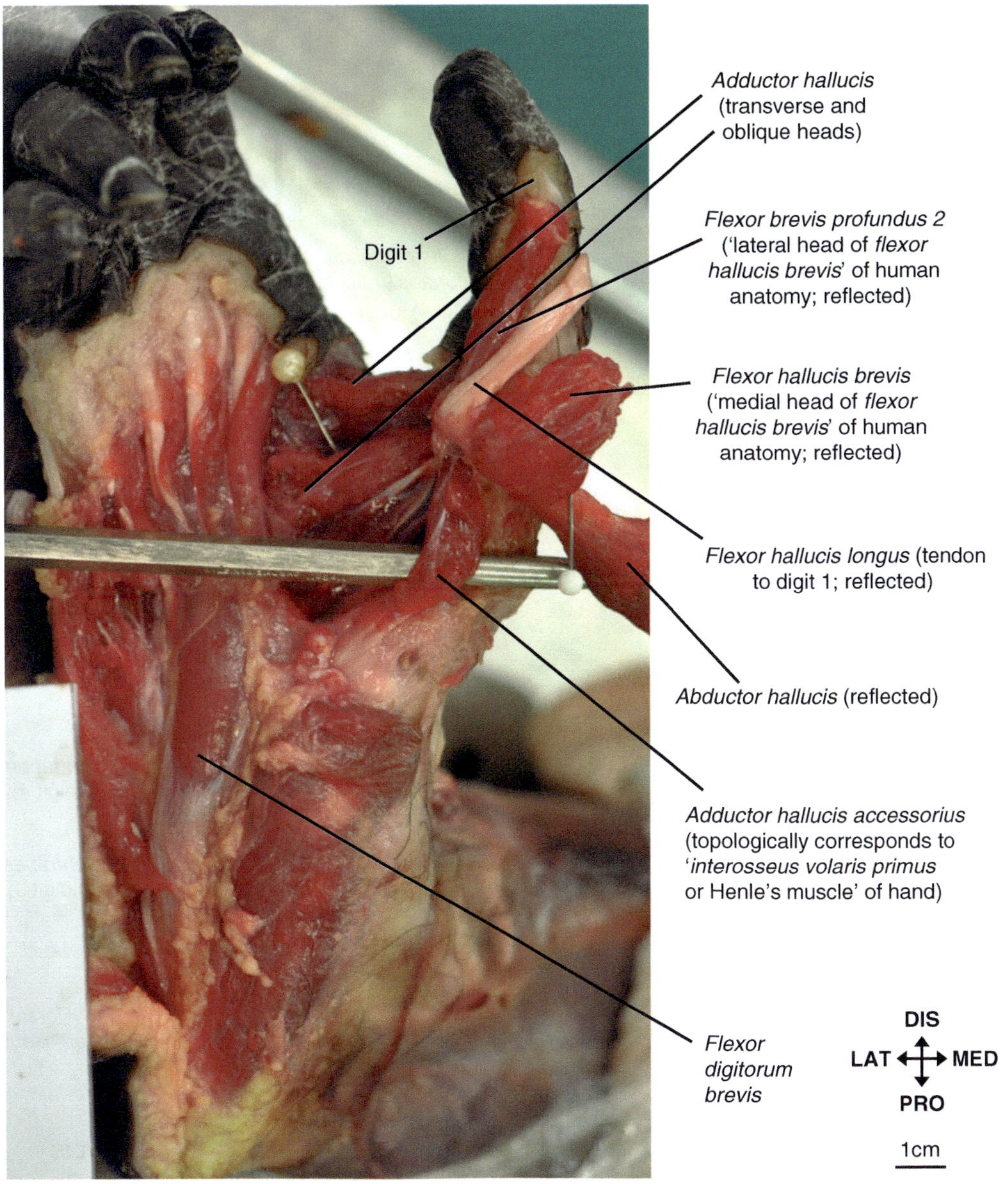


**Fig. 5.17** *Pan paniscus* (ZIMS 164047, Jasiri, adolescent female): plantar view of plantar muscles

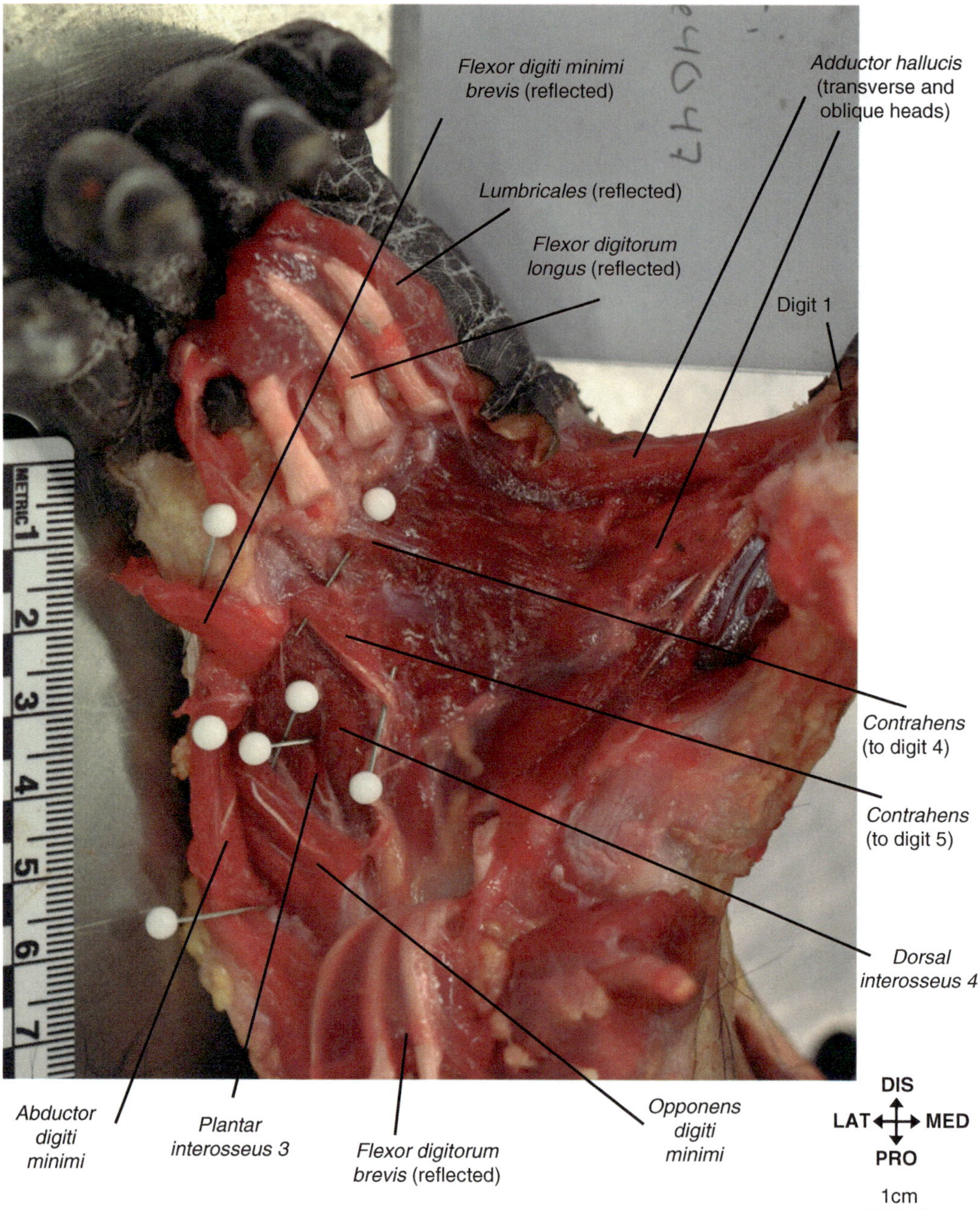


**Fig. 5.18** *Pan paniscus* (ZIMS 164047, Jasiri, adolescent female): plantar view of deeper plantar muscles

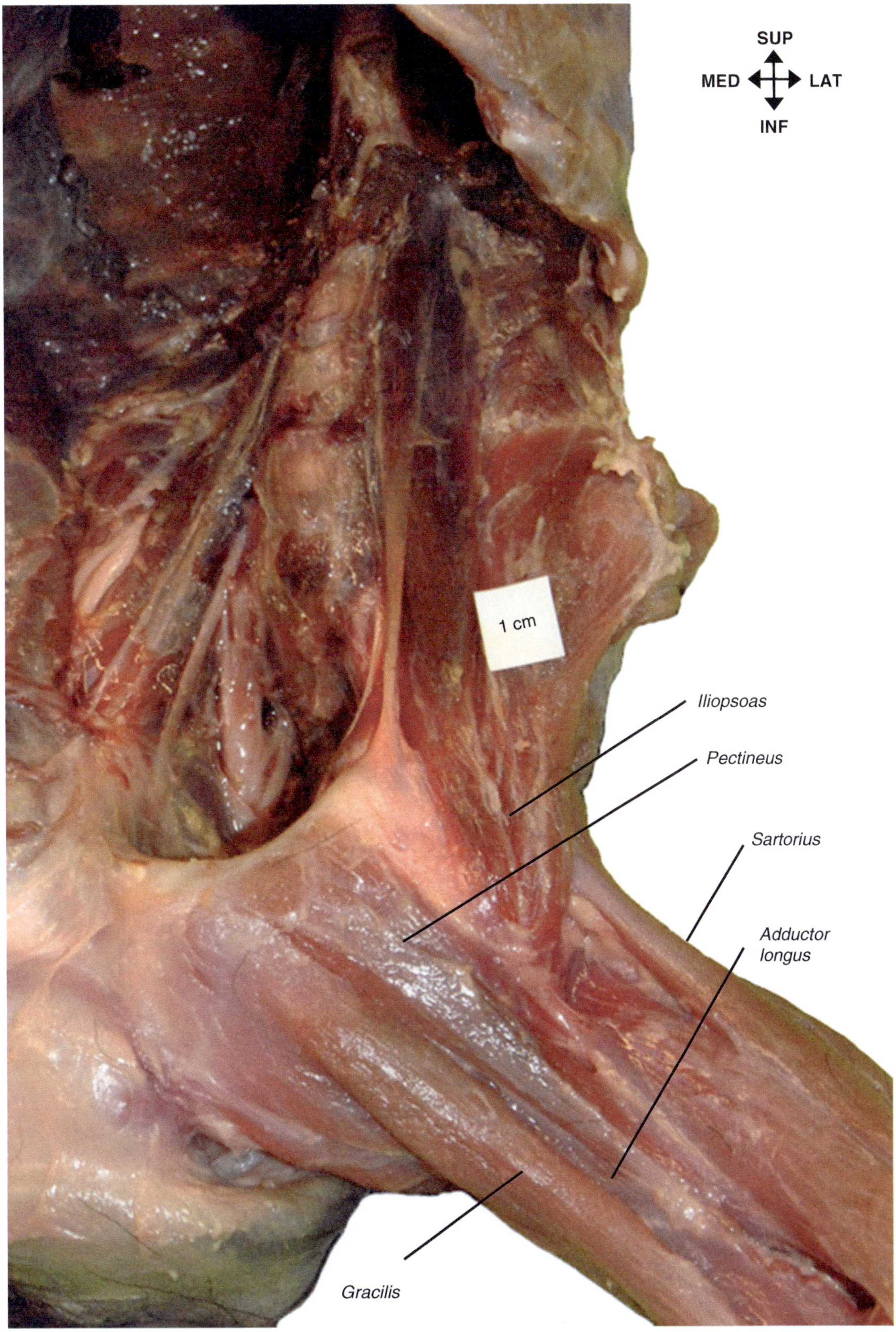


**Fig. 5.19** *Pan paniscus* (ZIMS 164041, Foyo, infant male): ventro-medial view of the proximal part of the left thigh

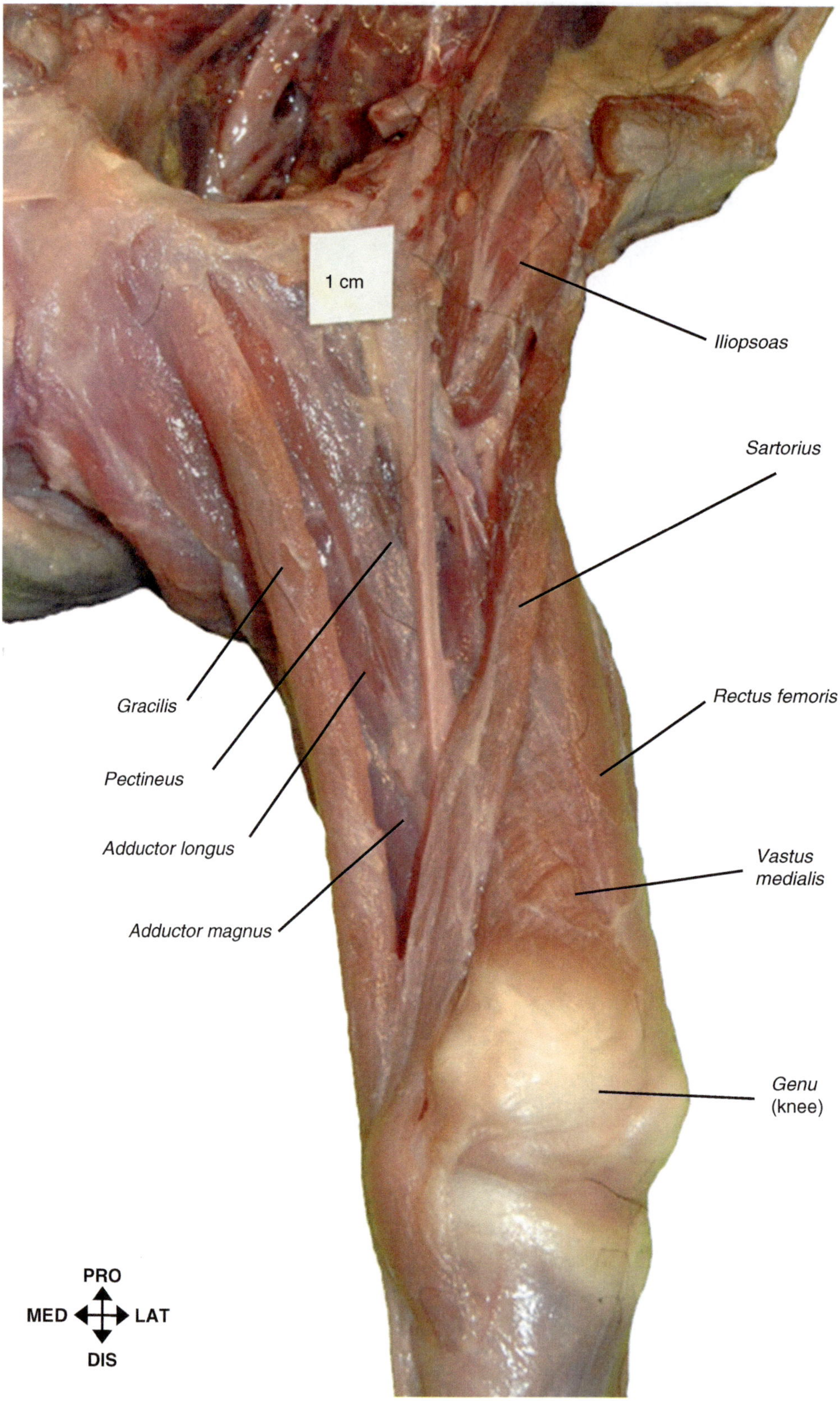


**Fig. 5.20** *Pan paniscus* (ZIMS 164041, Foyo, infant male): ventro-medial view of the muscles of the left thigh

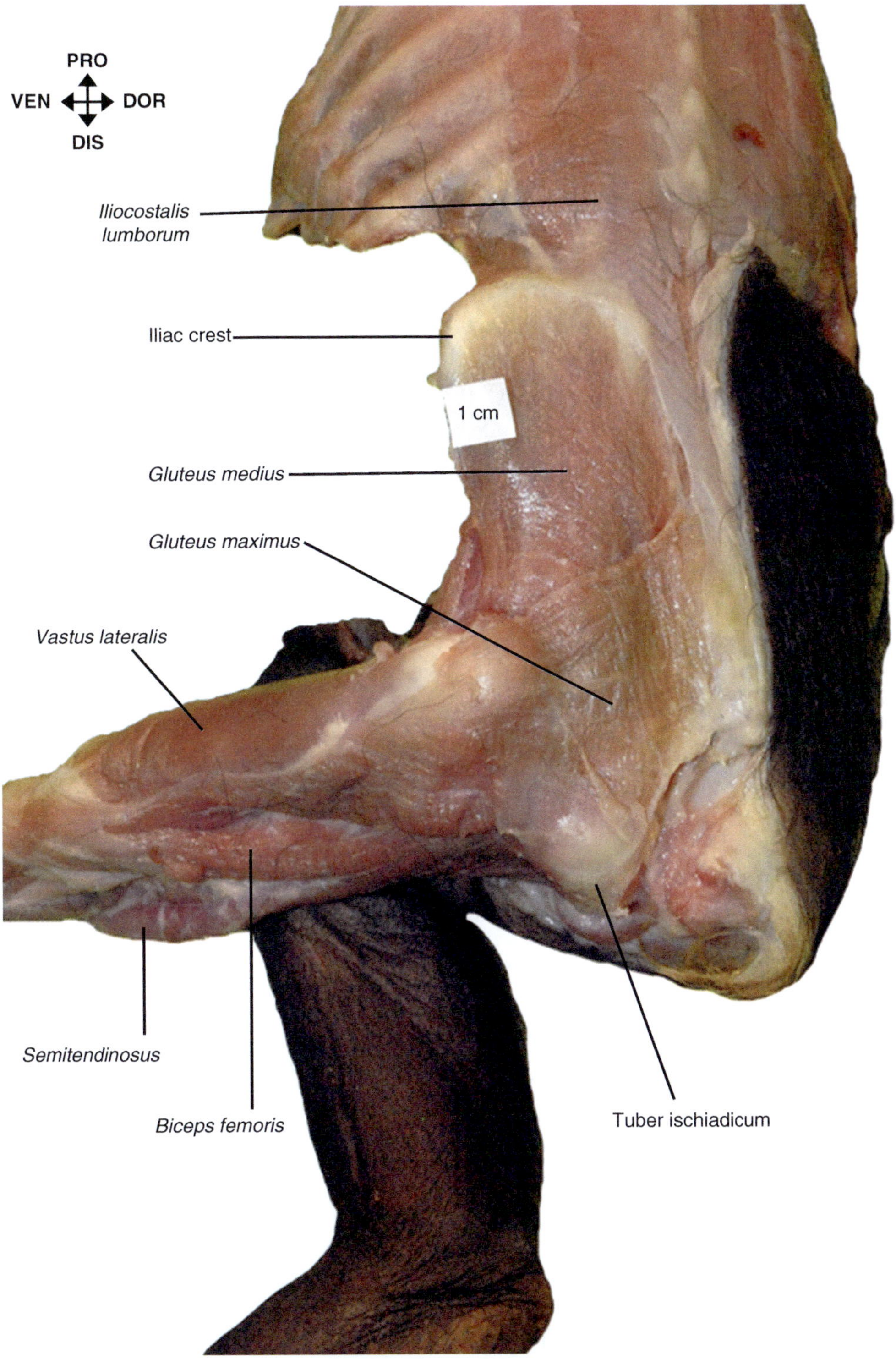


**Fig. 5.21** *Pan paniscus* (ZIMS 164041, Foyo, infant male): dorso-lateral view of the left thigh and gluteal musculature

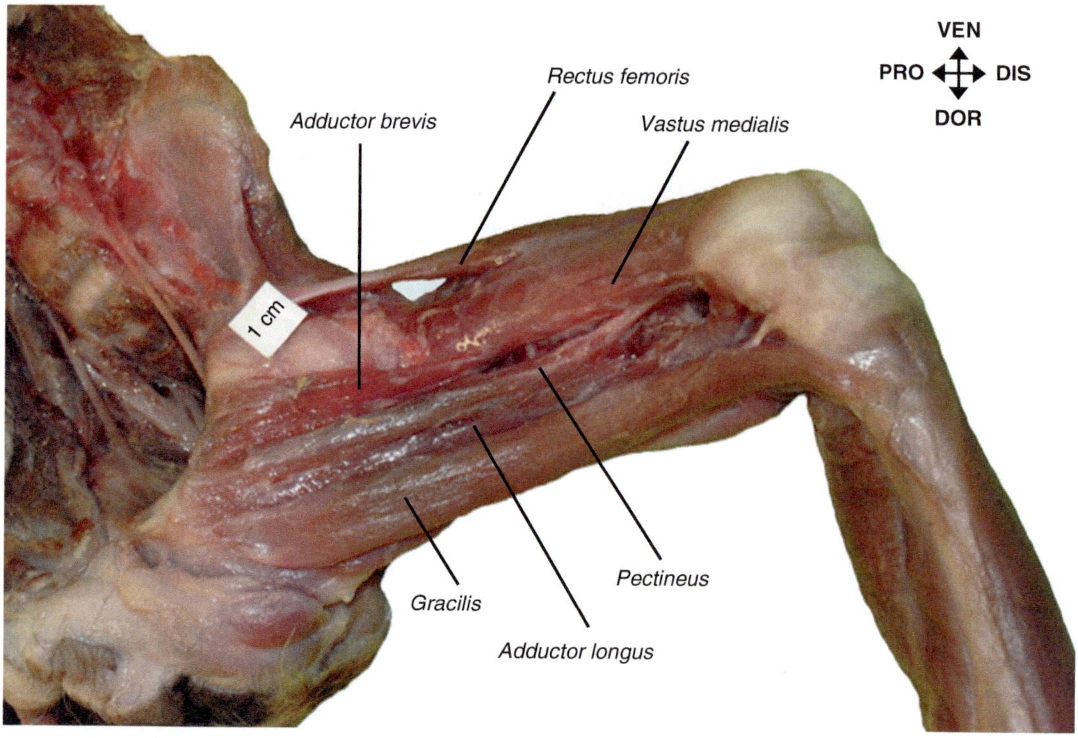


**Fig. 5.22** *Pan paniscus* (ZIMS 164041, Foyo, infant male): medial view of the thigh leg muscles, after removal of sartorius

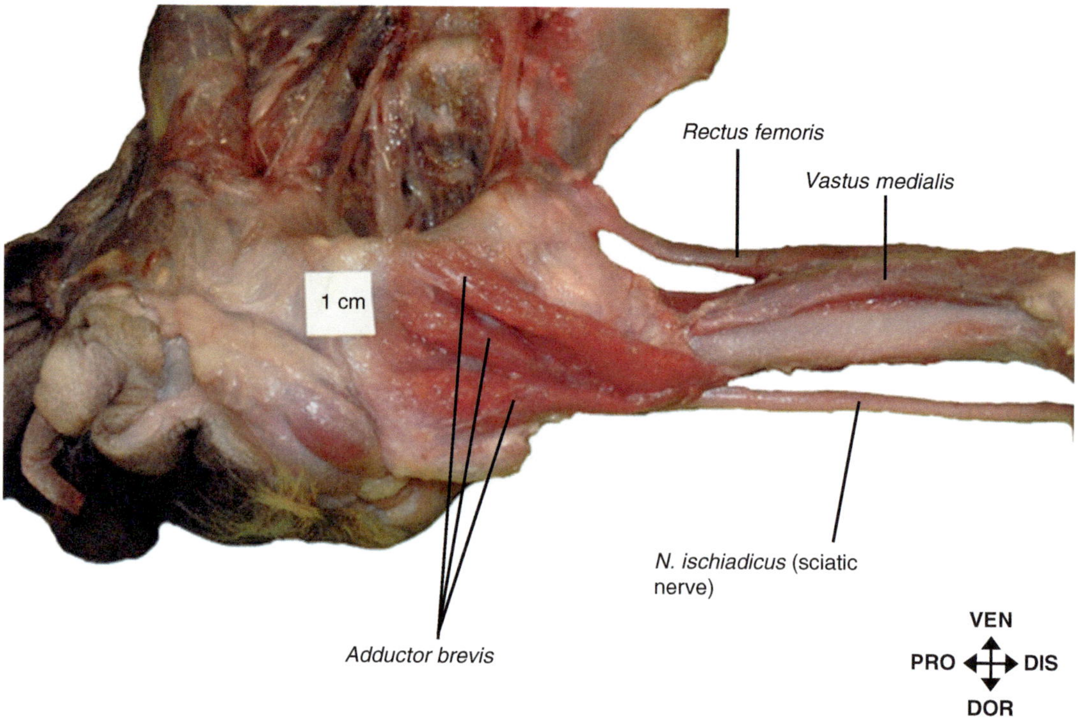


**Fig. 5.23** *Pan paniscus* (ZIMS 164041, Foyo, infant male): medial view of the deep muscles of the left thigh

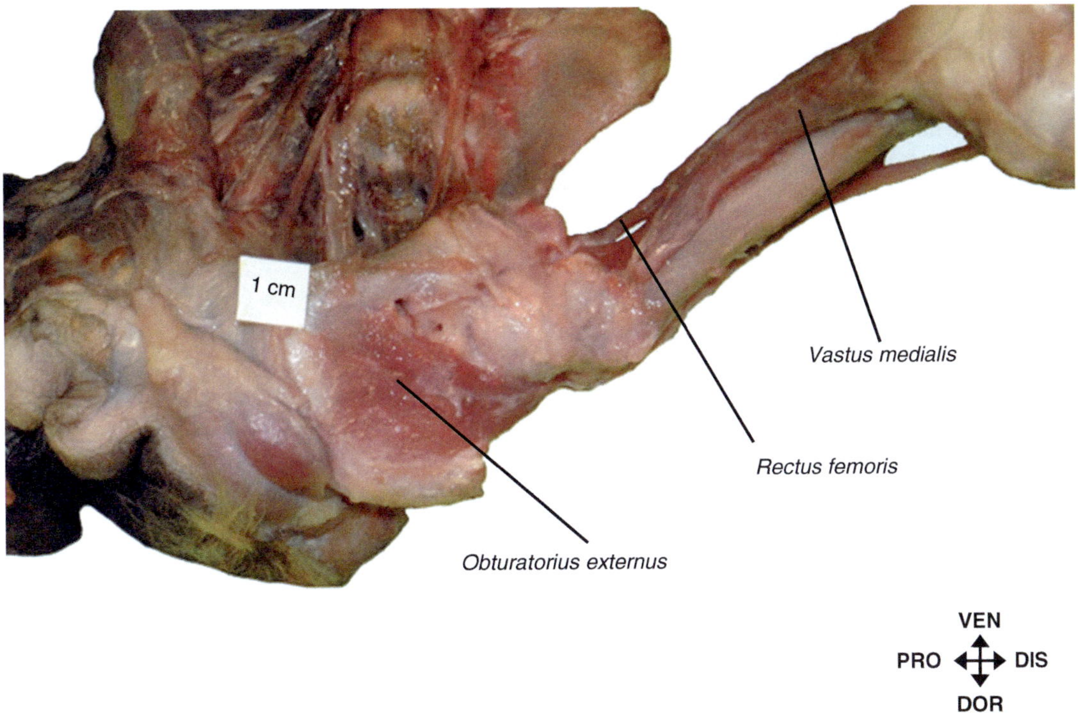


**Fig. 5.24** *Pan paniscus* (ZIMS 164041, Foyo, infant male): medial view of the deepest thigh muscles

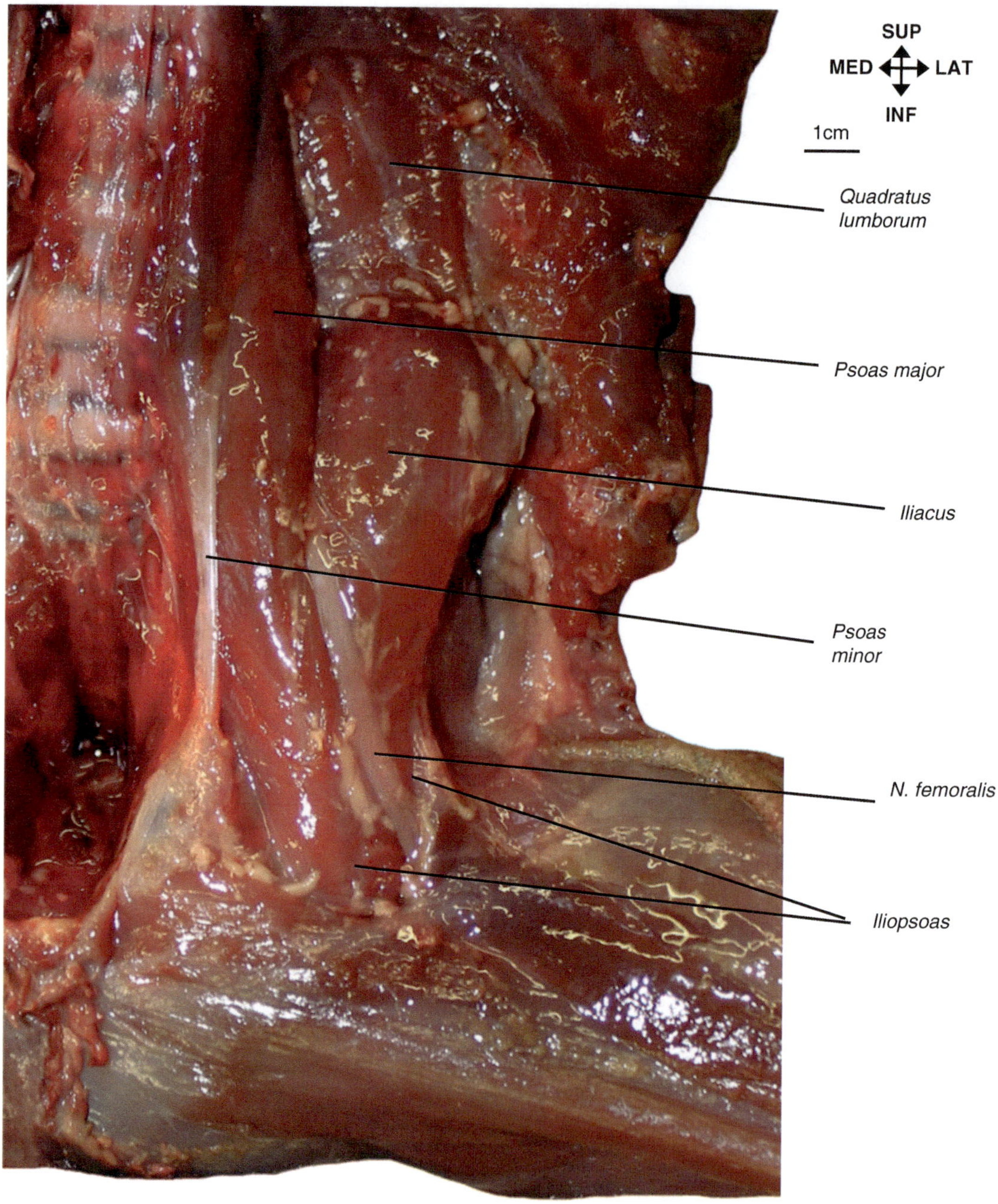


**Fig. 5.25** *Pan paniscus* (ZIMS 164052, Ano, fetus female): ventral view of the posterior abdominal wall

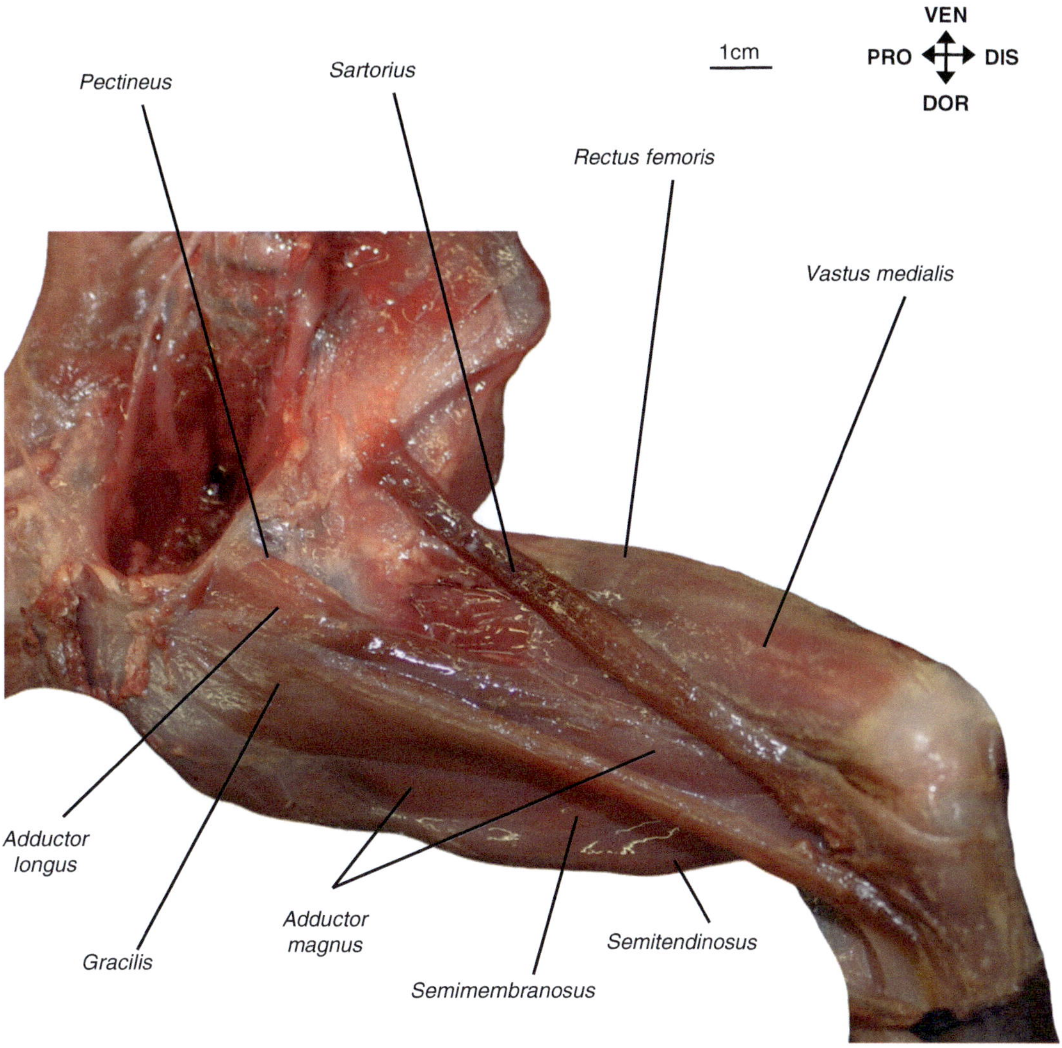


**Fig. 5.26** *Pan paniscus* (ZIMS 164052, Ano, fetus female): ventro-medial view of the muscles of the left thigh

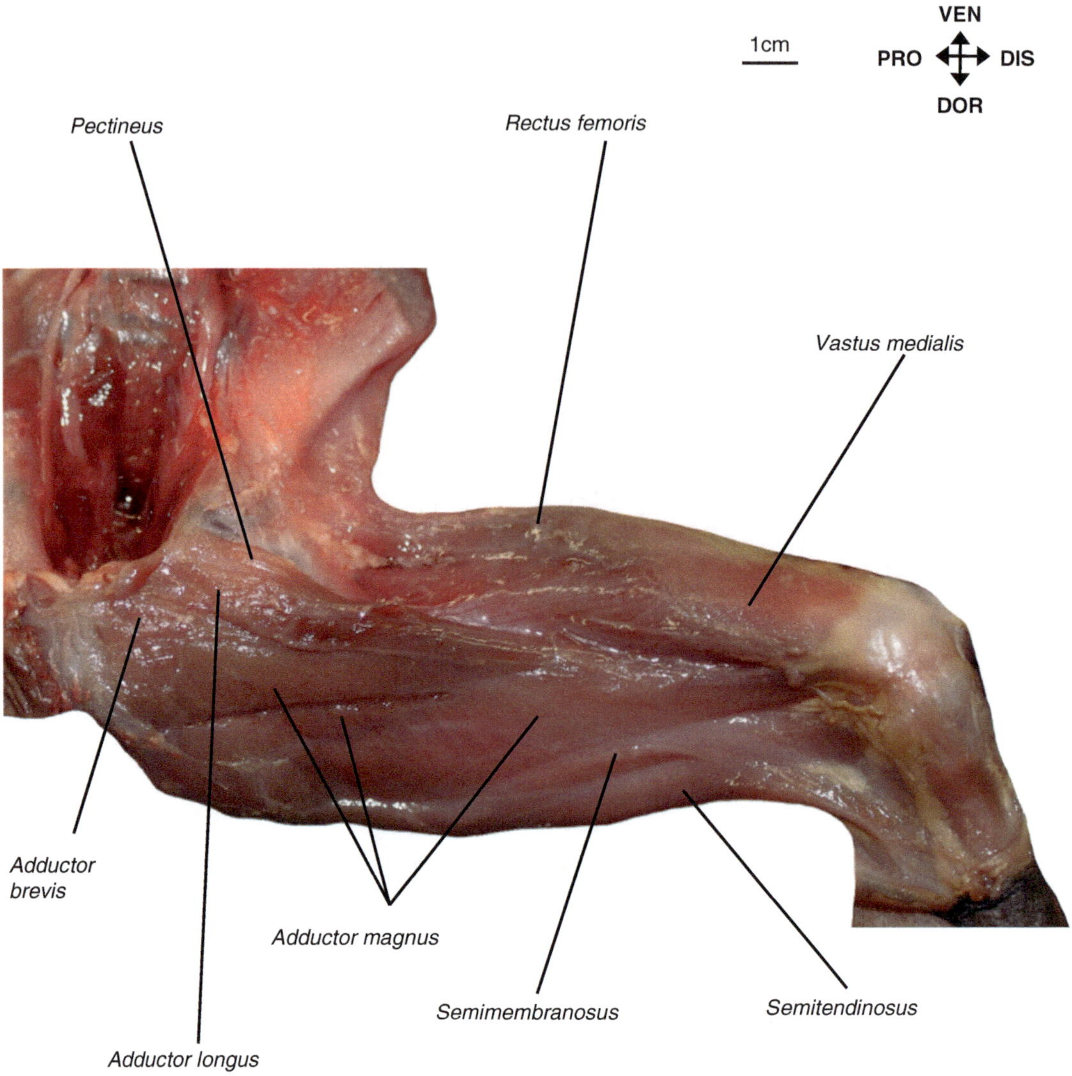


**Fig. 5.27** *Pan paniscus* (ZIMS 164052, Ano, fetus female): ventro-medial view of the deep muscles of the left thigh, after removal of sartorius and gracilis

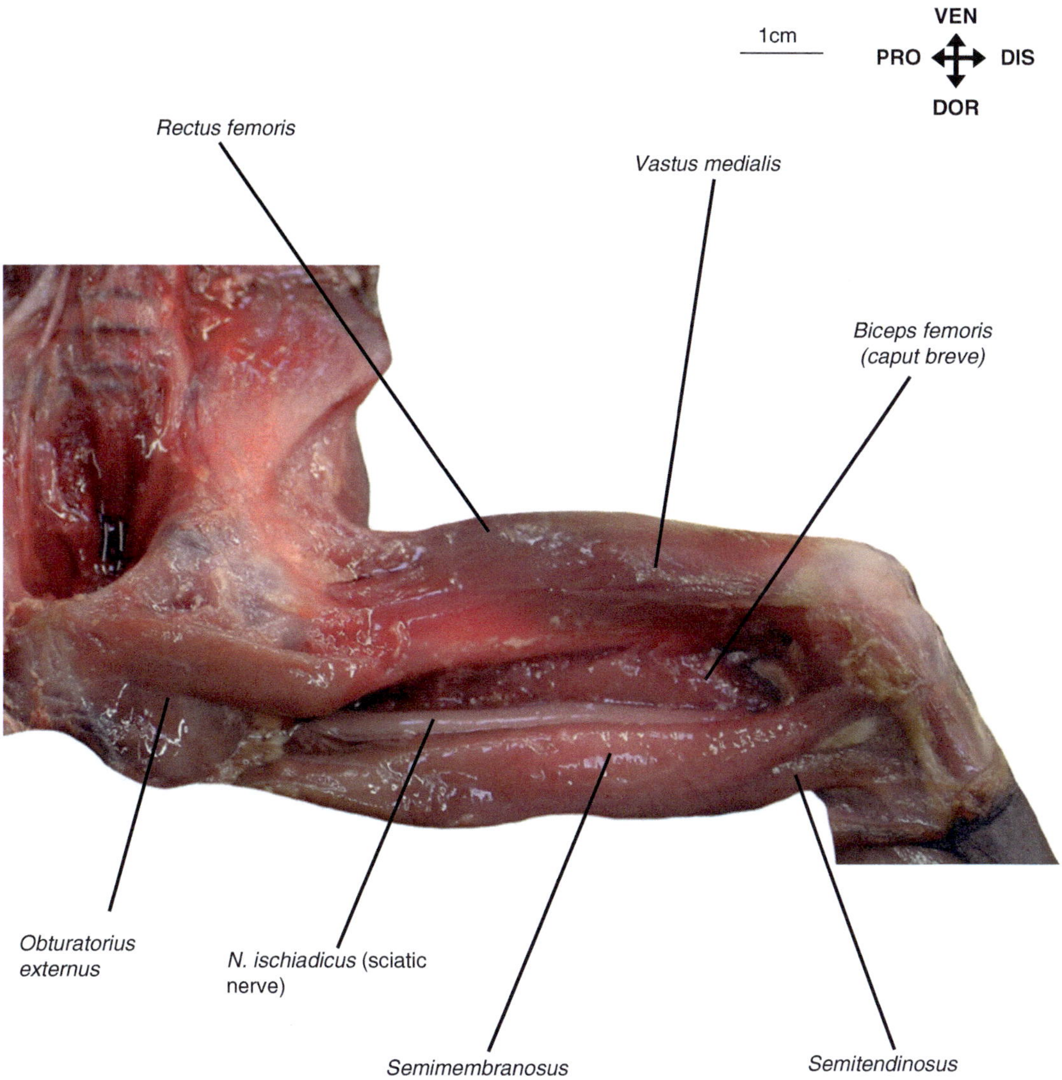


**Fig. 5.28** *Pan paniscus* (ZIMS 164052, Ano, fetus female): ventro-medial view of deep left thigh muscles after removal of pectineus, adductor longus, adductor brevis and adductor magnus

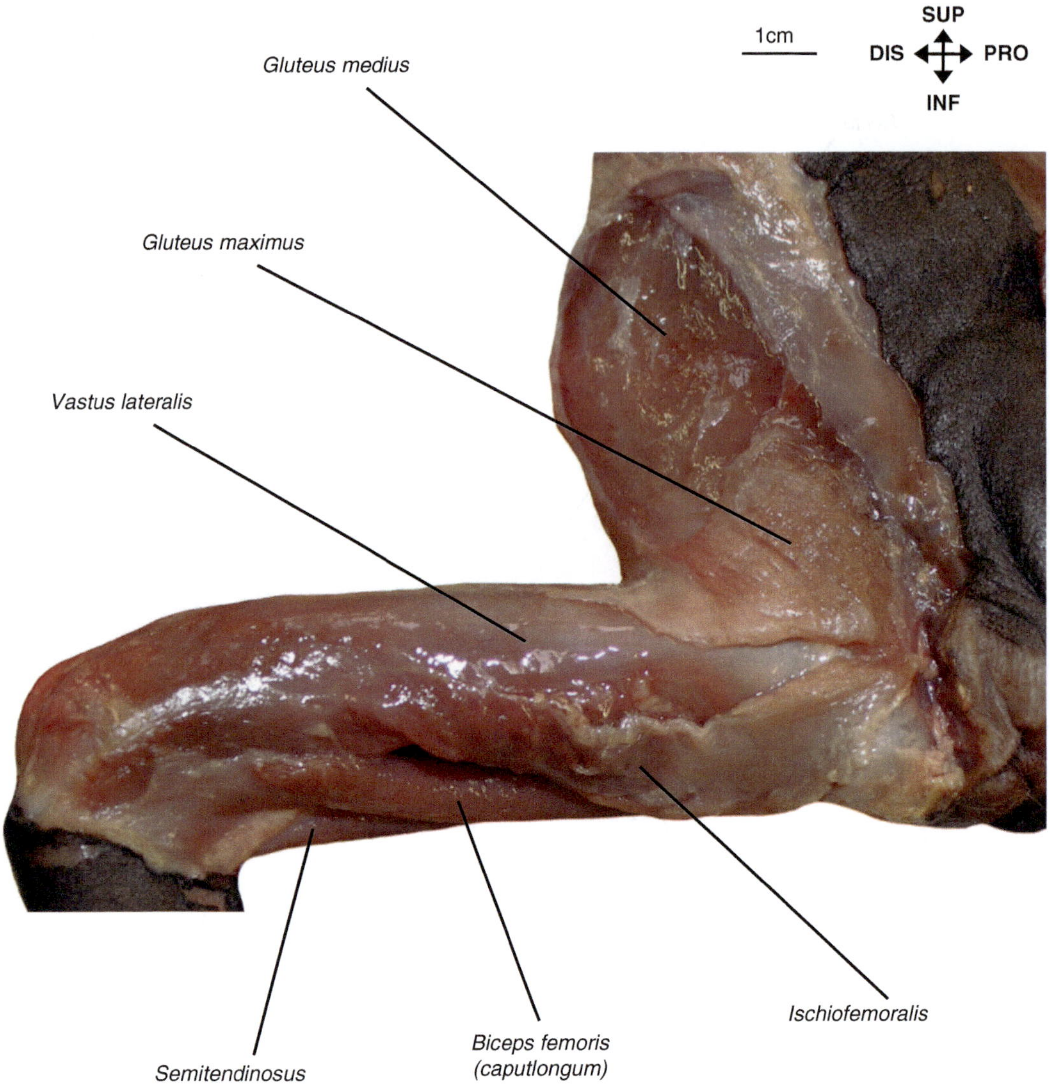


**Fig. 5.29** *Pan paniscus* (ZIMS 164052, Ano, fetus female): dorsal view of superficial gluteal and thigh muscles

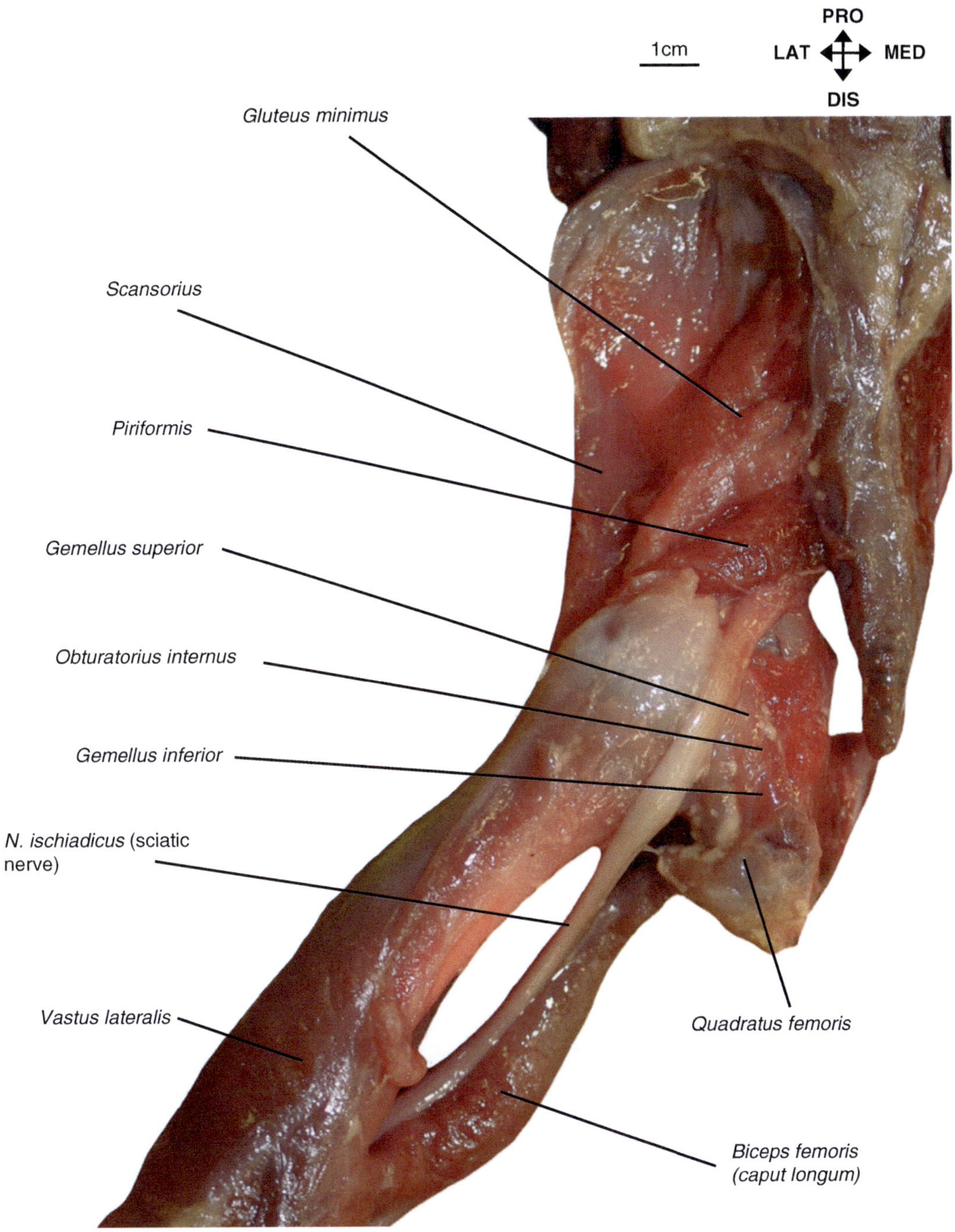


**Fig. 5.30** *Pan paniscus* (ZIMS 164052, Ano, fetus female): dorsal view of deep gluteal and thigh muscles after removal of gluteus maximus, gluteus medius and ischiofemoralis

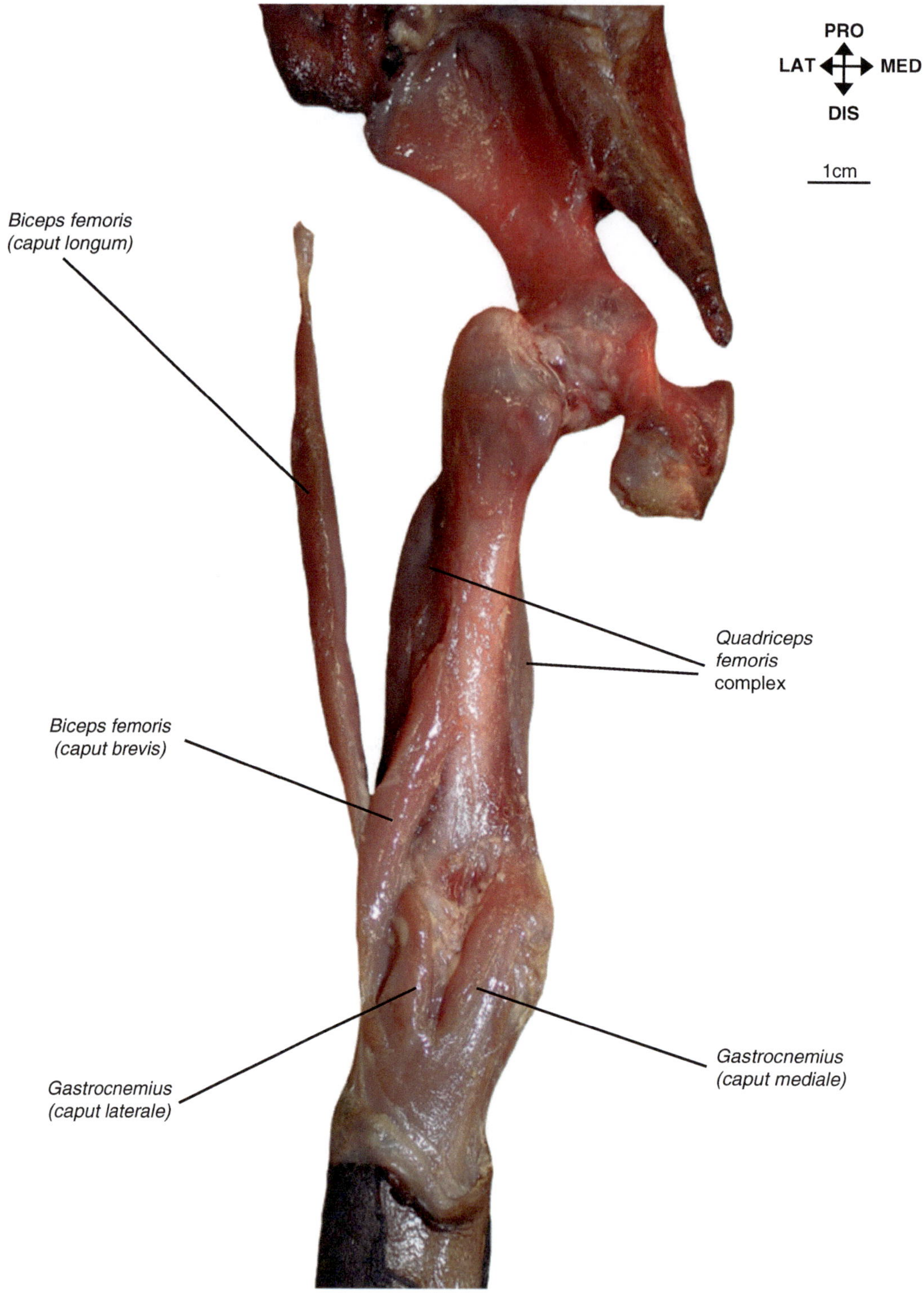


**Fig. 5.31** *Pan paniscus* (ZIMS 164052, Ano, fetus female): dorsal view of deep muscles of left thigh, after removal of semitendinosus and semimembranosus. The caput longum of biceps femoris is detached from its origin at the ischial tuberosity by dissection

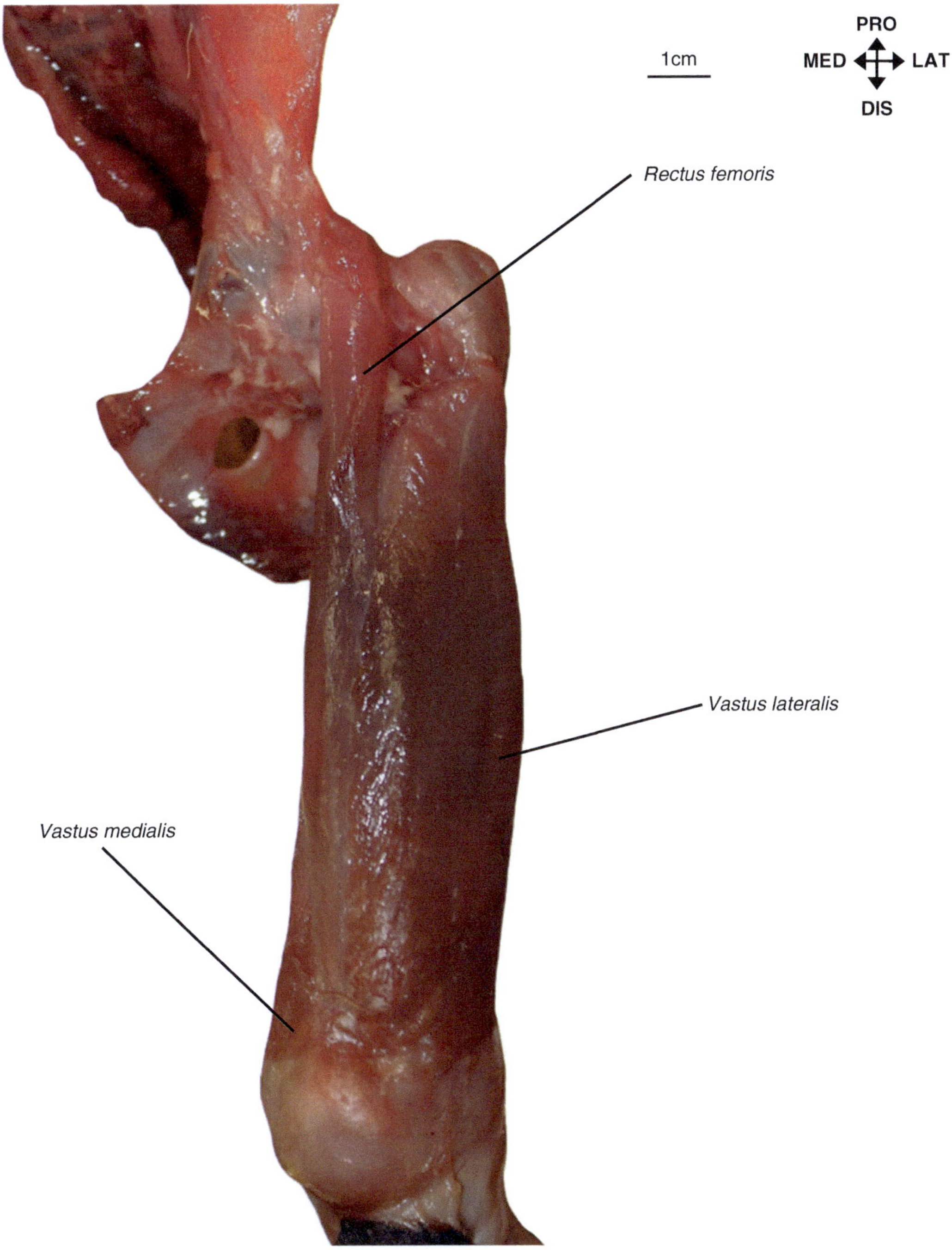


**Fig. 5.32** *Pan paniscus* (ZIMS 164052, Ano, fetus female): ventral view of superficial muscles of left thigh

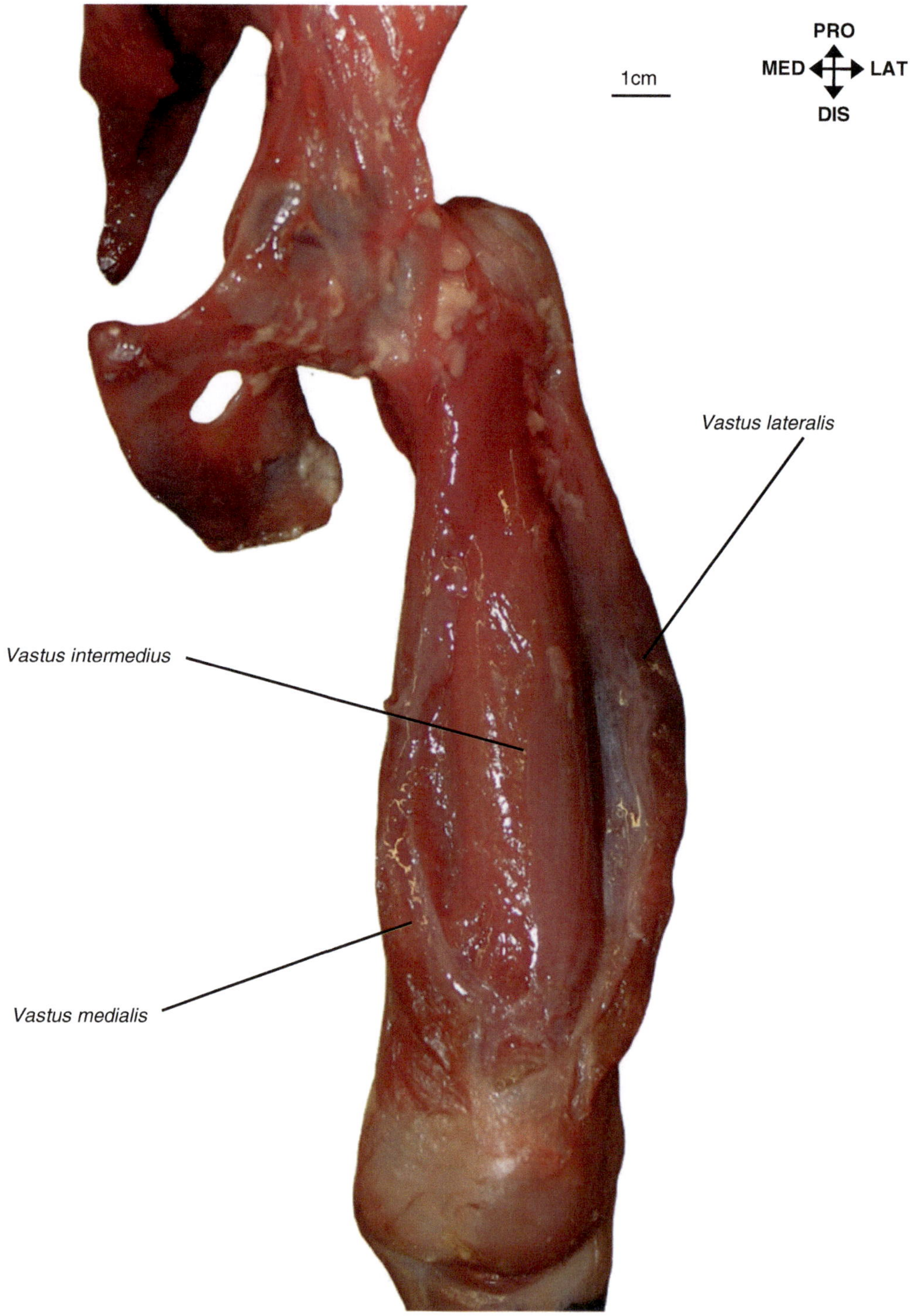


**Fig. 5.33** *Pan paniscus* (ZIMS 164052, Ano, fetus female): ventral view of deep muscles of left thigh after removal of rectus femoris

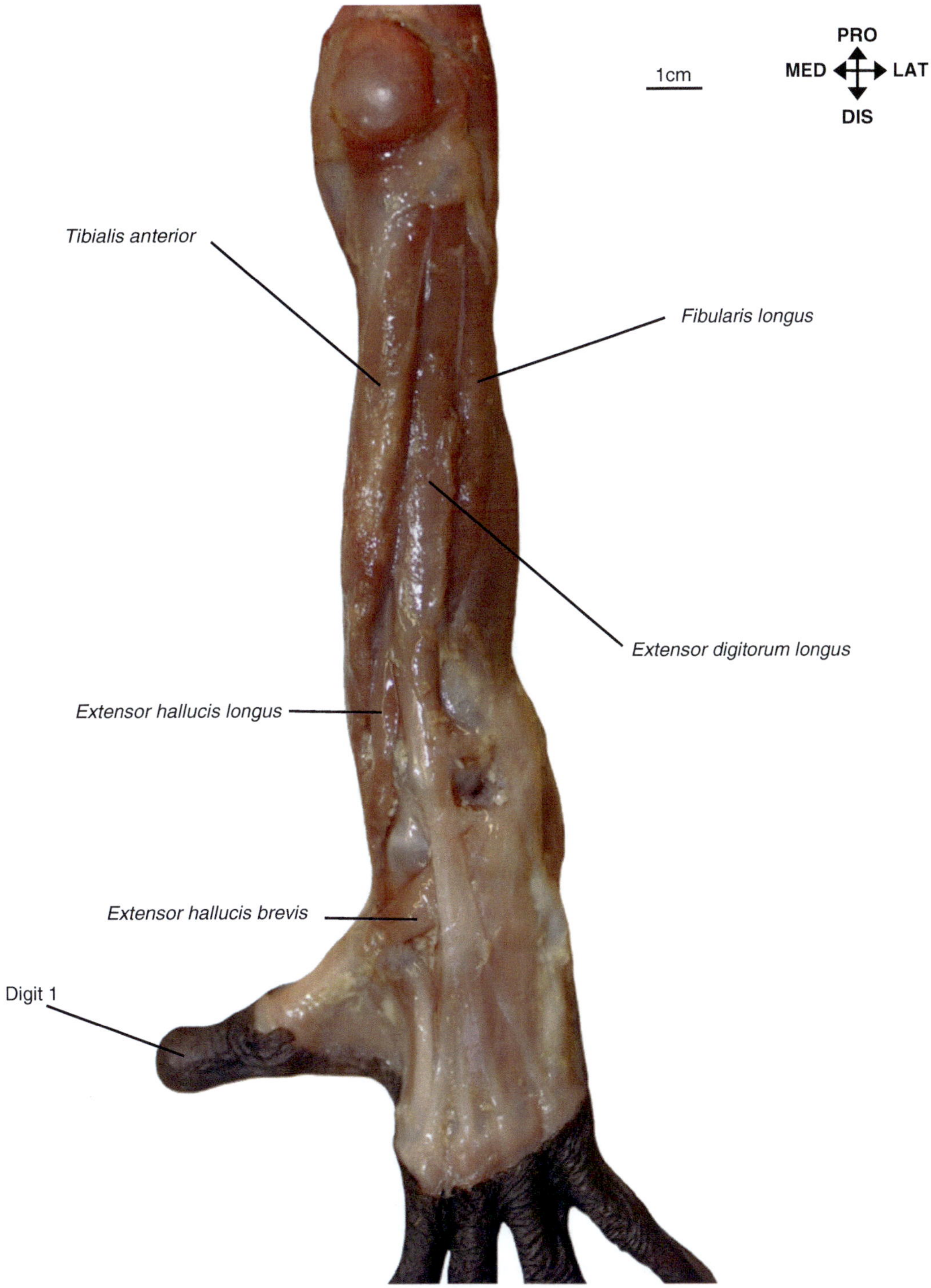


**Fig. 5.34** *Pan paniscus* (ZIMS 164052, Ano, fetus female): ventral view of the superficial muscles of the left leg

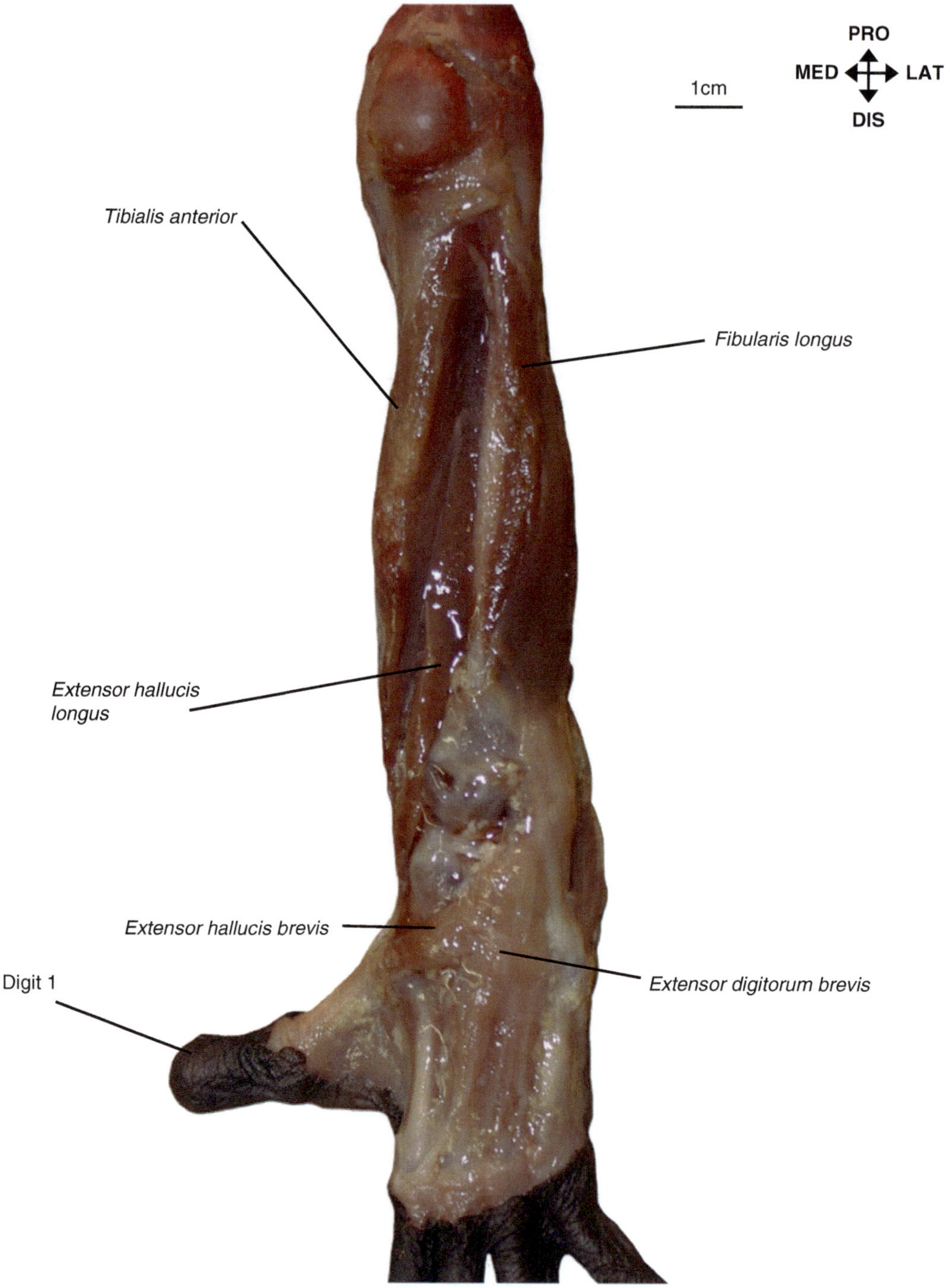


**Fig. 5.35** *Pan paniscus* (ZIMS 164052, Ano, fetus female): ventral view of the deep muscles of the left leg after removal of extensor digitorum longus

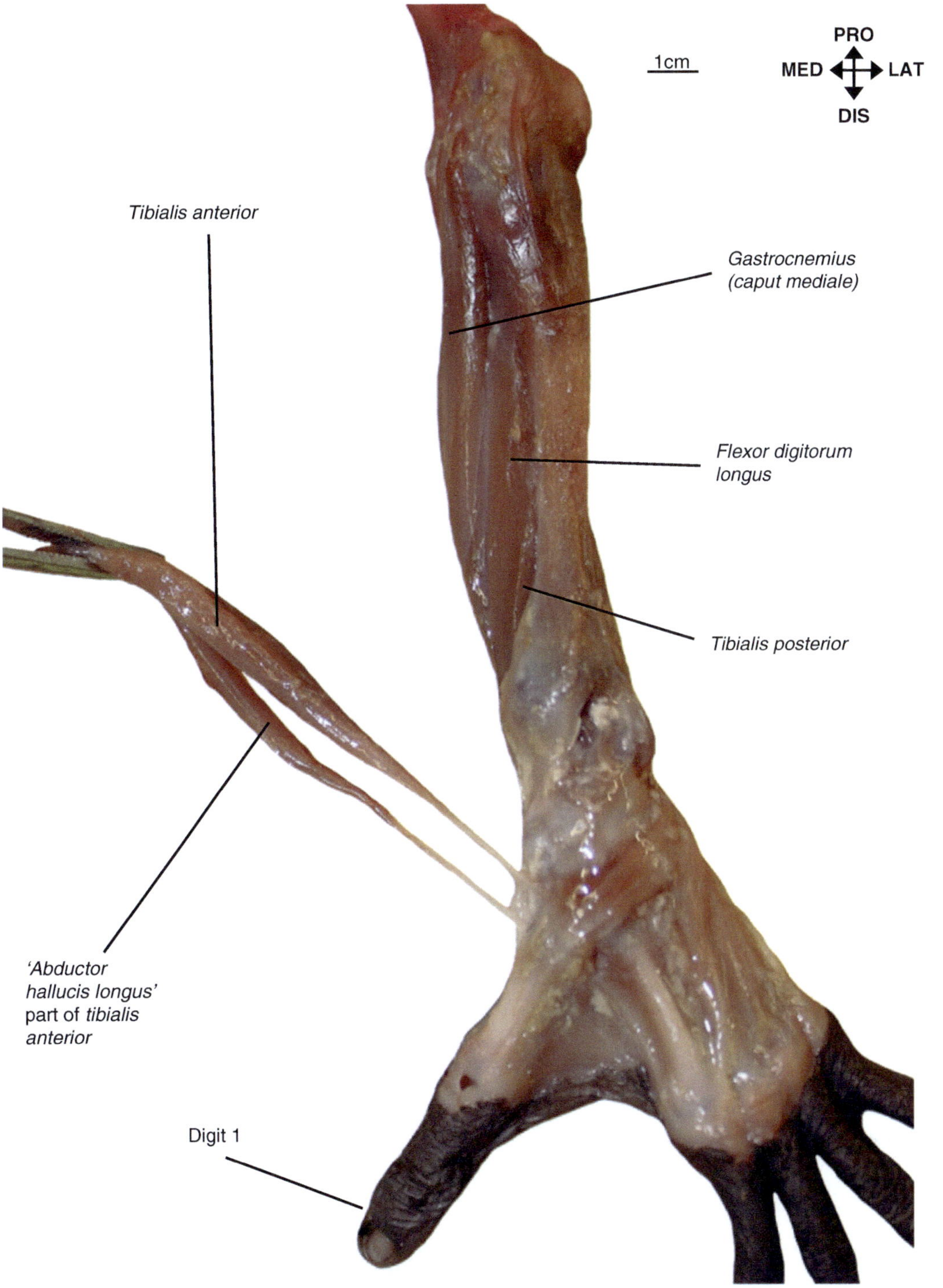


**Fig. 5.36** *Pan paniscus* (ZIMS 164052, Ano, fetus female): ventro-medial view of the muscles of the left leg after removal of extensor hallucis longus and extensor digitorum longus. Tibialis anterior has been detached showing a variation: its partially differentiated belly corresponding to an 'abductor hallucis longus' structure

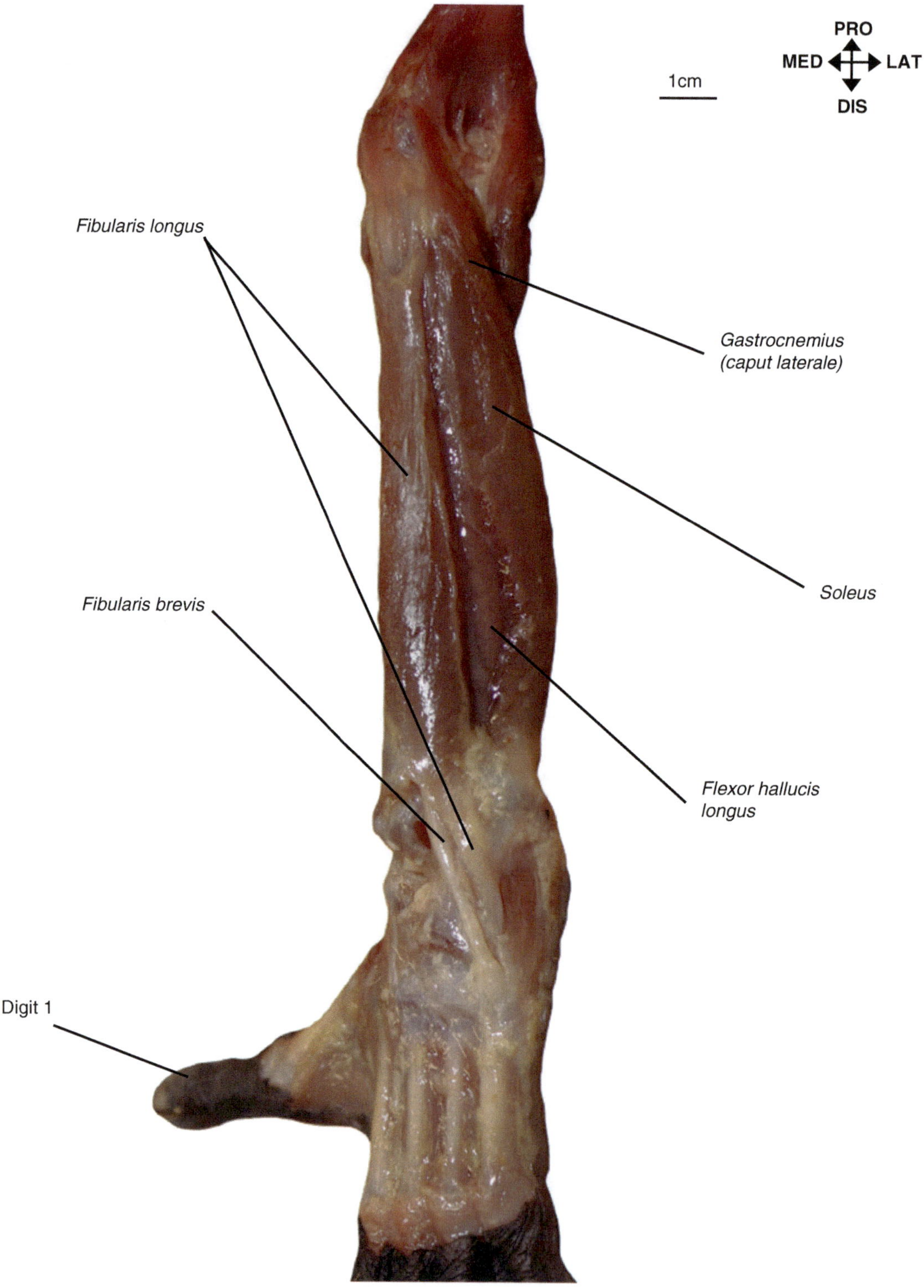


**Fig. 5.37** *Pan paniscus* (ZIMS 164052, Ano, fetus female): ventro-lateral view of the superficial muscles of the left leg

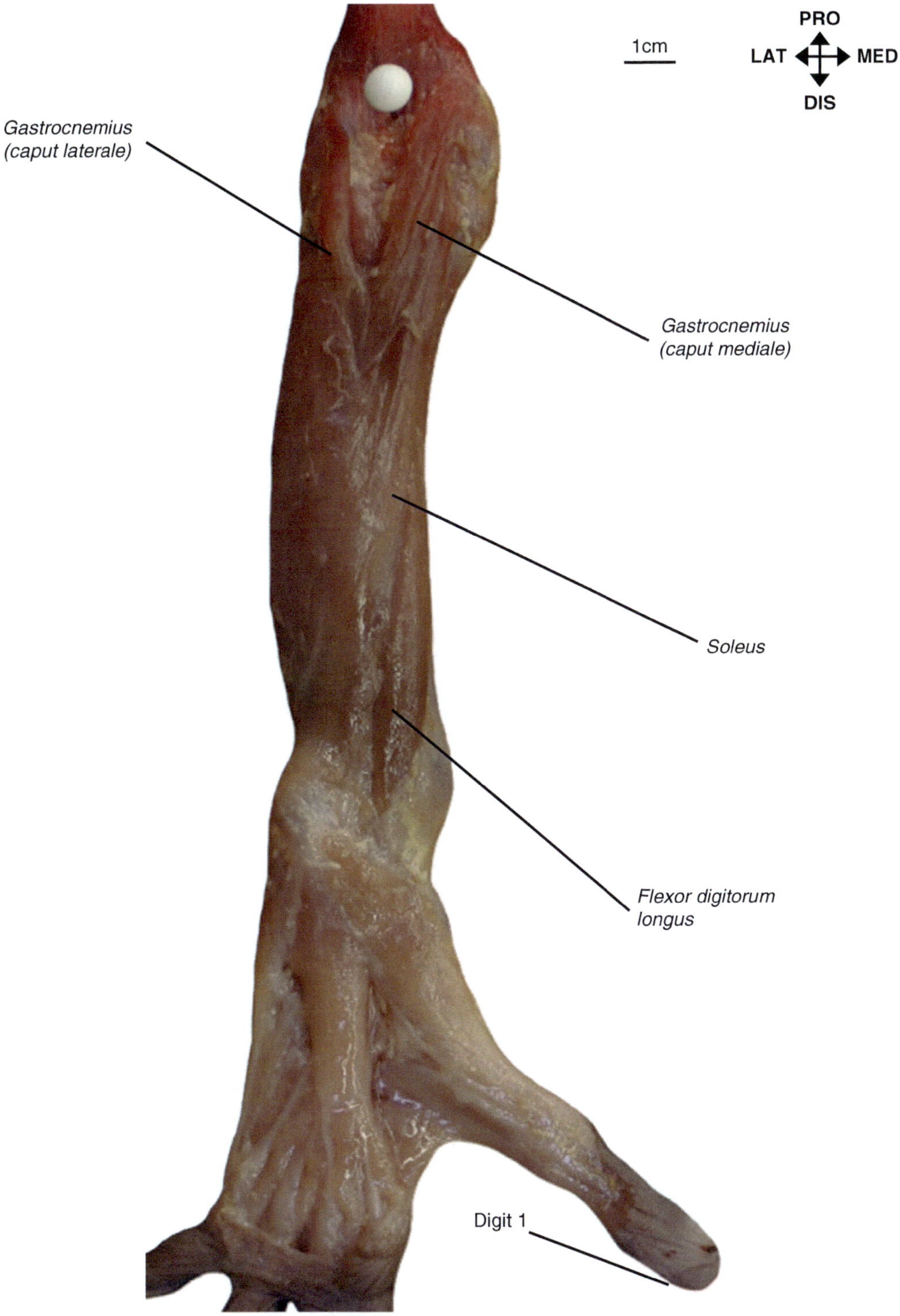


**Fig. 5.38** *Pan paniscus* (ZIMS 164052, Ano, fetus female): dorsal view of the superficial muscles of the left leg

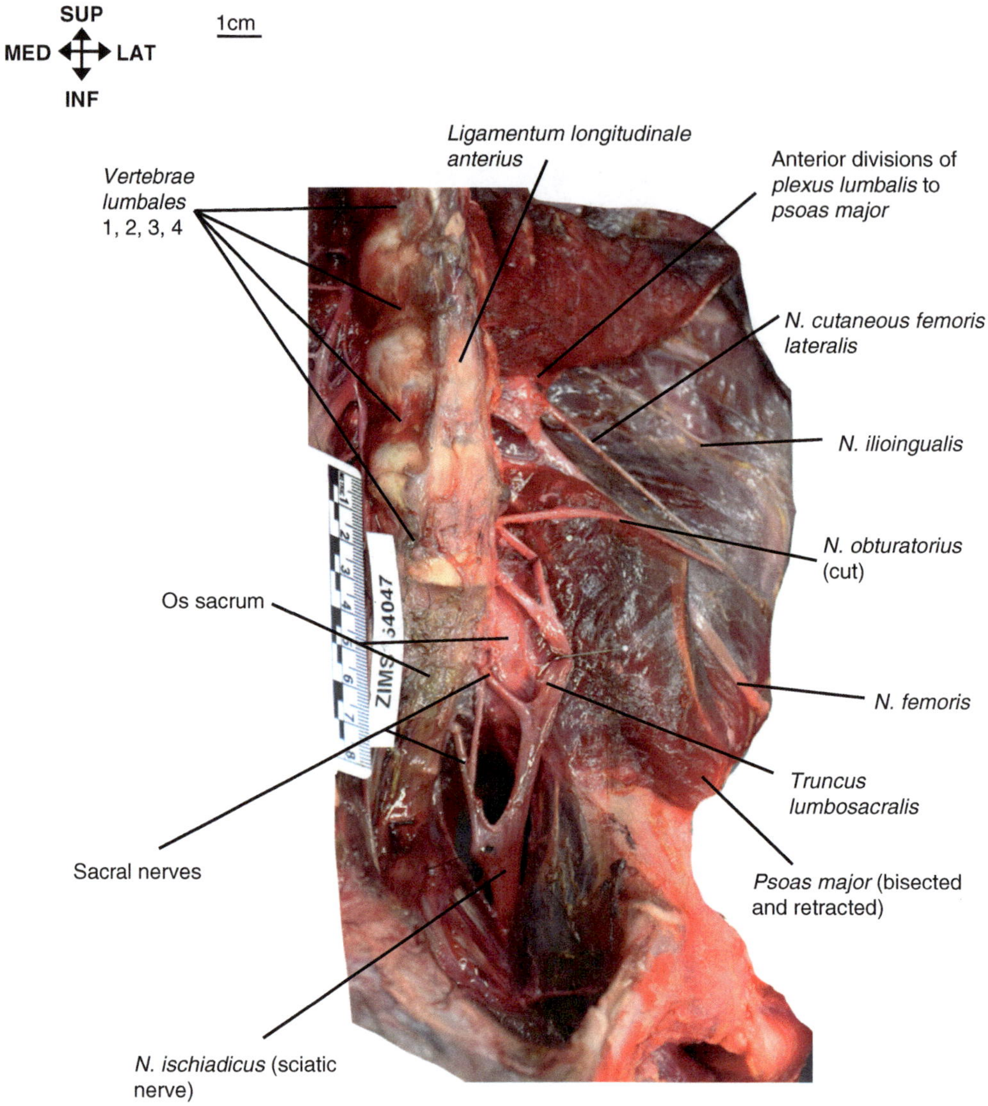


**Fig. 5.39** *Pan paniscus* (ZIMS 164047, Jasiri, adolescent female): ventral view of the left *plexus lumbosacralis*

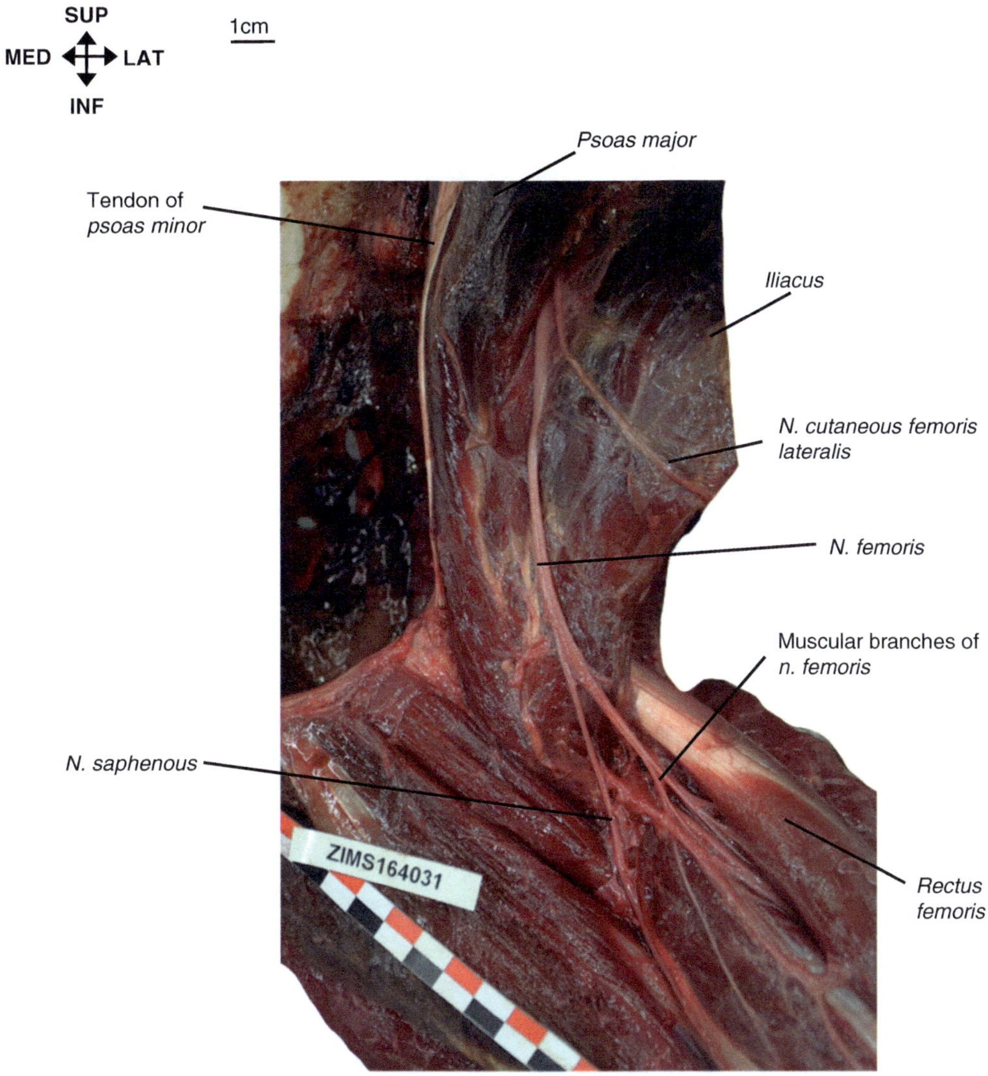


**Fig. 5.40** *Pan paniscus* (ZIMS 164031, Kidogo, adult male): ventral view of left *n. femoris* and its division distally to the *ligamentum inguinale* (not depicted). The sartorious has been reflected distally

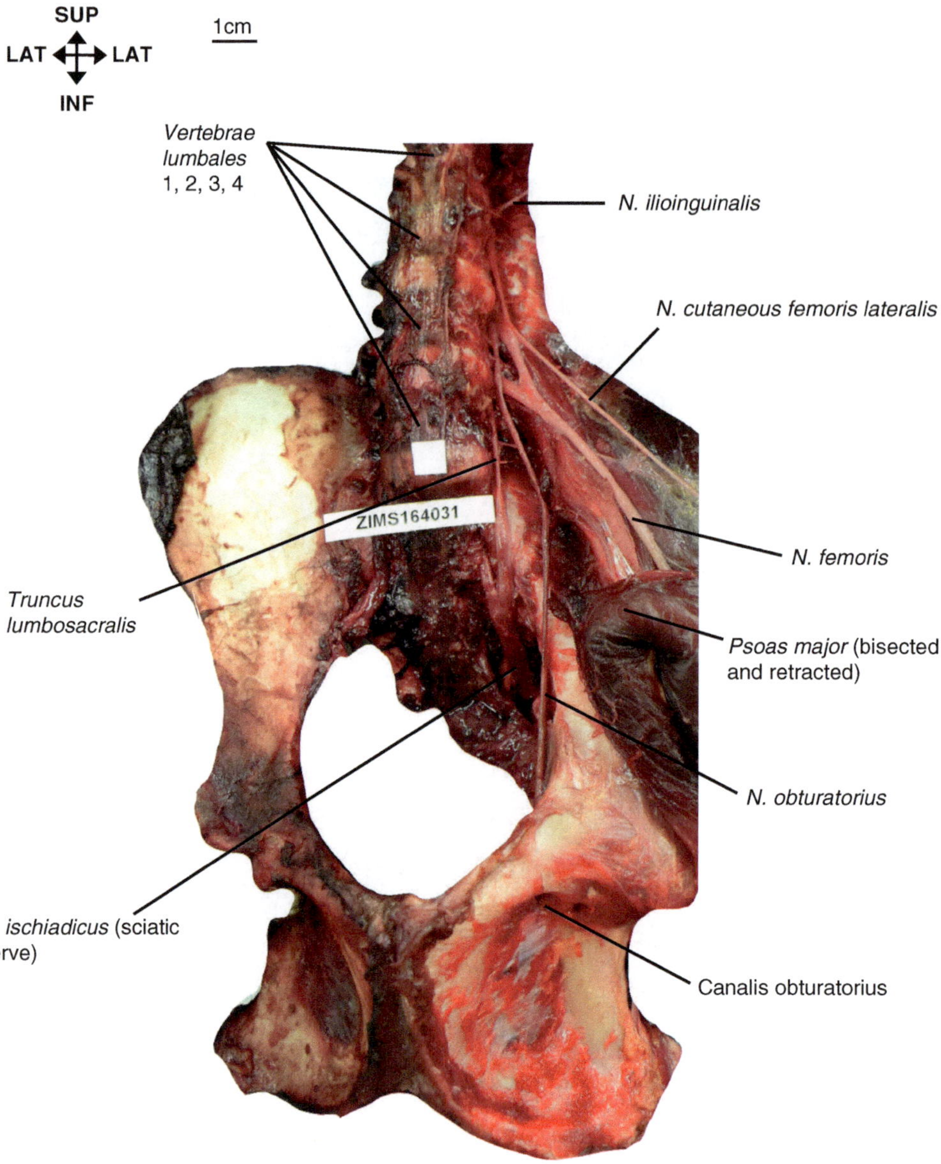


**Fig. 5.41** *Pan paniscus* (ZIMS 164031, Kidogo, adult male): ventral view of left *plexus lumbosacralis*; anterior pelvic and thigh musculature removed

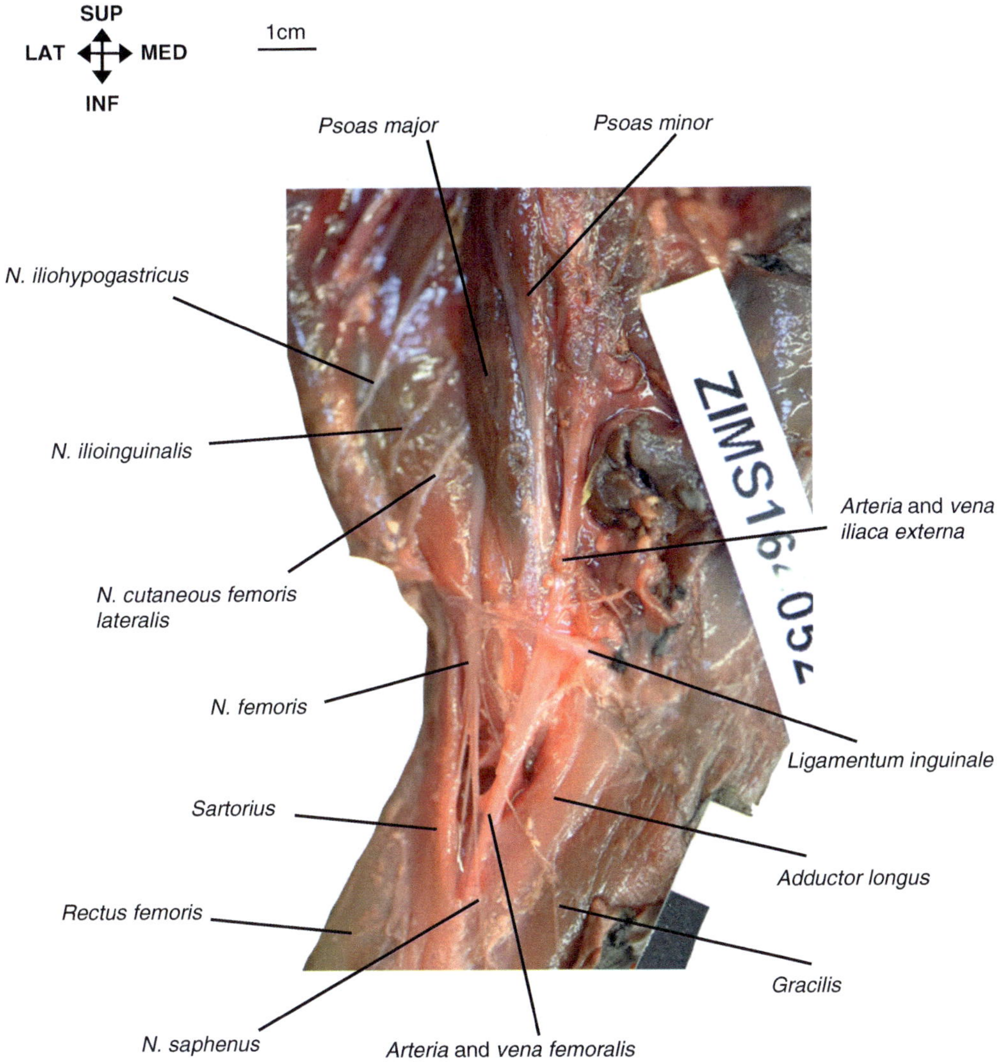


**Fig. 5.42** *Pan paniscus* (ZIMS 164052, Ano, female fetus): ventral view of the right *plexus lumbaris* and musculature of the anterior/medial compartment of the thigh

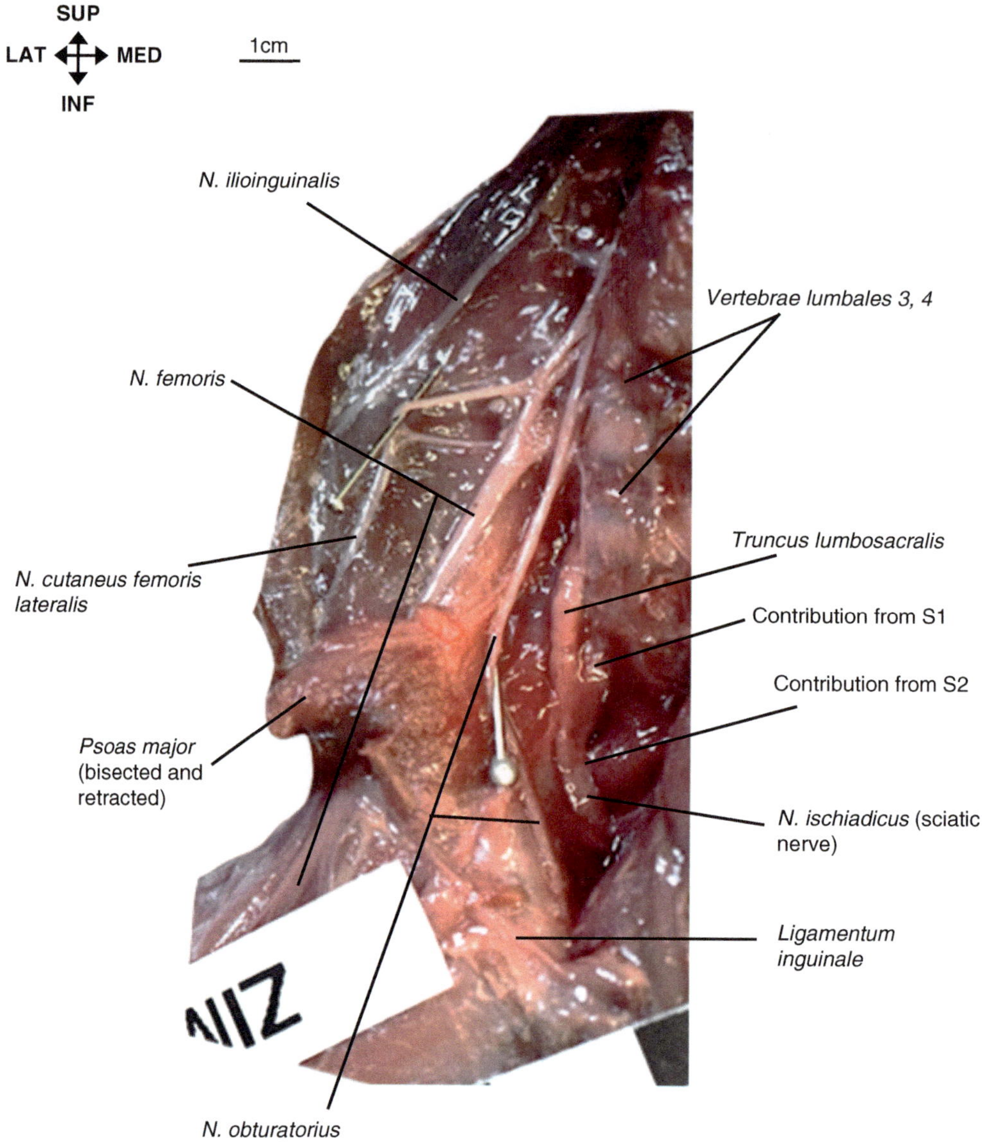


**Fig. 5.43** *Pan paniscus* (ZIMS 164052, Ano, female fetus): ventral view of the right *plexus lumbosacralis*; psoas major has been retracted

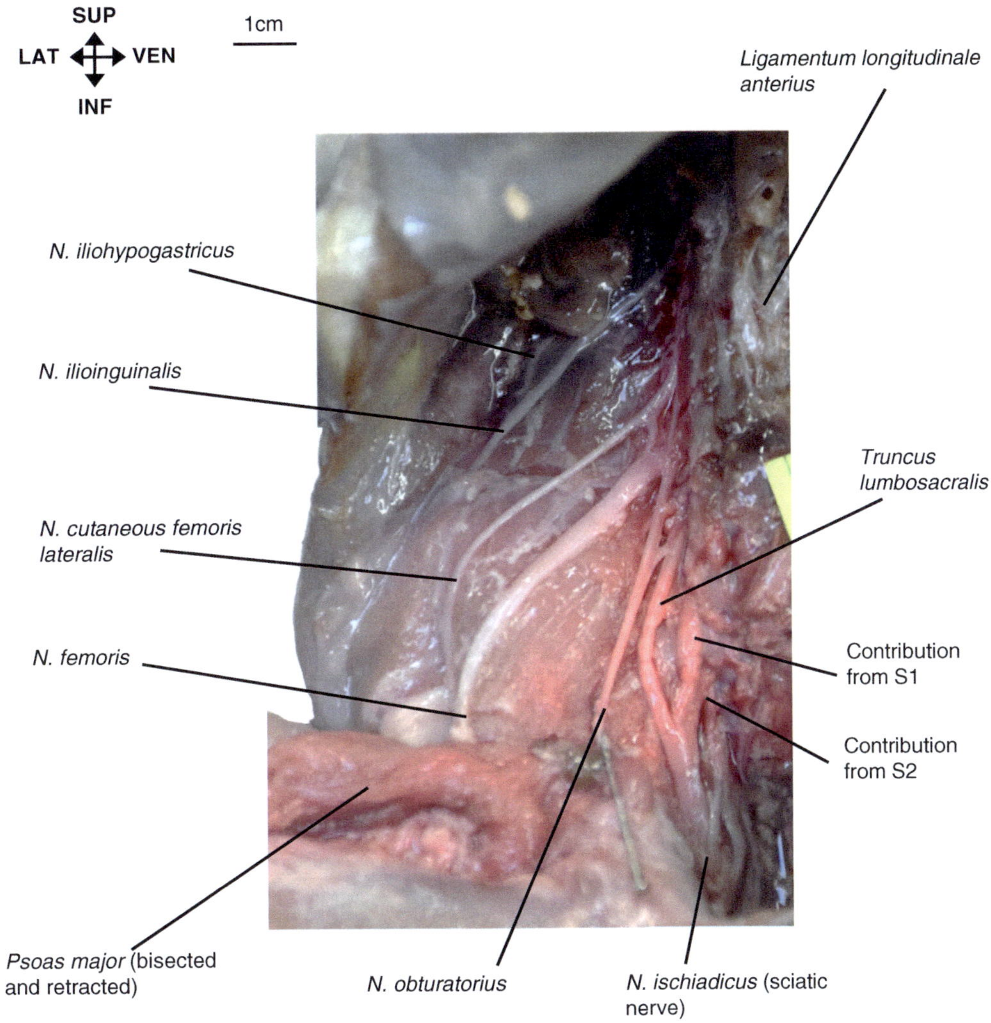


**Fig. 5.44** *Pan paniscus* (ZIMS 164040, Etje, infant male): ventral view of the right *plexus lumbosacralis*

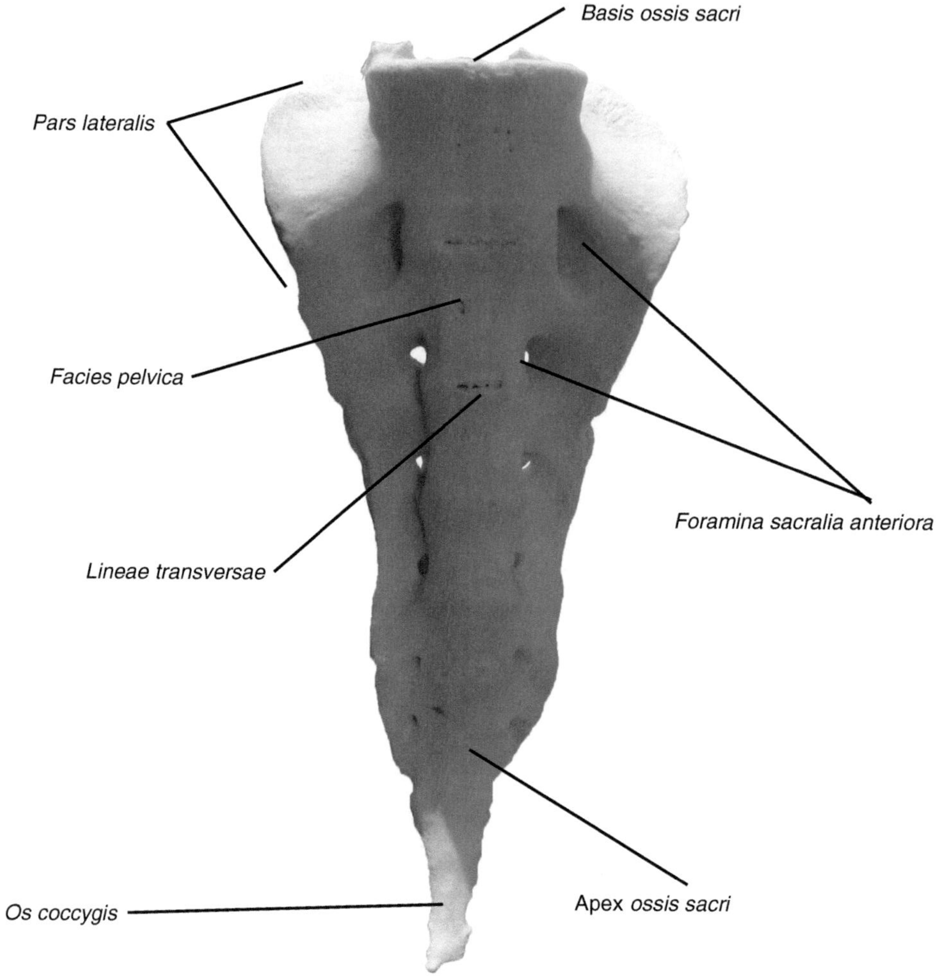


**Fig. 5.45** *Pan paniscus*: ventral (inferior in human anatomy) view of sacrum

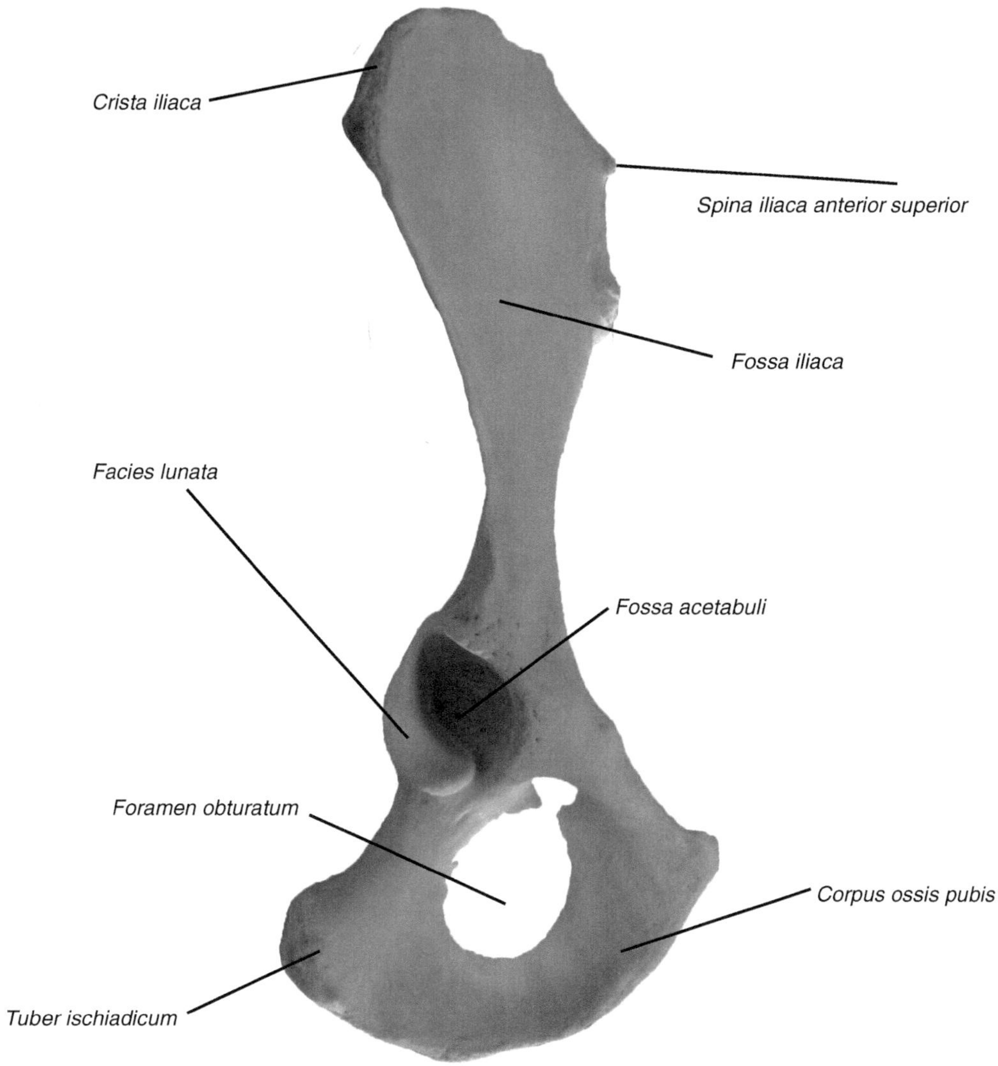


**Fig. 5.46** *Pan paniscus*: lateral view of hip bone

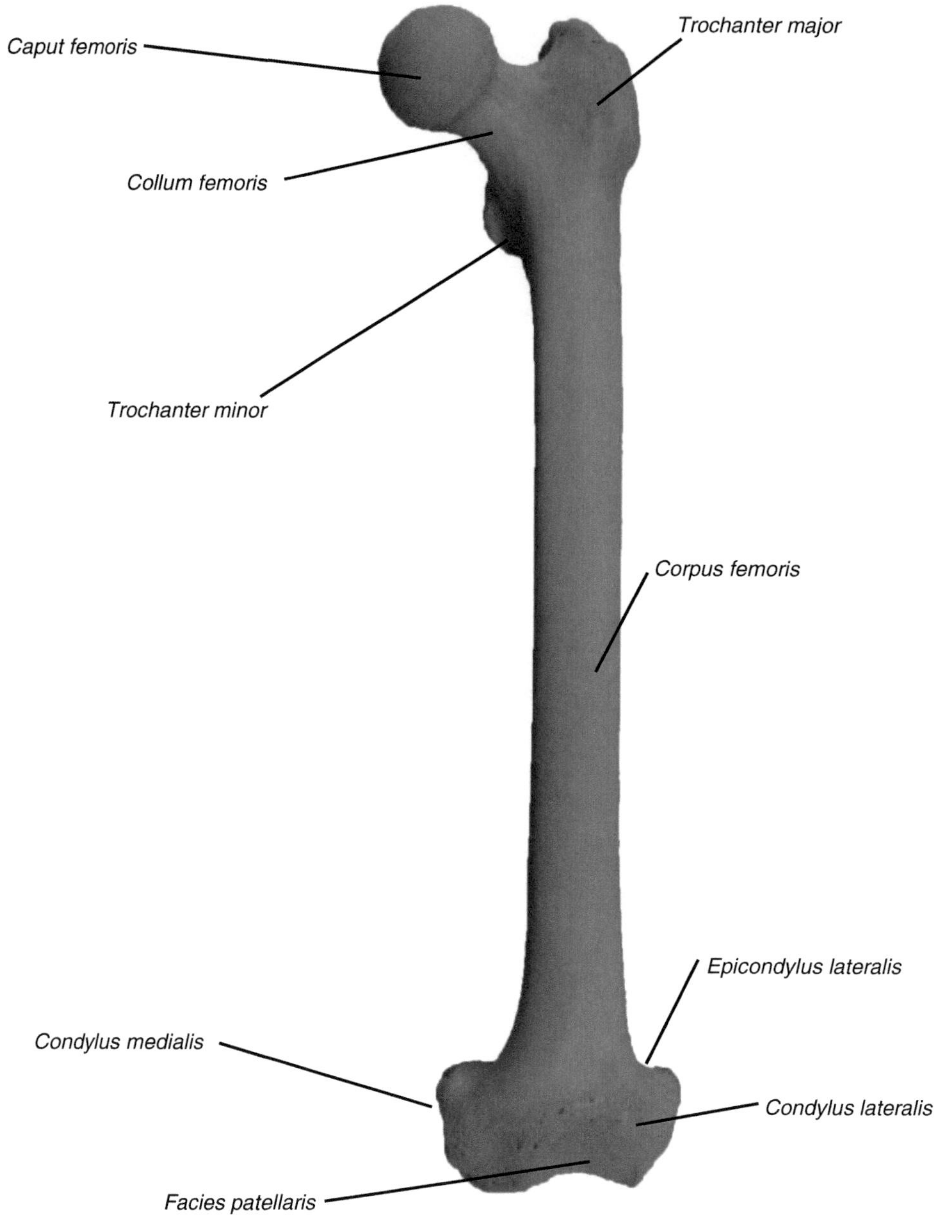


**Fig. 5.47** *Pan paniscus*: ventral (anterior in human anatomy) view of femur

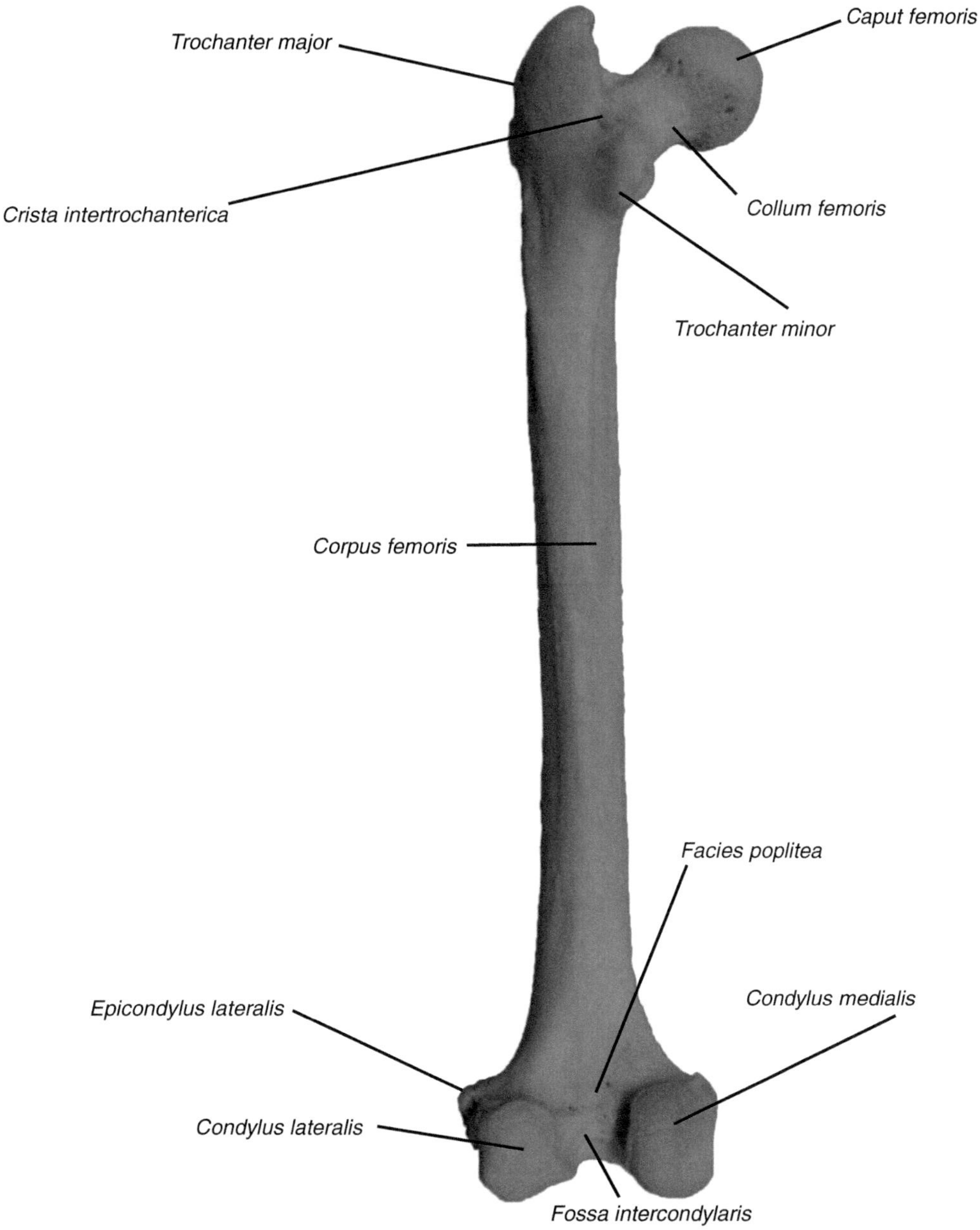


**Fig. 5.48** *Pan paniscus*: dorsal (posterior in human anatomy) view of femur

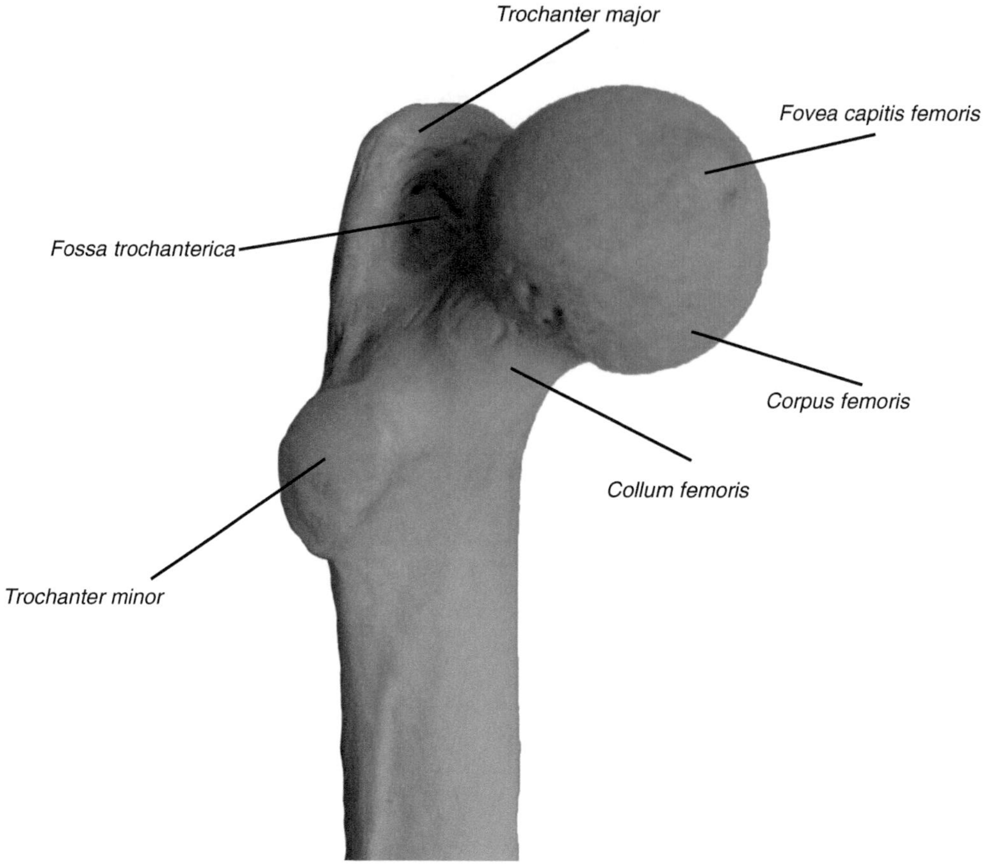


**Fig. 5.49** *Pan paniscus*: details of proximal femur

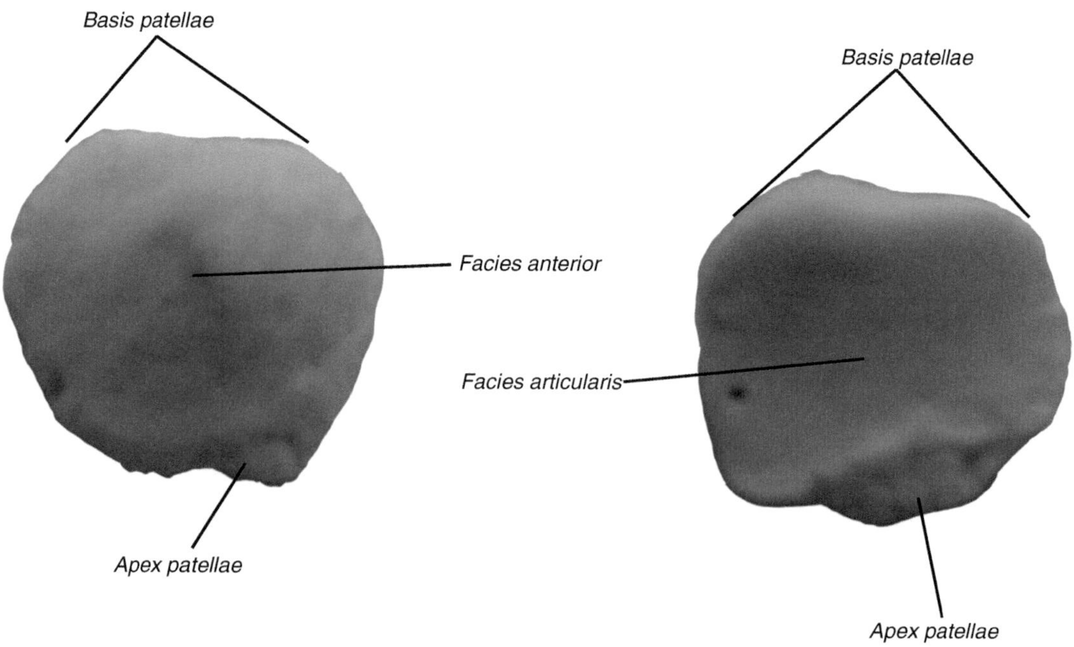


**Fig. 5.50** *Pan paniscus*: details of patella

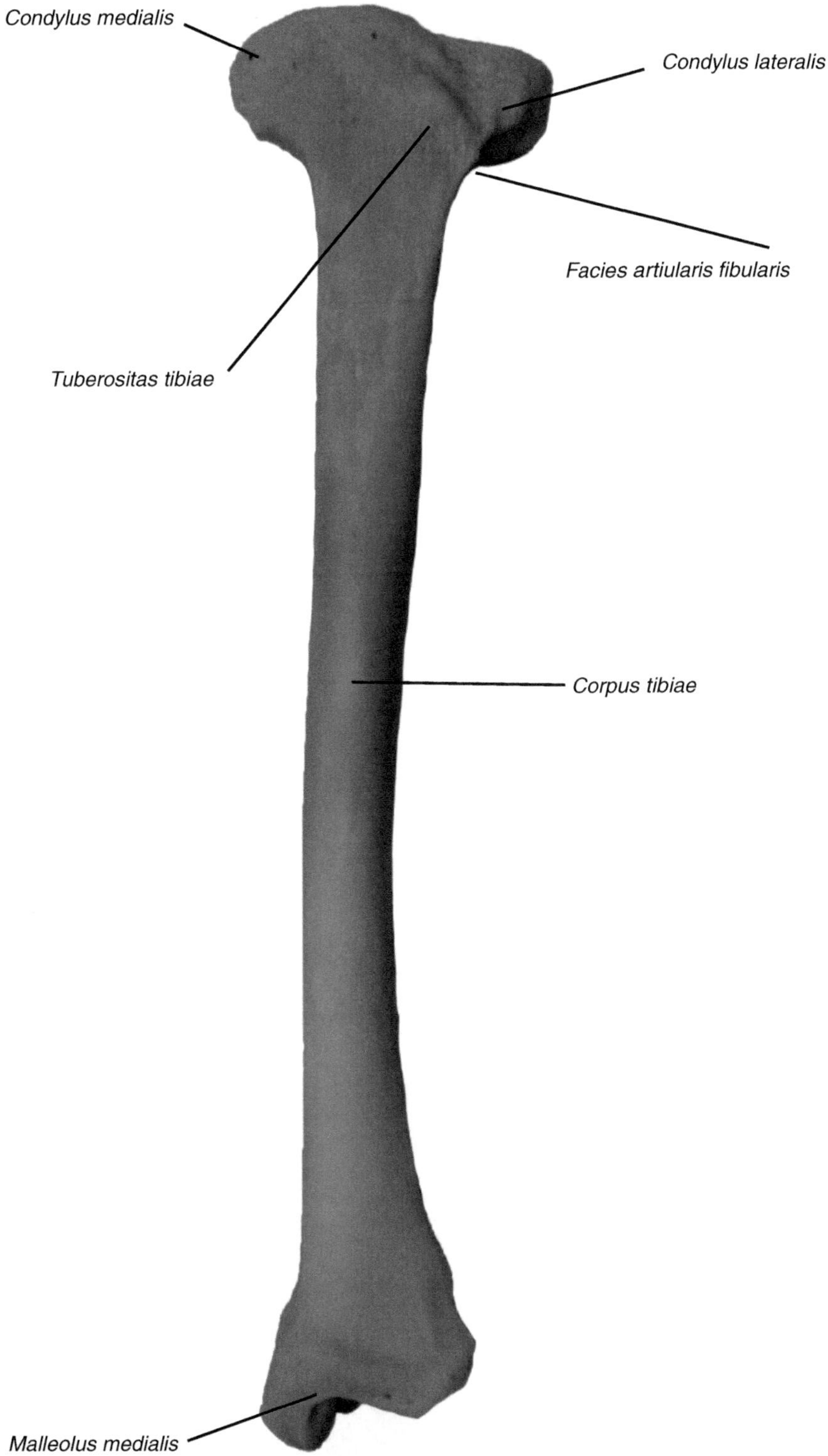


**Fig. 5.51** *Pan paniscus*: ventral (anterior in human anatomy) view of tibia

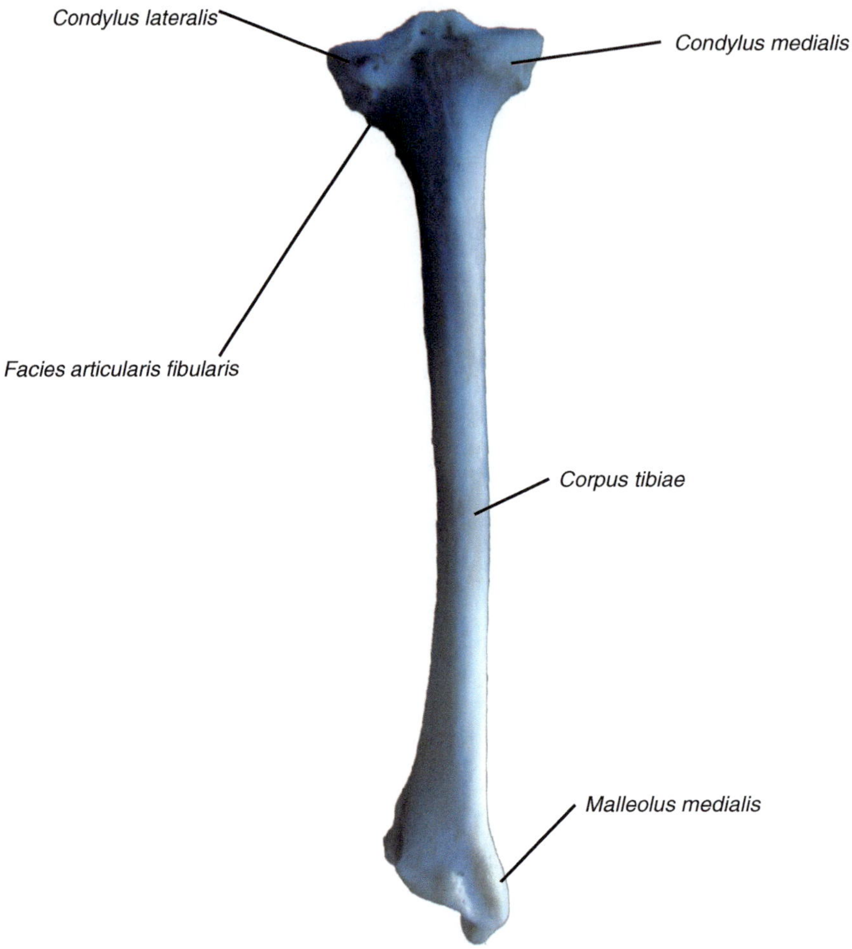


**Fig. 5.52** *Pan paniscus*: dorsal (posterior in human anatomy) view of tibia

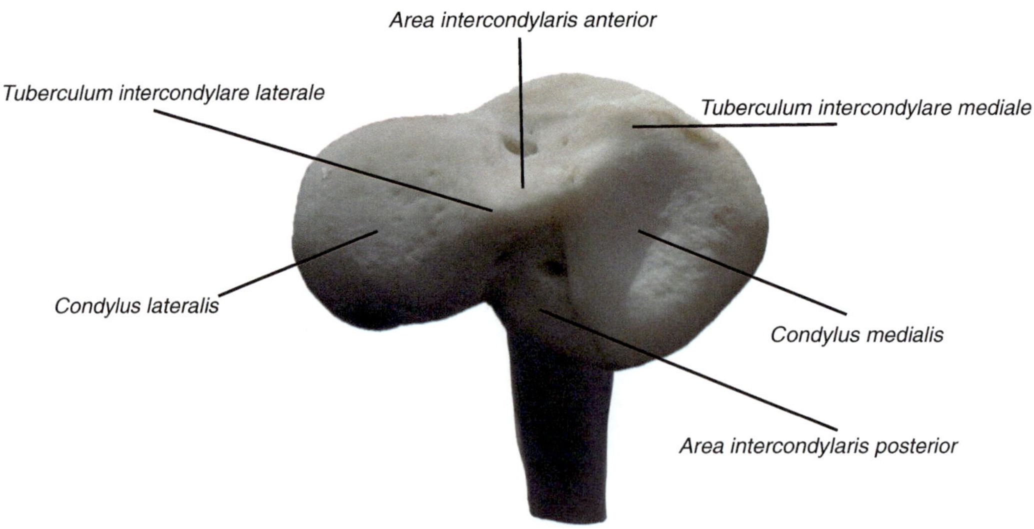


**Fig. 5.53** *Pan paniscus*: details of proximal tibia

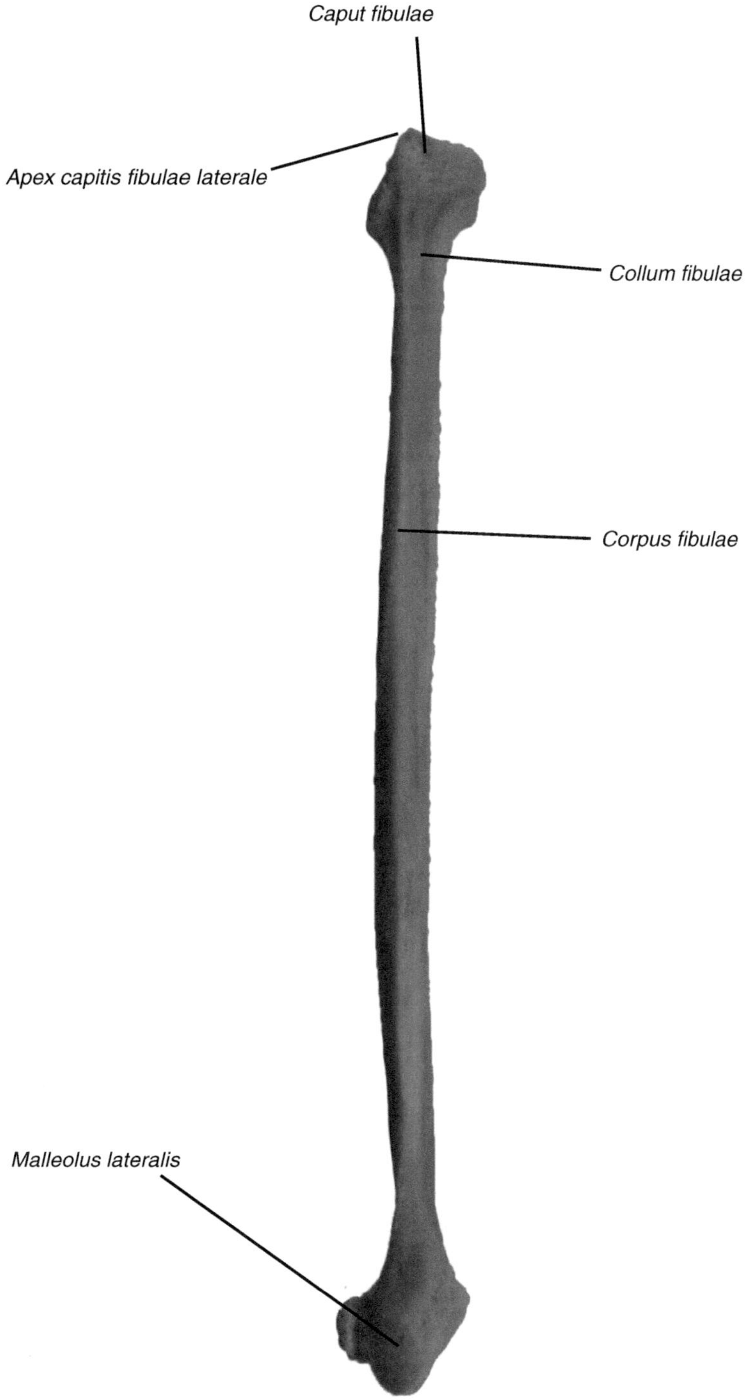


**Fig. 5.54** *Pan paniscus*: lateral view of fibula

**Fig. 5.55** *Pan paniscus*: details of proximal fibula

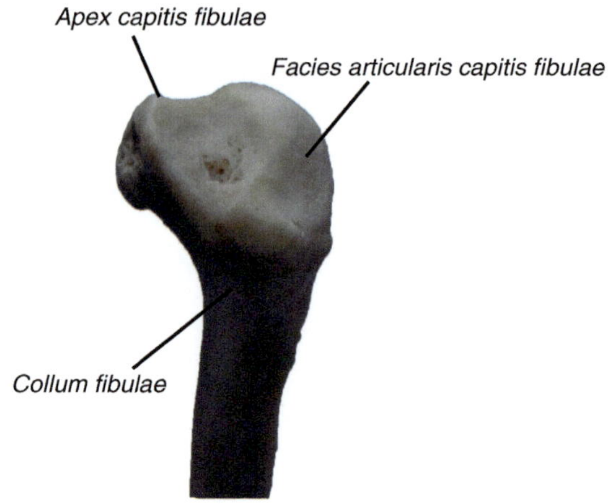


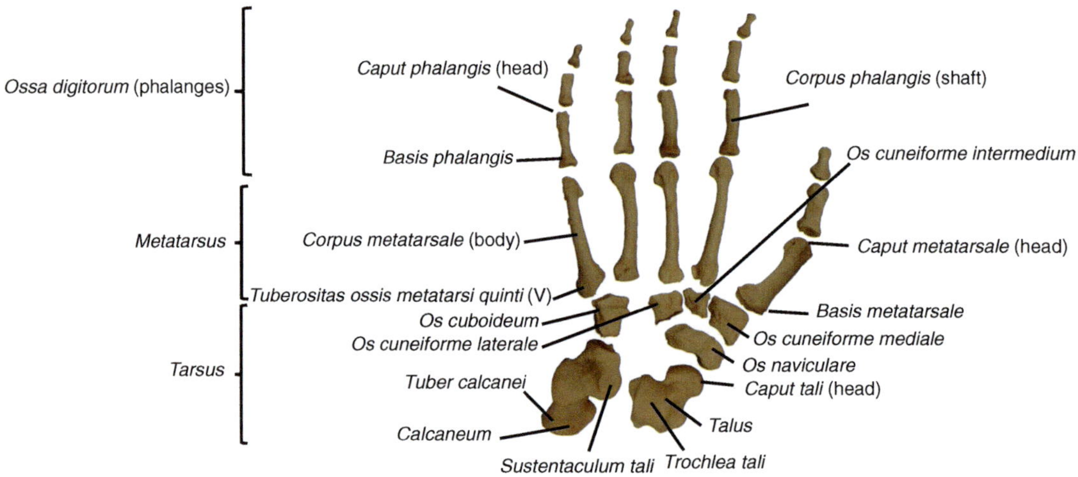


**Fig. 5.56** *Pan paniscus*: dorsal (superior in human anatomy) view of foot

# Bibliography

## Appendix I: Literature Including Information about the Muscles of Bonobos and Common Chimpanzees


Beddard FE (1893) Contributions to the anatomy of the anthropoid apes. Trans Zool Soc Lond 13:177–218

Blake ML (1976) The quantitative myology of the hind limb of Primates with special reference to their locomotor adaptations. Unpublished PhD Thesis, University of Cambridge, Cambridge

Boyer EL (1935) The musculature of the inferior extremity of the orang-utan, *Simia satyrus*. Am J Anat 56:192–256

Brooks HSJ (1887) On the short muscles of the pollex and hallux of the anthropoid apes, with special reference to the opponens hallucis. J Anat Physiol 22:78–95

Brown B (1983) An evaluation of primate caudal musculature in the identification of the ischiofemoralis muscle. Am J Phys Anthropol 60:177–178

Champneys F (1872) The muscles and nerves of a Chimpanzee (*Troglodytes Niger*) and a *Cynocephalus Anubis*. J Anat Physiol 6:176–211

Chapman HC (1879) On the structure of the chimpanzee. Proc Acad Nat Sci Phila 31:52–63

Cihak R (1972) Ontogenesis of the skeleton and intrinsic muscles of the human hand and foot. Adv Anat Embryol Cell Biol 46:1–194

Diogo R, Potau JM, Pastor JF, de Paz FJ, Ferrero EM, Bello G, Barbosa M, Aziz MA, Burrows AM, Arias-Martorell J, Wood BA (2013) Photographic and Descriptive Musculoskeletal Atlas of Chimpanzees - with notes on the attachments, variations, innervation, function and synonymy and weight of the muscles. Taylor and Francis, oxford

Duvernoy M (1855–1856) Des caracteres anatomiques de grands singes pseudoanthropomorphes anthropomorphes. Arch Mus Natl Hist Nat Paris 8:1–248

Dwight T (1895) Notes on the dissection and brain of the chimpanzee 'Gumbo'. Mem Boston Soc Nat Hist 5:31–51

Frey H (1913) Der Musculus triceps surae in der Primatenreihe. Morph Jahrb 47:1–192

Gibbs S (1999) Comparative soft tissue morphology of the extant Hominoidea, including Man. Unpublished PhD Thesis, The University of Liverpool, Liverpool

Hamada Y (1985) Primate hip and thigh muscles: comparative anatomy and dry weights. In: Kondo S (ed) Primate morphophysiology, locomotor analyses and human bipedalism. University of Tokyo Press, Tokyo, pp 131–152

Hepburn D (1892) The comparative anatomy of the muscles and nerves of the superior and inferior extremities of the anthropoid apes: I – myology of the superior extremity. J Anat Physiol 26:149–186

Humphry G (1867) On some points in the anatomy of the chimpanzee. J Anat Physiol 1:254–268

Kaplan EB (1958a) The iliotibial tract – clinical and morphological significance. J Bone Jt Surg 40A:817–832

Kaplan EB (1958b) Comparative anatomy of the extensor digitorum longus in relation to the knee joint. Anat Rec 131:129–149

Kaseda M, Nakamura M, Ischihara N, Hayakawa T, Asari M (2008) A macroscopic examination of M. Biceps Femoris and M. Gluteus Maximus in the Orangutan. J Vet Med Sci 70:217–222

Keith A (1894a) The myology of the Catarrhini: a study in evolution. Unpublished PhD thesis, University of Alberdeen, Alberdeen

Keith A (1894b) Notes on a theory to account for the various arrangements of the flexor profundus digitorum in the hand and foot of primates. J Anat Physiol 28:335–339

Kumakura H (1989) Functional analysis of the biceps femoris muscle during locomotor behavior in some primates. Am J Phys Anthropol 79:379–391

Lewis OJ (1962) The comparative morphology of M. flexor accessorius and the associated long flexor tendons. J Anat 96:321–333

Lewis OJ (1964) The evolution of the long flexor muscles of the leg and foot. In: Felts WJL, Harrison RJ (eds) International review of general and experimental zoology. Academic Press, New York, pp 165–185

Lewis OJ (1966) The phylogeny of the cruropedal extensor musculature with special reference to the primates. J Anat 100:865–880

Lewis OJ (1989) Functional morphology of the evolving hand and foot. Clarendon Press, Oxford

Macalister A (1871) On some points in the myology of The Chimpanzee and others of the primates. Ann Mag Nat Hist 7:341–351

MacDowell EC (1910) Notes on the myology of *Anthropopithecus niger* and *Papio-thoth ibeanus*. Am J Anat 10:431–460

Miller RA (1945) The ischial callosities of primates. Am J Anat 76:67–87

Miller RA (1952) The musculature of *Pan paniscus*. Am J Anat 91:182–232

Mysberg WA (1917) Über die Verbinderungen zwischen dem Sitzbeine und der Wirbelsäule bei den Säugetieren. Anat Hefte 54:641–668

Prejzner-Morawska A, Urbanowicz M (1971) The biceps femoris muscle in lemurs and monkeys. Folia Morphol 30:9465–9482

Ranke K (1897) Muskel-und Nervenvariationen der dorsalen elemente des Plexus ischiadicus der Primaten. Arch Anthropol 24:117–144

Robinson JT, Freedman L, Sigmon BA (1972) Some aspects of pongid and hominid bipedality. J Hum Evol 1:361–369

Ruge G (1878a) Untersuchung uber die Extensorengruppe aus Unterschenkel und Füsse der Säugethiere. Morphol Jahrb 4:592–643

Ruge G (1878b) Zur vergleichenden Anatomie der tiefen Muskeln in der Fusssohle. Morphol Jahrb 4:644–659

Sarmiento EE (1983) The significance of the heel process in anthropoids. Int J Primatol 4:127–152

Sigmon BA (1974) A functional analysis of pongid hip and thigh musculature. J Hum Evol 3:161–185
Sonntag CF (1923) On the anatomy, physiology, and pathology of the chimpanzee. Proc Zool Soc Lond 23:323–429
Sonntag CF (1924) The morphology and evolution of the apes and man. John Bale Sons and Danielsson, Ltd., London
Stern JT (1972) Anatomical and functional specializations of the human gluteus maximus. Am J Phys Anthropol 36:315–340
Sutton JB (1883) On some points in the anatomy of the chimpanzee (*Anthropopithecus troglodytes*). J Anat Physiol 18:66–85
Swindler DR, Wood CD (1973) An atlas of primate gross anatomy: baboon, chimpanzee and men. University of Washington Press, Seattle
Van den Broek AJP (1914) Studien zur Morphologie des Primatenbeckens. Morphol Jahrb 49:1–118
Vereecke EE, D'Aout K, Payne R, Aerts P (2005) Functional analysis of the foot and ankle myology of gibbons and bonobos. J Anat 206:453–476
Vrolik W (1841) Recherches d' anatornie comparé, sur le chimpanzé. Johannes Miller, Amsterdam
Wilder B (1862) Contributions to the comparative myology of the chimpanzee. Boston J Nat Hist 6:352–384
Yirga S (1987) Interrelation between ischium, thigh extending muscles and locomotion in some primates. Primates 28:79–86
Zihlman AL, Brunker L (1979) Hominid bipedalism: then and now. Yearbook Phys Anthropol 22:132–162

## Appendix II: Literature Cited, Not Including Information About the Muscles of Bonobos and Common Chimpanzees

Terminologia Anatomica (1998) Federative Committee on Anatomical Terminology. Georg Thieme, Stuttgart. New York: Columbia University Press

# Bibliography

## Appendix I: Literature Including Information about the Muscles of Bonobos and Common Chimpanzees


Abramowitz I (1955) On the existence of a palmar interosseous muscle in the thumb with particular reference to the Bantu-speaking Negro. South African J Sci 51:270–276

Aiello L, Dean C (1990) An introduction to human evolutionary anatomy. Academic Press, San Diego

Andrews P, Groves CP (1976) Gibbons and brachiation. In: Rumbaugh DM (ed) Gibbon and Siamang, vol 4. Karger, Basel, pp 167–218

Ashton EH, Oxnard CE (1963) The musculature of the primate shoulder. J Zool Soc Lond 29:553–650

Avril C (1963) Kehlkopf und Kehlsack des Schimpansen (*Pan troglodytes*). Gegenbaurs Morphol Jahrb 105: 74–129

Aziz MA, Dunlap SS (1986) The human extensor digitorum profundus muscle with comments on the evolution of the primate hand. Primates 27:293–319

Barnard WS (1875) Observations on the membral musculation of *Simia satyrus* (Orang) and the comparative myology of man and the apes. Proc Amer Assoc Adv Sci 24:112–144

Beddard FE (1893) Contributions to the anatomy of the anthropoid apes. Trans Zool Soc Lond 13:177–218

Blake ML (1976) The quantitative myology of the hind limb of Primates with special reference to their locomotor adaptations. Unpublished PhD Thesis, University of Cambridge, Cambridge

Blair DM (1923) A study of the central tendon of the diaphragm. J Anat Physiol 57:203–215

Bischoff TLW (1870) Beitrage zur Anatomie des *Hylobates leuciscus* and zueiner vergleichenden Anatomie der Muskeln der Affen und des Menschen. Abh Bayer Akad Wiss Miinchen Math Phys Kl 10:197–297

Bojsen-Møller F (1978) Extensor carpi radialis longus muscle and the evolution of the first intermetacarpal ligament. Am J Phys Anthropol 48:177–184

Bolk L (1902) Beiträge zur Affenanatomie, III, Der Plexus cervico-brachialis der Primaten. Petrus Campter 1:371–566

Boyer EL (1935) The musculature of the inferior extremity of the orang-utan, *Simia satyrus*. Am J Anat 56:192–256

Broca P (1869) L'ordre des primates – parallele anatomique de I'homme et des singes. Bull Soc Anthropol Paris 4:228–401

Brooks HSJ (1886) On the morphology of the intrinsic muscles of the little finger, with some observations on the ulnar head of the short flexor of the thumb. J Anat Physiol 20:644–661

Brooks HSJ (1887) On the short muscles of the pollex and hallux of the anthropoid apes, with special reference to the opponens hallucis. J Anat Physiol 22:78–95

Brown B (1983) An evaluation of primate caudal musculature in the identification of the ischiofemoralis muscle. Am J Phys Anthropol 60:177–178

Burrows AM (2008) The facial expression musculature in primates and its evolutionary significance. Bioessays 30:212–225

Burrows AM, Waller BM, Parr LA, Bonar CJ (2006) Muscles of facial expression in the chimpanzee (*Pan troglodytes*): descriptive, comparative and phylogenetic contexts. J Anat 208:153–167

Carlson KJ (2006) Muscle architecture of the common chimpanzee (*Pan troglodytes*): perspectives for investigating chimpanzee behavior. Primates 47:218–229

Champneys F (1872) The muscles and nerves of a Chimpanzee (*Troglodytes Niger*) and a *Cynocephalus Anubis*. J Anat Physiol 6:176–211

Chapman HC (1879) On the structure of the chimpanzee. Proc Acad Nat Sci Phila 31:52–63

Cihak R (1972) Ontogenesis of the skeleton and intrinsic muscles of the human hand and foot. Adv Anat Embryol Cell Biol 46:1–194

Day MH, Napier J (1963) The functional significance of the deep head of flexor pollicis brevis in primates. Folia Primatol 1:122–134

Dean MC (1985) Comparative myology of the hominoid cranial base, II, the muscles of the prevertebral and upper pharyngeal region. Folia Primatol 44:40–51



R. Diogo et al., *Photographic and Descriptive Musculoskeletal Atlas of Bonobos*,
DOI 10.1007/978-3-319-54106-8

Diogo R, Abdala V (2010) Muscles of vertebrates – comparative anatomy, evolution, homologies and development. Taylor & Francis, Oxford

Diogo R, Wood BA (2011) Soft-tissue anatomy of the primates: phylogenetic analyses based on the muscles of the head, neck, pectoral region and upper limb, with notes on the evolution of these muscles. J Anat 219:273–359

Diogo R, Wood BA (2012) Comparative anatomy and phylogeny of primate muscles and human evolution. Taylor & Francis, Oxford

Diogo R, Abdala V, Lonergan N, Wood BA (2008) From fish to modern humans – comparative anatomy, homologies and evolution of the head and neck musculature. J Anat 213:391–424

Diogo R, Abdala V, Aziz MA, Lonergan N, Wood BA (2009a) From fish to modern humans – comparative anatomy, homologies and evolution of the pectoral and forelimb musculature. J Anat 214:694–716

Diogo R, Wood BA, Aziz MA, Burrows A (2009b) On the origin, homologies and evolution of primate facial muscles, with a particular focus on hominoids and a suggested unifying nomenclature for the facial muscles of the Mammalia. J Anat 215:300–319

DuBrul EL (1958) Evolution of the speech apparatus. Thomas, Springfield

Duchin LE (1990) The evolution of human speech: comparative anatomy of the oral cavity in *Pan* and *Homo*. J Hum Evol 19:687–697

Duckworth WLH (1904) Studies from the anthropological laboratory, the Anatomy School, Cambridge. C. J. Clay & Sons, London

Duckworth WLH (1912) On some points in the anatomy of the plica vocalis. J Anat Physiol 47:80–115

Dunlap SS, Thorington RW, Aziz MA (1985) Forelimb anatomy of New World monkeys: myology and the interpretation of primitive anthropoid models. Am J Phys Anthropol 68:499–517

Duvernoy M (1855–1856) Des caracteres anatomiques de grands singes pseudoanthropomorphes anthropomorphes. Arch Mus Natl Hist Nat Paris 8:1–248

Dylevsky I (1967) Contribution to the ontogenesis of the flexor digitorum superficialis and the flexor digitorum profundus in man. Folia Morphol (Praha) 15:330–335

Dwight T (1895) Notes on the dissection and brain of the chimpanzee 'Gumbo'. Mem Boston Soc Nat Hist 5:31–51

Edgeworth FH (1935) The cranial muscles of vertebrates. Cambridge University Press, Cambridge

Elftman HO (1932) The evolution of the pelvic floor of primates. Am J Anat 51:307–346

Fick R (1925) Beobachtungen an den muskeln einger schimpansen. Z Anat Entwicklungsgesch 76:117–141

Frey H (1913) Der Musculus triceps surae in der Primatenreihe. Morph Jahrb 47:1–192

Giacomini C (1897) 'Plica semilunaris' et larynx chez les singes anthropomorphes. Arch Ital Biol 28:98–119

Gibbs S (1999) Comparative soft tissue morphology of the extant Hominoidea, including Man. Unpublished PhD Thesis, The University of Liverpool, Liverpool

Glidden EM, De Garis CF (1936) Arteries of the chimpanzee (*Pan* spec. ?), I, the aortic arch, II, arteries of the upper extremity, III, the descending aorta, IV, arteries of the lower extremity. Am J Anat 58:501–527

Göllner K (1982) Untersuchungen über die vom N. trigeminus innervierte Kiefermusculatur des Schimpansen (*Pan troglodytes*, Blumenbach 1799) und des Gorilla (*Gorilla gorilla gorilla*, Savage and Wyman 1847). Gegenbaurs Morphol Jahrb 128:851–903

Gratiolet LP, Alix PHE (1866) Recherches sur l'anatomie du *Troglodytes aubryi*. Nouv Arch Mus Hist Nat Paris 2:1–264

Grönroos H (1903) Die musculi biceps brachii und latissimocondyloideus bei der affengattung *Hylobates* im vergleich mit den ensprechenden gebilden der anthropoiden und des menschen. Abh Kön Preuss Akad Wiss Berlin 1903:1–102

Hamada Y (1985) Primate hip and thigh muscles: comparative anatomy and dry weights. In: Kondo S (ed) Primate morphophysiology, locomotor analyses and human bipedalism. University of Tokyo Press, Tokyo, pp 131–152

Harrison DFN (1995) The anatomy and physiology of the mammalian larynx. University Press, Cambridge

Hartmann R (1886) Anthropoid apes. Keegan, London

Hepburn D (1892) The comparative anatomy of the muscles and nerves of the superior and inferior extremities of the anthropoid apes: I – myology of the superior extremity. J Anat Physiol 26:149–186

Himmelreich HA (1971) M. levator veli palatini der Säugetiere. Gegenbaurs Morphol Jahrb 116:377–400

Himmelreich HA (1977) M. cephalopharyngeus der Säugetiere. Gegenbaurs Morphol Jahrb 123:556–588

Howell AB, Straus WL (1932) The brachial flexor muscles in primates. Proc US Natl Mus 80:1–31

Huber E (1930) Evolution of facial musculature and cutaneous field of trigeminus – Part II. Q Rev Biol 5:389–437

Huber E (1931) Evolution of facial musculature and expression. The Johns Hopkins University Press, Baltimore

Humphry G (1867) On some points in the anatomy of the chimpanzee. J Anat Physiol 1:254–268

Huntington GS (1903) Present problems of myological research and the significance and classification of muscular variations. Am J Anat 2:157–175

Huxley TH (1864) The structure and classification of the Mammalia. Med Times Gazette 1864:398–468

Jordan J (1971a) Studies on the structure of the organ of voice and vocalization in the chimpanzee, Part 1. Folia Morphol 30:99–117

Jordan J (1971b) Studies on the structure of the organ of voice and vocalization in the chimpanzee, Part 2. Folia Morphol 30:222–248

Jordan J (1971c) Studies on the structure of the organ of voice and vocalization in the chimpanzee, Part 3. Folia Morphol 30:323–340

Jouffroy FK (1971) Musculature des membres. In: Grassé PP (ed) Traité de Zoologie, XVI: 3 (Mammifères). Masson et Cie, Paris, pp 1–475

Jouffroy FK, Lessertisseur J (1959) Reflexions sur les muscles contracteurs des doigts et des orteils (contrahentes digitorum) chez les primates. Ann Sci Nat Zool, Ser 12(1):211–235

Jouffroy FK, Lessertisseur J (1960) Les spécialisations anatomiques de la main chez les singes à progression suspendue. Mammalia 24:93–151

Jouffroy FK, Saban R (1971) Musculature peaucière. In: Grassé PP (ed) Traité de Zoologie, XVI: 3 (Mammifères). Masson et Cie, Paris, pp 477–611

Juraniec J (1972) The aortic and esophageral hiatus in the diaphragm of primates. Folia Morphol 31:197–207

Juraniec J, Szostakiewicz-Sawicka H (1968) The central tendon of the diaphragm in primates. Folia Morphol 27:183–194

Kaneff A (1979) Évolution morphologique des musculi extensores digitorum et abductor pollicis longus chez l'Homme. I. Introduction, méthodologie, M. extensor digitorum. Gegenbaurs Morphol Jahrb 125:818–873

Kaneff A (1980a) Évolution morphologique des musculi extensores digitorum et abductor pollicis longus chez l'Homme. II. Évolution morphologique des m. extensor digiti minimi, abductor pollicis longus, extensor pollicis brevis et extensor pollicis longus chez l'homme. Gegenbaurs Morphol Jahrb 126:594–630

Kaneff A (1980b) Évolution morphologique des musculi extensores digitorum et abductor pollicis longus chez l'Homme. III. Évolution morphologique du m. extensor indicis chez l'homme, conclusion générale sur l 'évolution morphologique des musculi extensores digitorum et abductor pollicis longus chez l'homme. Gegenbaurs Morphol Jahrb 126:774–815

Kaplan EB (1958a) The iliotibial tract – clinical and morphological significance. J Bone Jt Surg 40A:817–832

Kaplan EB (1958b) Comparative anatomy of the extensor digitorum longus in relation to the knee joint. Anat Rec 131:129–149

Kaseda M, Nakamura M, Ischihara N, Hayakawa T, Asari M (2008) A macroscopic examination of M. Biceps Femoris and M. Gluteus Maximus in the Orangutan. J Vet Med Sci 70:217–222

Keith A (1894a) The myology of the Catarrhini: a study in evolution. Unpublished PhD thesis, University of Alberdeen, Alberdeen

Keith A (1894b) Notes on a theory to account for the various arrangements of the flexor profundus digitorum in the hand and foot of primates. J Anat Physiol 28:335–339

Keith A (1899) On the chimpanzees and their relationship to the gorilla. Proc Zool Soc Lond 1899:296–312

Kelemen G (1948) The anatomical basis of phonation in the chimpanzee. J Morphol 82:229–256

Kelemen G (1969) Anatomy of the larynx and the anatomical basis of vocal performance. In: Bourne GH (ed) The Chimpanzee, Anatomy, behaviour and diseases of chimpanzees, vol Vol 1. Karger, Basel, pp 35–186

Kikuchi Y (2010) Comparative analysis of muscle architecture in primate arm and forearm. Anat Histol Embryol 39:93–106

Kohlbrügge JHF (1896) Der larynx und die stimmbildung der Quadrumana. Natuurk T Ned Ind 55:157–175

Kohlbrügge JHF (1897) Muskeln und Periphere Nerven der Primaten, mit besonderer Berücksichtigung ihrer Anomalien. Verh K Akad Wet Amsterdam Sec 2(5):1–246

Koizumi M, Sakai T (1995) The nerve supply to coracobrachialis in apes. J Anat 186:395–403

Körner O (1884) Beiträge zur vergleichenden Anatomie und Physiologie des Kehlkopfes der Säugethiere und des Menschen. Abh Senckenb Naturforsch Ges 13:147–261

Körner F (1929) Vergleichend-anatomische Untersuchungen über den Faserverlauf der Pars lumbalis des Zwerchfells zur Begrenzung des Hiatus oesophageus bei Säugetieren. Morphol Jahrb 61:409–451

Kumakura H (1989) Functional analysis of the biceps femoris muscle during locomotor behavior in some primates. Am J Phys Anthropol 79:379–391

Lander KF (1918) The pectoralis minor: a morphological study. J Anat 52:292–318

Landsmeer JM (1986) A comparison of fingers and hand in *Varanus*, opossum and primates. Acta Morphol Neerl Scand 24:193–221

Larson SG, Stern JT (1986) EMG of scapulohumeral muscles in the chimpanzee during reaching and 'arboreal' locomotion. Am J Anat 176:171–190

Larson SG, Stern JT, Jungers WL (1991) EMG of serratus anterior and trapezius in the chimpanzee: scapular rotators revisited. Am J Phys Anthropol 85:71–84

Lewis OJ (1962) The comparative morphology of M. flexor accessorius and the associated long flexor tendons. J Anat 96:321–333

Lewis OJ (1964) The evolution of the long flexor muscles of the leg and foot. In: Felts WJL, Harrison RJ (eds) International review of general and experimental zoology. Academic Press, New York, pp 165–185

Lewis OJ (1966) The phylogeny of the cruropedal extensor musculature with special reference to the primates. J Anat 100:865–880

Lewis OJ (1989) Functional morphology of the evolving hand and foot. Clarendon Press, Oxford

Lightoller GHS (1925) Facial muscles – the modiolus and muscles surrounding the rima oris with some remarks about the panniculus adiposus. J Anat Physiol 60:1–85

Lightoller GS (1928) The facial muscles of three orang utans and two cercopithecidae. J Anat 63:19–81

Lightoller GS (1934) The facial musculature of some lesser primates and a *Tupaia*. Proc Zool Soc Lond 1934:259–309

Lightoller GS (1939) V. Probable homologues. A study of the comparative anatomy of the mandibular and hyoid arches and their musculature — Part I. Comparative myology. Trans Zool Soc Lond 24:349–382

Lightoller GS (1940a) The comparative morphology of the platysma: a comparative study of the sphincter colli profundus and the trachelo-platysma. J Anat 74:390–396

Lightoller GS (1940b) The comparative morphology of the m. caninus. J Anat 74:397–402

Lightoller GS (1942) Matrices of the facialis musculature: homologization of the musculature in monotremes with that of marsupials and placentals. J Anat 76:258–269

Loth E (1931) Anthropologie des parties molles (muscles, intestins, vaisseaux, nerfs peripheriques). Mianowski-Masson et Cie, Paris

Low A (1907) A note on the crura of the diaphragm and the muscle of Treitz. J Anat 42:93–96

Lunn HF (1948) The comparative anatomy of the inguinal ligament. Anat Phys 82:58–67

Lunn HF (1949) Observations on the mammalian inguinal region. Proc Zool Soc 118:345–355

Macalister A (1871) On some points in the myology of the chimpanzee and others of the primates. Ann Mag Nat Hist 7:341–351

MacDowell EC (1910) Notes on the myology of *Anthropopithecus niger* and *Papio-thoth ibeanus*. Am J Anat 10:431–460

Maier W (2008) Epitensoric position of the chorda tympani in Anthropoidea: a new synapomorphic character, with remarks on the fissura glaseri in Primates. In: Sargis EJ, Dagosto M (eds) Mammalian evolutionary morphology: a tribute to Frederick S. Szalay. Springer, Dordrecht, pp 339–352

Mijsberg WA (1923) Über den Bau des Urogenitalapparates bei den männlichen Primaten. Verh K Akad Wet Amsterdam 23:1–92

Miller RA (1932) Evolution of the pectoral girdle and forelimb in the primates. Amer J Phys Anthropol 17:1–56

Miller RA (1945) The ischial callosities of primates. Am J Anat 76:67–87

Miller RA (1947) The inguinal canal of primates. Am J Anat 90:117–142

Miller RA (1952) The musculature of *Pan paniscus*. Am J Anat 91:182–232

Mustafa MA (2006) Neuroanatomy. 10th Natl Congr Anat Bordum, Turkey, September 5:6–10

Myatt JP, Crompton RH, Payne RC, Vereecke EE, Isler K, Savage R, D'Août K, Günther MM, Thorpe SKS (2012) Functional adaptations in the forelimb muscles of non-human great apes. J Anat 220, 13–28.

Mysberg WA (1917) Über die Verbinderungen zwischen dem Sitzbeine und der Wirbelsäule bei den Säugetieren. Anat Hefte 54:641–668

Ogihara N, Kunai T, Nakatsukasa M (2005) Muscle dimensions in the chimpanzee hand. Primates 46:275–280

Oishi M, Ogihara N, Endo H, Ichihara N, Asari M (2009) Dimensions of forelimb muscles in orangutans and chimpanzees. J Anat 215:373–382

Owen R (1830–1831) On the anatomy of the orangutan (*Simia satyrus*, L.) Proc Zool Soc Lond 1:4–5, 9–10, 28–29, 66–72

Owen R (1868) The anatomy of vertebrates, Mammals, vol 3. Longmans, Green & Co., London

Parsons FG (1898a) The muscles of mammals, with special relation to human myology, Lecture 1, The skin muscles and muscles of the head and neck. J Anat Physiol 32:428–450

Parsons FG (1898b) The muscles of mammals, with special relation to human myology: a course of lectures delivered at the Royal College of Surgeons of England – lecture II, the muscles of the shoulder and forelimb. J Anat Physiol 32:721–752

Payne RC (2001) Musculoskeletal adaptations for climbing in hominoids and their role as exaptations for the acquisition of bipedalism. Unpublished PhD thesis, The University of Liverpool, Liverpool

Payne RC, Crompton RH, Isler K, Savage R, Vereecke EE, Günther MM, Thorpe SKS, D'Août K (2006a) Morphological analysis of the hindlimb in apes and humans. I. Muscle architecture. J Anat 208, 709–724.

Payne RC, Crompton RH, Isler K, Savage R, Vereecke EE, Günther MM, Thorpe SKS, D'Août K (2006b) Morphological analysis of the hindlimb in apes and humans. II. Moment arms. J Anat 208, 725–742.

Pellatt A (1979) The facial muscles of three African primates contrasted with those of *Papio ursinus*. S Afr J Sci 75:436–440

Potau JM, Bardina X, Ciurana N, Camprubi D, Pastor JF, De Paz F, Barbosa M (2009) Quantitative analysis of the deltoid and rotator cuff muscles in humans and great Apes. Int J Primatol 30:697–708

Prejzner-Morawska A, Urbanowicz M (1971) The biceps femoris muscle in lemurs and monkeys. Folia Morphol 30:9465–9482

Ranke K (1897) Muskel-und Nervenvariationen der dorsalen elemente des Plexus ischiadicus der Primaten. Arch Anthropol 24:117–144

Ribbing L, Hermansson K (1912) Kleinere muskelstudien, I11, die distale extremitatenmuskulatur eines schimpansen. Lunds Universitets Arsskrift NF Afd 2(8):1–10

Roberts JA, Seibold HR (1971) The histology of the primate urethra. Folia Primatol 14:59–69

Robinson JT, Freedman L, Sigmon BA (1972) Some aspects of pongid and hominid bipedality. J Hum Evol 1:361–369

Ruge G (1878a) Untersuchung uber die Extensorengruppe aus Unterschenkel und Füsse der Säugethiere. Morphol Jahrb 4:592–643

Ruge G (1878b) Zur vergleichenden Anatomie der tiefen Muskeln in der Fusssohle. Morphol Jahrb 4:644–659

Ruge G (1887a) Untersuchungen uber die Gesichtsmuskeln der Primaten. W. Engelmann, Leipzig

Ruge G (1887b) Die vom Facialis innervirten Muskeln des Halses, Nackens und des Schädels einen jungen *Gorilla*. Gegenbaurs Morphol Jahrb 12:459–529

Saban R (1968) Musculature de la tête. In: Grassé PP (ed) Traité de Zoologie, XVI: 3 (Mammifères). Masson et Cie, Paris, pp 229–472

Sarmiento EE (1983) The significance of the heel process in anthropoids. Int J Primatol 4:127–152

Schück AC (1913a) Beiträge zur Myologie der Primaten, I – der m. lat. dorsi und der m. latissimo-tricipitalis. Morphol Jahrb 45:267–294
Schück AC (1913b) Beiträge zur Myologie der Primaten, II – 1 die gruppe sterno-cleido-mastoideus, trapezius, omo-cervicalis, 2 die gruppe levator scapulae, rhomboides, serratus anticus. Morphol Jahrb 46:355–418
Schultz AH (1936) Characters common to higher primates and characters specific for man. Q Rev Biol 11(259–283):425–455
Seiler R (1970) Differences in the facial musculature of the nasal and upper-lip region in catarrhine primates and man. Z Morphol Anthropol 62:267–275
Seiler R (1971a) Facial musculature and its influence on the facial bones of catarrhine Primates, II. Morphol Jahrb 116:147–185
Seiler R (1971b) Facial musculature and its influence on the facial bones of catarrhine Primates, III. Morphol Jahrb 116:347–376
Seiler R (1974) Muscles of the external ear and their function in man, chimpanzees and *Macaca*. Morphol Jahrb 120:78–122
Seiler R (1976) Die Gesichtsmuskeln. In: Hofer H, Schultz AH, Starck D (eds) Primatologia, Handbuch der Primatenkunde, Bd. 4, Lieferung 6. Karger, Basel, pp 1–252
Seiler R (1979) Criteria of the homology and phylogeny of facial muscles in primates including man, II, Catarrhina. Morphol Jahrb 125:298–323
Sigmon BA (1974) A functional analysis of pongid hip and thigh musculature. J Hum Evol 3:161–185
Smith WC (1923) The levator ani muscle; its structure in man, and its comparative relationships. Anat Rec 26:175–204
Sonntag CF (1923) On the anatomy, physiology, and pathology of the chimpanzee. Proc Zool Soc Lond 23:323–429
Sonntag CF (1924) The morphology and evolution of the apes and man. John Bale Sons and Danielsson, Ltd., London
Starck D (1973) The skull of the fetal chimpanzee. In: Bourne GH (ed) The Chimpanzee, vol 6. Karger, Basel, pp 1–33
Starck D, Schneider R (1960) Respirationsorgane. In: Hofer H, Schultz AH, Starck D (eds) Primatologia III/2. Karger, Basel, pp 423–587
Stern JT (1972) Anatomical and functional specializations of the human gluteus maximus. Anta J Phys Anthropol 36:315–340
Stern JT, Larson SG (2001) Telemetered electromyography of the supinators and pronators of the forearm in gibbons and chimpanzees: implications for the fundamental positional adaptation of hominoids. Am J Phys Anthropol 115:253–268
Stern JT, Wells JP, Jungers WL, Vangor AK (1980) An electromyographic study of serratus anterior in atelines and *Alouatta*: implications for hominoid evolution. Am J Phys Anthropol 52:323–334
Stewart TD (1936) The musculature of the anthropoids, I, neck and trunk. Am J Phys Anthropol 21:141–204
Straus WL (1941a) The phylogeny of the human forearm extensors. Hum Biol 13:23–50
Straus WL (1941b) The phylogeny of the human forearm extensors (concluded). Hum Biol 13:203–238
Straus WL (1942a) The homologies of the forearm flexors: urodeles, lizards, mammals. Am J Anat 70:281–316
Straus WL (1942b) Rudimentary digits in primates. Q Rev Biol 17:228–243
Sullivan WE, Osgood CW (1925) The facialis musculature of the orang, *Simia satyrus*. Anat Rec 29:195–343
Susman RL, Stern JT (1980) EMG of the interosseous and lumbrical muscles in the chimpanzee (*Pan troglodytes*) hand during locomotion. Am J Anat 157:389–397
Susman RL, Nyati L, Jassal MS (1999) Observations on the pollical palmar interosseus muscle (of Henle). Anat Rec 254:159–165
Sutton JB (1883) On some points in the anatomy of the chimpanzee (*Anthropopithecus troglodytes*). J Anat Physiol 18:66–85
Swindler DR, Wood CD (1973) An atlas of primate gross anatomy: baboon, chimpanzee and men. University of Washington Press, Seattle
Testut L (1883) Le long fléchisseur propre du pouce chez l'homme et les singes. Bull Soc Zool Fr 8:164–185
Thompson P (1901) On the arrangement of the fasciae of the pelvis and their relationship to the levator ani. J Anat Phys 35:127–141
Tocheri MW, Orr CM, Jacofsky MC, Marzke MW (2008) The evolutionary history of the hominin hand since the last common ancestor of *Pan* and *Homo*. J Anat 212:544–562
Tuttle RH (1969) Quantitative and functional studies on the hands of the Anthropoidea, I, the Hominoidea. J Morphol 128:309–363
Tuttle RH (1970) Postural, propulsive and prehensile capabilities in the cheiridia of chimpanzees and other great apes. In: Bourne GH (ed) The Chimpanzee, vol 2. Karger, Basel, pp 167–263
Tuttle RH, Basmajian JV (1976) Electromyography of pongid shoulder muscles and hominoid evolution I – retractors of the humerus and rotators of the scapula. Yearbook Phys Anthropol 20:491–497
Tuttle RH, Basmajian JV (1978a) Electromyography of pongid shoulder muscles II – deltoid, rhomboid and "rotator cuff". Am J Phys Anthropol 49: 47–56
Tuttle RH, Basmajian JV (1978b) Electromyography of pongid shoulder muscles III – quadrupedal positional behavior. Am J Phys Anthropol 49:57–70
Tyson E (1699) Orang-Outang sive Homo sylvestris, or the anatomy of a pygmie compared to that of a monkey, an ape and a man. T. Bennett and D. Brown, London
Van den Broek AJP (1908) Ueber einige anatomische Merkmale von *Ateles* in Zusammenhang mit der Anatomie der Platyrrhinin. Anat Anz 33:111–124
Van den Broek AJP (1914) Studien zur Morphologie des Primatenbeckens. Morphol Jahrb 49:1–118

Vereecke EE, D'Aout K, Payne R, Aerts P (2005) Functional analysis of the foot and ankle myology of gibbons and bonobos. J Anat 206:453–476
Virchow H (1915) Gesichtsmuskeln des Schimpansen. Abh Preuss Akad Wiss 1:l–81
Vrolik W (1841) Recherches d' anatornie comparé, sur le chimpanzé. Johannes Miller, Amsterdam
Waller BM, Vick SJ, Parr LA, Bard KA, Smith Pasqualini MC, Gothard K, Fuglevand A (2006) Intramuscular electrical stimulation of facial muscles in humans and chimpanzees: Duchenne revisited and extended. Emotion 6:367–382
Waterman HC (1929) Studies on the evolution of the pelvis of man and other primates. Bull Am Mus Nat Hist 55:585–642
Wilder B (1862) Contributions to the comparative myology of the chimpanzee. Boston J Nat Hist 6: 352–384
Winckler G (1947) Recherches sur le systéme musculaire du transverso-spinalis. Arch Anat Histol Embryol 30:151–231
Winckler G (1950) Contribution a I'étude des muscles larges de la paroi abdominale, Étude d'anatomie comparée. Arch Anat Histol Embryol 33:157–228
Wood Jones F (1920) The principles of anatomy as seen in the hand. J. & A Churchill, London
Wyman J (1855) An account of the dissection of a black chimpanzee. Proc Boston Soc Nat Hist 5:270–275
Yirga S (1987) Interrelation between ischium, thigh extending muscles and locomotion in some primates. Primates 28:79–86
Ziegler AC (1964) Brachiating adaptations of chimpanzee upper limb musculature. Am J Phys Anthropol 22:15–32
Zihlman AL, Brunker L (1979) Hominid bipedalism: then and now. Yearbook Phys Anthropol 22:132–162

## Appendix II: Literature Cited, Not Including Information About the Muscles of Bonobos and Common Chimpanzees

Diogo R (2004a) Morphological evolution, aptations, homoplasies, constraints, and evolutionary trends: catfishes as a case study on general phylogeny and macroevolution. Science Publishers, Enfield
Diogo R (2004b) Muscles versus bones: catfishes as a case study for an analysis on the contribution of myological and osteological structures in phylogenetic reconstructions. Anim Biotechnol 54:373–391
Diogo R (2007) On the origin and evolution of higher-clades: osteology, myology, phylogeny and macroevolution of bony fishes and the rise of tetrapods. Science Publishers, Enfield
Diogo R (2008) Comparative anatomy, homologies and evolution of the mandibular, hyoid and hypobranchial muscles of bony fish and tetrapods: a new insight. Anim Biotechnol 58:123–172
Diogo R (2009) The head musculature of the Philippine colugo (Dermoptera: *Cynocephalus volans*), with a comparison to tree-shrews, primates and other mammals. J Morphol 270:14–51
Diogo R, Abdala V (2007) Comparative anatomy, homologies and evolution of the pectoral muscles of bony fish and tetrapods: a new insight. J Morphol 268: 504–517
Diogo R, Potau JM, Pastor JF, de Paz FJ, Ferrero EM, Bello G, Barbosa M, Wood B (2010) Photographic and Descriptive Musculoskeletal Atlas of *Gorilla* – with notes on the attachments, variations, innervation, synonymy and weight of the muscles. Taylor & Francis, Oxford
Diogo R, Potau JM, Pastor JF, de Paz FJ, Ferrero EM, Bello G, Barbosa M, Aziz MA, Burrows AM, Arias-Martorell J, Wood B (2012) Photographic and Descriptive Musculoskeletal Atlas of Gibbons and Siamangs (*Hylobates*) – with notes on the attachments, variations, innervation, synonymy and weight of the muscles. Taylor & Francis, Oxford
Diogo R, Potau JM, Pastor JF, de Paz FJ, Ferrero EM, Bello G, Barbosa M, Aziz MA, Burrows AM, Arias-Martorell J, Wood B (2013) Photographic and Descriptive Musculoskeletal Atlas of Chimpanzees – with notes on the attachments, variations, innervation, function and synonymy and weight of the muscles. Taylor & Francis, Oxford
Netter FH (2006) Atlas of human anatomy, 4th edn. Saunders, Philadelphia
Terminologia Anatomica (1998) Federative Committee on Anatomical Terminology. Georg Thieme, Stuttgart. New York: Columbia University Press
Wood J (1870) On a group of varieties of the muscles of the human neck, shoulder, and chest, with their transitional forms and homologies in the Mammalia. Philos Trans R Soc Lond 160:83–116

# Index




R. Diogo et al., *Photographic and Descriptive Musculoskeletal Atlas of Bonobos*,
DOI 10.1007/978-3-319-54106-8

E



F



G



H



I



J



L

GPSR Compliance

*The European Union's (EU) General Product Safety Regulation (GPSR) is a set of rules that requires consumer products to be safe and our obligations to ensure this.*

*If you have any concerns about our products, you can contact us on ProductSafety@springernature.com*

In case Publisher is established outside the EU, the EU authorized representative is:

Springer Nature Customer Service Center GmbH
Europaplatz 3
69115 Heidelberg, Germany

**Batch number: 10684391**

Printed by Printforce, the Netherlands